The
COSMIC PULSE
OF LIFE

The Revolutionary Biological Power Behind UFOs

By
Trevor James Constable

THE BOOK TREE
San Diego, California

The Cosmic Pulse of Life
© 1976, 2008
by Trevor James Constable

Revised and Updated Fourth Edition

ISBN 978-1-58509-115-7

Cover layout & design: Atulya
Interior layout and design: Atulya
Editor: Paul Tice

Printed in USA on Acid-Free Paper

Published by
The Book Tree
P O Box 16476
San Diego, CA 92176
www.thebooktree.com
We provide fascinating and educational products to help awaken the public to new ideas and information that would not be available otherwise.
Call 1 (800) 700-8733 for our *FREE BOOK TREE CATALOG*.

For

GLORIA
who brought me
happiness and peace

CONTENTS

A Note from the Original Publisher vii

Publisher's Foreword ix

Preface xi

Preface to Fourth Edition xv

Chapter 1 The Great Impasse 1

Chapter 2 Dimensions Unmeasurable..14

Chapter 3 Finding a New Pathway 26

Chapter 4 Probing the New Reality 36

Chapter 5 Eye Into the Ethers...46

Chapter 6 Ether Ships 57

Chapter 7 The Case for the Critters 67

Chapter 8 Expanding the Case for the Critters...............................77

Chapter 9 The Mask of Officialdom..87

Chapter 10 Steiner—Prodigal Titan 99

Chapter 11 Cosmic Blueprint 111

Chapter 12 Ethereans 128

Chapter 13 The Boys Downstairs 138

Chapter 14 Criminal or Genius...151

Chapter 15 Cosmic Electronics...167

Chapter 16 Cosmic Electronics in Action..179

Chapter 17 Avatar Extraordinary—Wilhelm Reich 192

Chapter 18 From Orgasm to UFOs..202

Chapter 19 Cosmic Breakthrough..214

Chapter 20 The New Knowledge at Work 227

Chapter 21 The Battle for the Earth 243

Appendix I: Photographs of Invisible UFOs 258

Appendix II: The Work of Luciano Boccone 293

Appenidix III: The Tether Incident 313

Appendix IV: The Critters in History, by Larry Arnold 318

Index 339

A NOTE FROM THE ORIGINAL PUBLISHER

Opposition arose in the major New York publishing houses to the publication of this book, despite the international success of the author's major works in the aviation field. His flying books are all classics. The world public has now also accorded classic status to his first UFO book, *They Live in the Sky*, even though it created a furor when published in 1958.

Resistance to the publication of *The Cosmic Pulse of Life*, his masterwork, through the channels that would be normal for such an author, seemed to stem from irrational fear. The real possibility arose that the major work of his life would be suppressed, simply because of the author's criticism of official science. He was told several times that his book would do a lot of harm in the world.

In my view, the author has said that formal science has been unable to make a single significant finding on UFOs since the beginning of the modern UFO era in 1947. Not only is that a simple statement of fact, but the author draws attention to functional weaknesses in the formal approach. He shows how these problems may be surmounted. He provides objective evidence of the truth of what he says. Furthermore, the author is no longer alone in obtaining such evidence. Since *The Cosmic Pulse of Life* was originally published, groups of engineers and technicians in both Italy and Rumania who did not know Mr. Constable even existed, have obtained shoals of infrared photographs indistinguishable from those first obtained by the author in the 1950's and beyond. The correlation is shown in this revised and updated book.

Harm cannot possibly result from anything Mr. Constable has to say. On the contrary, a new beginning could result, and the author has blazed a trail—now widened by independent confirmatory work in Europe and elsewhere.

What the author has learned from his labors is passed on to us all, whether we are scientists or just plain citizens. From first hand, I know something of the terrible and even tragic price that this man has paid for what he freely and openly gives to his fellows. The idea that a work of this scope and significance could be kept from the world public was not acceptable to me back in 1975, and I therefore brought out the original Merlin Press edition of this book. Once again, classic status was achieved, and *Pulse* became known as "the underground classic."

Since the original publication, during which the entire first printing was destroyed at completion in a mysterious printing plant fire, I have become aware of the worldly machinations of the hidden powers described in this book. In my opinion, the author is correct when he writes of the subversion of evolution. He has linked certain categories of UFO intelligences directly to large-scale, real-world events. No other writer has had the courage, insight or practical experience to perceive these connections, which are cunningly masked. Nowhere is this more evident than in the systematic destruction of the industrial, financial, moral and spiritual fabric of the USA. Only an unthinking person could believe that such an immense and rapid disintegration is a random happening, rather than closely planned—an engineering job.

Whatever the ultimate reality of the UFO mystery, never can it reach mankind through an establishment steeped in duplicity, increasingly under psychic control, and terrified by truth and light. We are wise to listen instead to those individuals whose sacrifices in the service of truth qualify them to be heard.

Once again therefore, *The Cosmic Pulse of Life* appears, revised and enlarged with the latest information. Verified forcefully by the march of events, this book will inspire and help guide those who want a better tomorrow for mankind, and who will work and fight for that tomorrow. As I said in my foreword to the original 1975 version of this book, there is no fear on my part that anything between these covers is going to hurt the world.

Irwin Trent
Merlin Press

Special note: Irwin Trent has since passed on, in 2002. This newer version is even more updated than he mentions above. A debt of gratitude is owed to Mr. Trent. We gratefully acknowledge his original recognition of this book's value, and his courage to publish it.

PUBLISHER'S FOREWORD

This is one of the strangest, fascinating and most ground-breaking books one will ever find. Even today, Trevor James Constable remains ahead of his time with this pioneering work. As its current publisher I am confident that it has found a safe home, despite the continuing saga—or one could even call legendary events—that come with it.

There are those who have been threatened by this work in the past, refusing to publish it because one can be overwhelmed by the implications it presents. People become fearful when presented with evidence that clearly shows that an entire ocean of beings exist unseen around us. This time it is an ocean of invisible air, and many of the inhabitants are smarter than those found in our watery depths. Although many of these beings are peaceful and would prefer to be left alone, some of them may not have our better interests in mind. Whatever their true agenda, *anything* that can operate unseen around us has a tremendous and frightening advantage over our lives. We are at their complete mercy if they wish to exercise such an advantage.

Do we want to live our lives this way? Apparently so. People are naturally afraid of the unknown, so it is deemed better by our established, societal "gurus" to adhere to the "out of sight, out of mind" philosophy. If you cannot see it, it does not exist. Right?... Wrong! This book *shows you* that it does exist. *They* exist, in an unseen reality.

John Keel and others have echoed Charles Fort's cry that "we are property." The world may be in the mess that it's in because we have been manipulated all this time by unseen forces. Forces that prey off negative energies and suffering, as long as it can continue to be generated. Most of the people you know and will ever meet are kind-hearted, caring souls. So why is the world in such turmoil? Why do thousands of innocent children starve to death every day, along with all the other cruelties that we allow or inflict on others? Are we really that terrible as a race?

In some ways we are, but it may not be all our doing. Yes, this book really can cause trouble, but even more so for *them*. They want to remain hidden, but this book shines a spotlight on them for the very first time. As for us, and for the way we are approaching this evidence, it should be remembered that you cannot see the fruits of the valley without climbing the mountain. As a culture we must climb, or we will simply remain as "property".

As we move forward into the exploration of these other invisible life forms that share our planet, we do need to keep in mind the point of view of what could be the majority of these hidden ones. They may be peaceful just like us, and want to be left alone to live a natural, unencumbered life. They may wish to be hidden not so much for nefarious reasons, but for self-preservation alone. Anything humankind can see, it has killed, exploited or depleted to the point of extinction. We have raped the planet and they, on the other hand, may not want to be our next victims. Can you blame them? Champions of animal rights or anti-pollution could present much of this argument better than I could, but I am sure you get the point.

To those who will go out with their cameras as a result of this work and explore the infrared realms out of curiosity or for research, I applaud you. However, I ask that you be *respectful*, as well. If you harass these creatures there is a price to pay, as evidenced by some clear examples in this book. This carefully researched book is not an invitation for "fools to rush in". Again, I cannot stress strongly enough for you being respectful toward other life forms if you should choose to proceed. There is a balance in nature that should not be disrupted any further than it already is, so such work should be done by professional researchers or conscientious people who are acutely aware of the science and energies involved.

Before choosing to publish this book, the author warned me of the implications ahead of time, more than once. He explained the previous fire and complete destruction of the publishing company and the entire first printing of this book, followed by the explosion of a tanker near his home and the loss of more books when they were reprinted. The reader may think these events were mere coincidence, or paranoia on the part of the author. But until you come into contact with these entities or their energies on a daily basis, you should not judge. They are real, and they have intelligence. This book is important because it brings them out into the open for the very first time, and shows you how to observe them.

I told Trevor that yes, we would publish and that there was nothing to worry about. This time the books would be printed digitally and kept on disk out of state, so we'd be immune to these sorts of shenanigans. My father also lives out of state. His house caught fire yesterday, as of this writing, just as we were about to go to press. While his wife was out he was napping on the couch and woke up choking on smoke. Huge flames greeted him when he opened the door to the garage. He was able to put out most of the flames with a fire extinguisher, and was then helped by the fire department. If he had not been home, the house would be gone. If he had not awakened in time, he would be gone. They had to rip the ceiling out of the garage, where it had started, and many items were lost. The cause? The garage door opener had mysteriously burst into flames. By itself.

Do these events, going back to the other printings, form a pattern? Or is all this merely coincidence? You decide.

Paul Tice
Publisher
The Book Tree

PREFACE

Had official science risen to its responsibilities in connection with UFO phenomena—responsibilities that it owes to the human race—this book would never have been necessary.
There would have been no more need for me to research, investigate and write about UFOs than there is for me to write texts on medicine, electronics or chemistry. Science is professionally engaged in those areas.

History clearly records that official science *defaulted* in the UFO field by evading its responsibilities at the modern inception of UFO phenomena in the late 1940's. In the interim, official science has proved itself incapable of much beyond organized obscurantism and ridicule of those who sought to assume the responsibility for UFOs from which it has recoiled.

The book will show that UFOs are the death knell of the old order in science—mechanism—and that they simultaneously herald a new epoch in science and culture. In this coming era, much of the old knowledge that is now gospelized and crammed by force into unwilling young heads will be either unmasked as apocryphal or be subject to extensive review and revision. All disciplines of science will be revivified.

This coming, life-positive science will largely shape the future of mankind in accordance with living needs, giving man back his humanity. Mechanistic science deserves full respect for its vast achievements, and for the way it has clarified thinking. Claims that it is an adequate tool in its present form for cosmic investigation must, however, be regarded with unrelenting suspicion.

There is something fundamentally wrong with a mode of scientific cognition that can neither face nor cope with the irruption of the cosmic mainstream into human consciousness, leaving the field to amateurs like myself.

Official science in the UFO field has become a priesthood defending itself against cosmic truth, rather than a brotherhood of free men bound by their common pursuit of truth. A definite process of evasion has been operative in science ever since UFOs burst in upon mankind during and following the Second World War. Even the external, superficial aspects of UFOs have not been honorably or honestly tackled. There has been no penetrating investigation in the main direction in which UFO phenomena lead human inquiry by the very mode of their manifestation. That main direction is *into the invisible.*

In the supreme adventure of my life, I followed stumblingly in that direction. I found strange lifeforms and objects in a dynamic, unseen physical borderland from which scientific attention has been largely diverted. This book is an account of those adventures and concomitant findings, written as nontechnically as possible, so that any intelligent layman may follow me.

Forced to work on my own because I was rejected as "too far out" even within the UFO field, aided by only one faithful companion, financing everything personally and at every stage breaking new theoretical and experimental ground, this project has been relentlessly demanding. Inadequacies accordingly exist in my work that would be easily rectifiable with resources, time and assistance. Further delay in disseminating what has been learned nevertheless cannot be justified on that account. The discharge of this work without obligation to any outside source on this earth is offered as a compensating factor.

Today's truth is that the UFO subject is so staggeringly vast and complex as to defy full elucidation at this time. This book is no more than the end of the beginning. I believe its general outline will prove valid and enduring, especially my conviction that Dr. Wilhelm Reich's discovery of the orgone energy is the master technical key to the mystery.

Young, aggressive and truthful men and women in the science of the future will establish the full determinism of the strange biological phenomena that I have recorded. This will bring them both satisfaction and renown. I have neither the drive nor the resources to tackle such a project, for I am a generalist rather than a specialist, and an adventurer rather than a scientist. Within a relatively short time, however, energetic young people are going to plunge into this profound cosmic mystery in large numbers, and they will get more and better evidence than I have been able to get. The rational securing of the New Knowledge is the only way to end the Old Order, because we must have hold of the new before we may let go of the old. I wish these bright young people of tomorrow the best.

Because of my faith in the new humanity now coming on the Earth, I point to my pictures and say to those people of the past who perennially seek to block the New Knowledge:

"Here is what you have to investigate and deal with in your space age, if not now, then tomorrow. Here are some of the lifeforms that your grandchildren and their grandchildren will know all about, in a better world when you have left the Earth. This is my gift to my fellow humans—specimens brought back from the upper borderland of physical nature, by an adventurer who dared to do while you burrowed your heads in slaughterous filth. An ordinary man pulled these things out of space around you, while you reached out uncomprehendingly for the corpse of the moon. *Of course UFOs are alive!* In your deadness and sterility you cannot grasp that. Living unseen beings, living in an ocean of living energy; living unseen macrobacteria; etheric fauna feeding on living energy; and *constructs*—ships driven through space unseen by the living energy that fills all space. This life panorama drops sporadically into human sight nowadays by processes of etherian physics that we are just beginning to understand. When this happens, you can neither bear to look at it, nor can you comprehend it, because it goes to the bioenergetic roots of your own existence, turning you into irrational and frightened primitives."

Had it not been for four outstanding human beings, I might well have been among the millions who have no option but to accept the jabberwocky dispensed by official science on the subject of UFOs. These four humans contributed decisively to my personal development, and to my understanding of cosmic processes as they present themselves in UFO phenomena. Each of these four avatars entered my life in turn and at the right time, and without them, this book could not have been written.

Dr. Franklin Thomas, an optometrist, Doctor of Divinity, master printer and publisher, was also an Adept and Master Teacher. He is responsible for setting my feet on the pathway that led to this book.

Dr. Rudolf Steiner, Austrian philosopher and founder of the Anthroposophical Society, provided the human race with the spiritual-scientific conceptions necessary for the renewal of human culture, and for a healthy, foolproof understanding of the laws and forces operative in such cosmic phenomena as UFOs. Directed to Steiner by Franklin Thomas, I found my whole life changed as a result.

Dr. Ruth B. Drown was a pioneer inventor and New Age physician whom I was privileged to serve and love in her final years. She provided the fundamental techniques for manipulating the cosmic life energy, more commonly known as orgone energy, in medical diagnosis, therapy and medical photography, the latter self-validating.

Dr. Drown was also responsible for the first application of such radionic knowledge to the UFO field, where its implications are endless. She was unmercifully smeared, persecuted and pilloried by uncomprehending contemporaries. Her work is central to the forthcoming cosmic communications break-in.

Dr. Wilhelm Reich, distinguished protégé of Sigmund Freud and a pioneer in psychoanalysis and scientific sexology, was the discoverer in the 20th century of the cosmic life energy or orgone energy. This discovery will someday be recognized as the most important technical discovery in the history of mankind. Dr. Reich's discovery at last permits otherwise impenetrable UFO phenomena to be understood as fundamentally *bioenergetic* manifestations. He died in a federal prison in 1957, and U.S. agents under court orders put his scientific works to the torch in a spectacle seen nowhere in the world since Nazi times. I freely acknowledge the magnitude of my debt to Wilhelm Reich, the most gifted natural scientist of the 20th century and one of humanity's greatest benefactors.

I stood on the shoulders of these four avatars in sketching my outline of the greatest mystery of all time. Without them I would have had no adequate perspective, no insight to the New Knowledge, and no means of charting my course.

To all who may find it incredible that an ordinary man could come in under the learned nose of official science and do what is herein recounted, I can say only that my four Titans were with me all the way. They taught me that the cosmos speaks to him who asks and seeks lawfully and truthfully, and never to him who turns away. As to how it was possible to do all these things, the answer lies in my activism, which at all times overwhelmed quibbles concerning classical methodology that is already obsolescent.

I have been able to present this work because I am for Life and Love, and reached out into the cosmos with that in mind and heart. I expected to find the *pulse of life*—and I did. So will all others who do likewise.

TREVOR JAMES CONSTABLE
San Pedro, California 2002

PREFACE TO FOURTH EDITION

This book has a checkered history, beginning with its rejection by all the publishers who had done well with my books on ace fighter pilots. One publisher said I would "hurt the world" with what was in the book. Eventually, rejected everywhere, the decision to self-publish was made, and my dearest friend, the late Irv Trent, came up with the funds. The printer of the book in Santa Fe Springs, California, was warned to take every possible precaution against sabotage, a warning he took lightly.

Six weeks later, the book had been printed and sat in the printer's warehouse awaiting pickup the next day to go to the bindery. This was the last step before the public got the book. By dawn the next day, the first printing of the book was ash, the black residuum of a mysterious printing plant fire which gutted the place. Back to square one.

Books of the #2 printing were stored on the seaward wall of my apartment garage in San Pedro, overlooking Los Angeles harbor. While I was absent on sea duty, the tanker Sansinena blew up about 700 yards from my place, heavily damaging many buildings and homes in San Pedro, including mine. The books were scattered all over the garage, and we lost a lot of them.

All in all, we had a heck of a ride with these books, the original Merlin Press edition. Soon widely known via the underground, its main "distribution" outlet, the Merlin Press edition eventually sold out. Again, we were back on square one, in need of a publisher. None came with the gumption to risk putting the book before the public. All of them began well, but on actually reading the book, declined to publish *Cosmic Pulse*. A strong man was required.

Tom Brown, then Director of Borderland Sciences Research Foundation, stepped up to the plate and produced an updated 1990 edition of *Cosmic Pulse*. That is essentially this book, to which supporting, independent photographs have been added from two other countries. The late Luciano Boccone in Genoa, Italy, and Florin Georghitza in Romania agree essentially with my interpretation of the UFOs, even though they had never heard of me or my work when pioneering their own. This circumstance proves that significant advances in human knowledge usually occur simultaneously, at places and in cultures well separated from each other.

When Mr. Paul Tice of The Book Tree in San Diego undertook to publish this book, we discussed whether we should issue a complete rewrite of *Cosmic Pulse*. Certainly it would be justified. We nevertheless decided against it, because so much that is in the book has come within reach of public recognition and acceptance. That was not true with earlier editions.

In this connection, today's UFO enthusiast cannot consider himself fully informed unless he has read and studied the late Colonel Philip Corso's brilliant effort *The Day After Roswell*. This book is essentially a U.S. government confessional on the UFO subject. The reasons for government silence, restrictions on information and the plain fact that the UFOs are *hostile* are all covered. They coincide basically with what I have presented in *The Cosmic Pulse of Life*, which you have in your hands. Originally advanced by me in a most stumbling fashion in my 1958 opus, *They Live in the Sky*, those basic ideas got me banned as a lecturer on UFOs, for ten years or more, while the world public slowly grew into those same ideas.

Colonel Corso tells you in his book of his adventures in farming out UFO devices to U.S. industrial giants. This was material retrieved by the Army from a UFO crash near Roswell, N.M. in 1947. Exciting and enlightening information, without which you are ill-informed on UFOs until reading it. Corso was in charge of the U.S. Army's Foreign Technology Division at the Pentagon. Colonel Corso was assistant to Lt. General Arthur Trudeau, Chief of U.S. Army Research and Development. I recommend Corso's book because he, like myself, was unafraid to take a leap into an area deemed "questionable"—yet, over time, the truth does come out.

The Author
February, 2008

Chapter One

THE GREAT IMPASSE

There is no proof. There are no authorities whatever. No president, academy, court of law, congress or senate on this earth has the power or the knowledge to decide what will be the knowledge of tomorrow.
—WILHELM REICH

Mankind has been on a collision course with other-world intelligences ever since the Wright Brothers lifted clear of the earth, shortly after the turn of the 20th century. Science and technology moved farther and faster between Kittyhawk and Hiroshima than in all the rest of human history combined. During this brief, forty-year tick of the evolutionary time clock, civilization developed a technical cleverness beyond the prior conception even of visionaries, but in the more vital art of understanding himself, man made no comparable progress.

Beneath man's glittering technical veneer, and behind his social facade, there still throbs the black force of barbarism. Scientist and savage are separated by a paper-thin veil that is easily punctured. Atomic energy as the major technical achievement of the 20th century was unlocked for the specific purpose of incinerating scores of thousands of innocents. The Second World War, and the mechanized slaughter of millions—including women, children and the aged—verified at mid-century man's parlous inadequacy in handling himself, knowing himself and facing himself.

At this point in history, when men with the rockets born of war talked of conquering space and other planets, initial encounters with nonhuman intelligences were already occurring. The latter years of the war are rich in pioneer UFO observations by both visual and electronic means. Observations of UFO phenomena multiplied all over the world during the ensuing quarter century, but mechanistic science could contribute *not one significant discovery* pertaining to these objects. What was official science actually doing during this unprecedented period, when technical progress continued pell-mell, and man was openly preparing for the conquest of space?

Between 1946 and the present time, billions of dollars have been literally *dumped* into thousands of alleged "research" projects, undertaken by official science and its minions. Professors and doctors and chemists, and universities whose scientific resources dwarf those of whole nations in bygone times, have been investigating everything from the fat layering of female Korean divers to the sex life of Arkansas wart hogs. Federal information storage facilities are bulging with this monstrous pedantry, this fantastic plethora of inconsequential minutiae, palmed off as "science" at the dawn of the Cosmic Age. Official science would look anywhere, at anything no matter how ridiculous or insignificant, provided it could look away from pressing cosmic phenomena.

Amid this evasive scientism, UFOs have been manifesting throughout the world, successfully resisting all official and sub rosa efforts by mechanistic science to penetrate their secrets. Open-minded individual scientists who have tackled UFOs avocationally have been frustrated in their efforts to establish any deterministic guidelines for these multiform phenomena, *without departing from the officially sanctioned basis of their disciplines.*

Mechanistic science has suffered an unmitigated defeat on the UFO problem, even as Mechanism reached its peak in human influence. The history of science may be searched in vain for a comparable example of an established system of cognition reaching such a total impasse. Official science is bankrupt on UFOs—methodologically, ethically and emotionally bankrupt. We are at the end of an era. For his Space Age, man must have a new thinking— a *new mode of mentation*—and he must have the New Knowledge, a small portion of which is being related in this book to the crucial and otherwise insolvable problems created by the advent of the UFOs.

All who have enshrined Mechanism as the foundation of their world conception, in or out of science, can maintain their illusion of emotional security in these times only by retreating within the armored structure of Mechanism for a neurotic counterpart of Custer's Last Stand. The handwriting is on the wall. Cosmic sights and sounds are pouring in on man from space, as the high priests of official science struggle to shut them out, to evade and avoid these manifestations.

This evasion is responsible for the staggering resources available in America today for mechanistic scientific ventures devoid of cosmic implications, while comprehensive, honest and honorable investigation of UFOs is economically throttled by a frightened Establishment, whose power is under attack at its roots. UFOs have brought with them a new source of energy, a new mode of propulsion and the conquest of gravity. In this book we shall penetrate the basic determinism of this energy and demonstrate why it is that the social and scientific reaction to its discovery and development has been irrational, repressive and antagonistic. The energy cannot be put into a wire or a tank and *sold.*

For a period of over sixty years, the majority of the world's scientists have avoided the clear responsibility they bear to the human race on the UFO question. All too many have used their professional standing and prestige to ridicule the work of those who were forced, by the conceptual paralysis of mechanistic science, to investigate UFOs on their own initiative. This anti-knowledge attitude came to full expression in the so-called Condon Report, spawned by Colorado University.

Space around the earth has been ignored. Strange aeroforms inexplicable to the technology we think of as advanced continue to manifest, and their implications for world thought and human knowledge have been deliberately evaded. Hope is endemic among classical mechanists that UFOs will just go away. Death-oriented mechanistic science reaches out instead for the corpse of the moon because the character structure of its devotees is anti-life. Every manifestation connected with UFOs, when submitted to authority of this kind goes through the same process: the life is crushed out of it. So-called "scientific controls" are widely misused to this end, and already they are old history.

The discovery of the orgone energy[1] by Dr. Wilhelm Reich, the energy that will be shown in this book to be the technical and functional principle behind UFOs, invalidates and nullifies all mechanistic conceptions of control that are not predicated upon its presence. That presence is universal and demonstrable. The orgone energy is the overriding new factor in science.

"Authority" in science has not yet recognized the discovery of the orgone energy. "Authority" in science today also does not stand for the free and honest investigation of these strange, transcendental and ubiquitous phenomena subsumed under the term UFOs. "Authority" stands for a sinister, modern obscurantism that is little different from that imposed on struggling mankind by generations of vicious theocrats. Science once had to fight clear of the same kind of reactionary control that its high priests are now using to defend their own world conception.

The Great Impasse in which mechanistic science finds itself arises from the irreconcilability between *living phenomena* and a mode of cognition anchored in sterility and deadness. Science is in danger of losing its ideals of serving life and raising knowledge to ever higher levels because of the control exercised over its reception systems by life-hating, life-killing individuals. They rule what is accepted. Their characteristic mode of reacting to living phenomena—their terror of life and movement—make up the psychic anatomy of the Great Impasse. A powerful effort by younger, more vitally alive men and women will be needed to rescue science from their dominance.

The scope and reality of the Great Impasse can be demonstrated readily by reviewing the unique nexus of events amid which UFOs first impinged on mankind in the modern period, together with the resultant riddles. The most important factor attending the modern advent of UFOs is the development of radar. Mechanistic science has been unable thus far to face the devastating UFO evidence that this instrument has provided.

During the Second World War, radar was developed more rapidly and to a higher degree of refinement than any other electrical device in history. Prior to the war, radar was essentially experimental, but in the six years of 1939-45 the technique was perfected, and by war's end scores of thousands of radar sets were functioning on land, aboard ships and in aircraft. The period of 1939-45 is by evolutionary reckoning no more than an instant.

At this instant, planet Earth suddenly became a veritable electromagnetic hedgehog.

If seen from outer space with eyes sensitive to microwave radiation, the earth would appear to suddenly produce thousands upon thousands of jets of high-powered energy—lancing, blasting, squirting into an ether heretofore tranquil at those wavelengths. This brand new fact of earth life coincides historically with the modern advent of UFOs. Thus far, nobody appears able to discern its implications correctly.

This radiation, sudden and massive, differed from the prior electrical works of man in several fundamental ways. Radars radiated electromagnetic energy at the unprecedented power levels made possible by the development of the cavity magnetron. This British invention became the key to modern, high-powered radar, which began its role in terrestrial existence barely half a century ago.

PLANET EARTH—PRE-RADAR

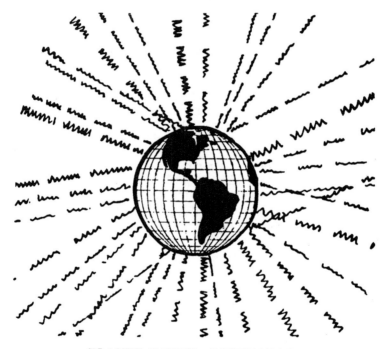

PLANET EARTH—POST-RADAR

Entire planetary environment permeated with pulsed
electromagnetic radiation—an unprecedented condition.

Radar also uses microwaves, a portion of the electromagnetic spectrum not previously utilized in this fashion, or for that matter, in any widespread way, prior to the Second World War. Most radars work in a band of frequencies lying between those used for television broadcasting and those of visible light as perceived by the unaided human eye. Frequencies used for radar adjoin those of the *infrared,* the electromagnetic radiation lying before the visible red and invisible to the unaided human eye. Every UFO photograph I have ever made utilized the infrared in some way.

The most important new aspect to the development of radar—from the point of view of understanding UFOs—was that this radar energy was *pulsed,* chopped into hundreds of short bursts each second. Scientists in the medical field have long known that chopping or pulsing electrical energies used for tissue stimulation increases their effects. In the portion of the electromagnetic spectrum used for radar, modern spectrum engineering has already determined that electromagnetic energy directly affects the human body through tissue penetration. Heating of the skin and nerve endings is created in this way, together with eye damage and coagulation of body proteins.

There is thus an *interface* between man's microwave works and his own organism, as well as with the organisms of animals and insects. Coagulation of proteins is one known effect. At least 50 years of systematic work will be required to understand the full extent and implications of this interface between man's use of the spectrum and natural phenomena—especially biological phenomena.[2]

The sudden, widespread pulsed-wave activity that commenced with wartime radar did not take place in a vacuum, but in the ether. Rational scientists postulate an ether because there cannot be waves without a medium for them to wave in. Amid the triumphs of modern physics and electronics, we do well to observe that the ether itself and its qualities and properties— or even its very existence for that matter—continue to bemuse the scientific fraternity. Numerous scientists are conceptually and philosophically unhappy with Professor Einstein's etherless universe, for example.

Official science being largely supportive of lifeless and sterile conceptions wherever possible, the tendency therein is to endow the ether with stationary characteristics. Physicists like the late Carl F. Krafft, who saw the ether dynamically and differently, have been denied recognition. Official scientific sequestration of his book forced Krafft to publish his theories himself, which he did in *The Ether and its Vortices.* This rare title has been reprinted by Borderland Sciences Research Foundation. Essays from Krafft's book were published in the appendix to W. Gordon Allen's *Spacecraft From Beyond Three Dimensions* (Exposition Press, Jericho, N.Y. 1959). Those willing to concede that there may be radical advances in 21st century science should acquaint themselves with Krafft's little known work.

The rapid and prolific development of radar disturbed the ether as it had never been disturbed before in the terrestrial environment. This monumental fact has been ignored in ufology, and not connected to the concomitant advent of UFOs. Repeatedly we shall be forced to return to this signal historical event as our investigations lead us to the nature of the ether. We shall find in due course that the ether itself has been discovered as a physical natural force and presence, and named *orgone energy* by Dr. Wilhelm Reich.

A working knowledge of radar is indispensable in ufology. Most laymen still do not know how a radar set operates. Many scientists qualified in non-electrical disciplines also lack this understanding. Let us quickly describe how a radar set works.

A radar functions by sending out hundreds of mighty bursts of electro-magnetic, microwave energy every second. These bursts of energy strike and are reflected from objects remote from the radar set. The amount of energy reflected is exceedingly small. The "echo" that returns to the radar set is thus only an infinitesimal portion of the energy originally radiated. The time that the pulse of energy requires to travel from the radar antenna to the target and back—at 186,000 miles per second—is accurately measured. This gives the range of the target. The actual physical location of the target relative to the radar antenna is displayed on cathode ray screens in various orientations. From these screens, data such as the position, speed, altitude and course of the target may be read directly.

Since only a miniscule fraction of the radiated energy is reflected back to the radar set to be measured and displayed, enormous radiated power is required. Many thousands of times more power is needed to detect an aircraft by radar than is needed to talk to the same aircraft by radiotelephone. Only a few broadcasting stations are licensed today for 50,000 watts, but such a power output is common in military and naval radars.

Radar differs from radio in the way in which the energy is radiated into the ether. Radio signals are focused into narrow beams only in specialized communications, but radar depends upon a focused beam of energy. The radiation of the energy from a radar antenna becomes like a long, electro-magnetic finger, which men are able to stick up into the air or extend across the surface of the sea. With this finger they are able to *feel electronically for objects that they cannot see with their eyes.* The electronic finger penetrates fog, mist and cloud. Radar is thus at once an extension of man's sight and an extension of his sense of touch.

Developed primarily for long range detection of enemy aircraft, the crude radars of 1939-40 were quickly superseded by units of ever-increasing complexity, efficiency and capability. When these sensitive electronic feelers were directed into the air, they detected aircraft with great precision over previously unattainable distances. The military learned to control the aerial defense of large territories with the aid of radar.

All this new-type "feeling around", however, led to the detection of other things—not aircraft—and in the process landed official science in a highly embarrassing jackpot. To this day, science has been unable either to escape from, or to conceptually confront these strange "other things" that were not—and are not—aircraft. Many of the objects detected with this all-new instrument proved to be *invisible to the eye even when within known visual range.*

In his book *Flying Saucer Conspiracy,* published by Henry Holt and Company of New York, Major Donald Keyhoe recounted what is probably the most significant of all wartime radar detections of UFOs. There were between 200 and 300 objects involved. All of them were *ocularly invisible.* Called the Nansei-shoto case, because of its occurrence near the islands south of Okinawa, the event is one instance worth a thousand. The essential elements are sufficient for us here.

The story was given to Major Keyhoe by a U. S. Navy officer, pseudonymously identified as James Dawson, who had served as a Combat Information Center Officer aboard an aircraft carrier during the incident. Patrolling the area south of Okinawa in company with another aircraft carrier, Dawson's ship detected with radar a huge force of enemy aircraft approaching from the northeast.

Initial radar contact was 120 miles. The "blip" or echo returning from the incoming force was very large, supporting an estimate of 200 to 300 aircraft. At 100 miles range, their speed was determined to be 650 knots—nearly 700 mph. No known aircraft in the world at that time could attain such a speed. At eighty miles and at an altitude of 12,000 feet, the incoming force of aircraft began to spread out in two formations from the main body, as though preparing to assault the whole American task force.

Only twelve aircraft were available as air cover for the U.S. ships, all other American aircraft being absent on strikes against Japanese positions in the Ryukyus. These twelve fighters were scrambled and vectored by Dawson towards the attacking force. In bright weather, with only scattered clouds at 5,000 feet, the American fighter pilots enjoyed unrestricted visibility. At 15,000 feet on intercept they could see fifty miles *but they never did see the incoming 200 to 300 enemy aircraft.*

Directed accurately from the carrier to intercept the "enemy", the veteran Navy fighter pilots could not see the attacking force even when directly above them, as seen on the carrier radar. The "enemy" nevertheless kept coming towards the task force, now at general quarters for the impending attack. Dawson warned the bridge by telephone that the enemy was almost upon them. The skipper told him he was crazy, that there was nothing to be seen in the sky. Radar still showed the horde of attackers. The incredulous Dawson went on deck himself and saw—*nothing.*

The fighter director on the other carrier had also picked up the enemy machines on his radar. According to Major Keyhoe, the U.S. Navy acknowledged after the war that there had been this inexplicable "ghost of Nansei-shoto." There have been numerous neurotic attempts to verbalize away this significant incident, although there have been numerous subsequent incidents of similar character that have not been publicized.

The facts to be faced are:

1. UFOs were detected at speeds exceeding those attainable by enemy aircraft in that period and theater of war.

2. There was simultaneous detection by two independently operating radars aboard two different ships.

3. UFOs were invisible to twelve experienced Navy pilots at the peak of their powers, as well as to the shipboard personnel, all of them being trained to observe and deal with attack from the air. All of them were in a position to make a visual sighting, but were unable to do so.

When man stuck his pulsing electronic finger into the heavens, he was not prepared either to find, or to accept, *anything other than what he was specif-*

ically seeking. Revolution was upon him. The instruments fashioned by his own cleverness had led him to an impasse from which only a new mode of thinking would provide release.

Similar cases to the Nansei-shoto incident have occurred all over the world since. The stratagem of pigeonholing such unwelcome, unwanted and mechanistically incomprehensible evidence, has failed. The manifestations continue as this is written. Official science is conceptually strapped. Having devised an exquisite instrument to locate invisible airborne things, surely it is downright queer that men of science recoil from that instrument's evidential yield.

Radar designers did not anticipate that their devices would detect objects that were *completely invisible*. The Great Impasse has developed its bulk and immobility out of this fact. No mechanist will ever get around that impasse with the old thinking. New mental resources are needed to deal with what is *demonstrably supersensible, yet capable of firm objectification*.

Radar is telling people of science now, as it has told them times without number since the war—pressed down and running over—that there are at work in the atmosphere of the earth and beyond, invisible objects that are nevertheless physical. Giving them names such as "propagation anomalies" does nothing but divert attention from the matter. Calling them "angels" provides only uneasy chuckles, unworthy of men of truth.

Evidence of the invisible reality around us points to the collapse of the mechanico-mystical conceptions of the cosmos, the neurotic *weltanschauung* that is the emotional underpinning of both official science and organized religion. The choice for man is not between finding a new way or hanging on to the old, but one of how to find the new way in full health and clarity of mind. *Invisible yet physical!* Whether or not you can live with such a concept concerning UFOs, and accept its consequences for your own world view, will largely determine your posture in the Space Age.

There is an old, moribund and fossilized order in science—the strictly mechanistic priesthood that has won great benefits for mankind—and the new scientists who pursue scientific truth wherever it leads. The old order still winds the clock, and will continue to do so for perhaps another fifty years, or longer. Then the new people will leave behind the small behavior of the old timers toward the great drama of cosmic commerce that is now opening, and renew the earth through the immense catharsis of that drama.

The high priests of official science cannot be converted. The task is comparable to converting a wheelchair arthritic into a two-rounds-a-day golfer, or to getting people under thirty years of age into the U.S. Congress. The rigid older minds will rule for the time being. Under their political and scientific suzerainty, opinions inimical to further investigation of UFOs in their invisible aspects can be purchased from official science by a doomed establishment. This process is no more difficult than purchasing potatoes at a supermarket, just considerably more elaborate and costly.

The analysis of new facts with old thinking, or with underdeveloped thinking, has caused the Great Impasse to extend itself even into ufology. UFO literature cites abundant cases of invisibility, and even more instances of objects making *transitions* from visibility into invisibility. Dutifully recorded, these dynamic happenings and their implications have been largely evaded by ufology. Invisible UFOs are seemingly a source of disquiet and even annoyance to most ufologists bent upon mechanistic theories and their proof.

Attention is directed preferentially upon cases where objects deemed to be craft have been observed with the naked eye in aerial maneuver or in landings. Objects of this type do not stretch the rigidized mind farther than from A to B. Mechanical concepts compatible with mechanistic science can be comfortably applied, albeit indecisively, to objects that appear to be "solid". *Emotional security*—which is threatened and jeopardized by anything implying intradimensionality or lack of fixity within the physical sphere of observation—is thereby maintained.

Conceiving of beings on other planets as building spaceships and flying across space to Earth does not disturb the mechanist emotionally, because such thinking is linear, i.e., it goes in a straight line. Objects that are physical but invisible have, by contrast, enormous power to disturb. Routine, linear thinking cannot cope with such radical concepts because they are destructive of the mechanistic rubric. The living, functional elements immanent everywhere in UFO phenomena elude the linear mind, and cannot be handled by engineering principles, since they are biological and bioenergetic. Empirical approaches to UFOs based on Life instead of sterility, automatically violate mechanistic method and outrage the linear-minded mechanist. Ships-from-other-planets is the limit of his tolerance.

Only one firm instance of a phenomenon is sufficient to justify scientific investigation. Firm instances of radar sightings unconfirmed by visual means, when such confirmation should have been obtainable under known laws of physics and optics, number into the hundreds. Involving service aircraft and highly-trained personnel, they are high-quality observations but are suppressed by Department of Defense security.

Interceptions have been made by radar-equipped fighters, and the interceptions confirmed from ground radar. While in some cases UFOs have been machine-gunned and piece shot off for later recovery and analysis,[3] the crucial cases are those where the intercepting aircraft finds *nothing* in the air at the point of interception and is seen by ground radar to *fly right through an apparently massive airborne object*. These events have taken place repeatedly, at the dawn of the Space Age, while the aerospace industry has been plundering the treasury for stupendous amounts of money for *space research*.

Whatever it is in space around us, picked up on radars, sighted by our best pilots, photographed by our astronauts, and easily outperforming the most advanced products of earthly engineering, has never been considered worthy of thorough investigation. This irrational attitude condemns itself. The Croesus-rich American research foundations, swollen with tax-sheltered funds, the U.S. Government, our great universities and the aerospace industry have all avoided investigating propulsion of this type in favor of juvenile fireworks for adults at Cape Kennedy. The ultimate in mechanistic crudity and mind-numbing complexity, these operations consume billions of dollars and delude us into thinking we are smart.

That large numbers of UFOs are invisible is a fact beyond rational dispute, and it is the terrors attendant on this fact that have provoked the irrational in man. One of the few certainties in ufology is that mankind became aware of these invisible objects only with the advent of radar. From this it may be fairly inferred that *such objects have always been present—unseen—in the terrestrial environment.*

A high probability exists that these objects are an integral part of the life of the earth. That is why they must be faced in full scientific honesty, and their determinism established. Mechanistic science always casts its evolutionary vote for accident and coincidence, rather than letting phenomena speak in their own way to modern consciousness. The realm of the microbe existed for scores of millennia as a condition of terrestrial life and as integral with and essential to that life as the sun. Microbes only entered human consciousness when the microscope removed man's optical limitation.

Radar has performed a similar function macrotellurically, extending man's sense of touch into an adjacent yet invisible and seemingly interpenetrating realm. Radar has opened to human consciousness the signal possibility of transitions from one state of tangibility to another, supported by those numerous UFO sightings where visible objects literally disappear while in full view. These cases also have involved highly-trained, experienced personnel using modern equipment.

A classic encounter of this type took place in March 1953, when UFOs were detected by radar over Detroit, Michigan. The case involved USAF F-94B jets under the command of Lt. Col. Howard C. Strand. This officer's story, as told to UFO investigators Brad Steiger and Joan Whritenour, was published in *Saga* magazine in August, 1968:

"In February of this year, the authors were in Chicago to tape the pilot film of a new television series, *UFOs—Here* and *Now*. At this time it was our pleasure to meet and spend some time with Lt. Col. Howard C. Strand, Base Commander of the Detroit Air National Guard. Lt. Col. Strand has over 7,000 hours of military flying time, more than half in jets. Strand is an honest, straight-from-the-shoulder military man. He is soft-spoken, a gentleman-officer of the old school. He most certainly is not the sort of man to fabricate a story to bring attention to himself.

"On a clear spring day in 1953, Lt. Col. Strand encountered a number of UFOs while flying over Detroit. At that time, he was on active duty in the Air Force, flying F94-B aircraft, and was stationed at Selfridge AFB, Michigan. He had not been a 'believer' in flying saucers prior to that sighting, and even today he devotes no time to ufology, other than to do selective reading on the subject. Lt. Col. Strand has only had that single experience in 1953, but it is a particularly impressive sighting made by a highly qualified observer. Here is Lt. Col. Strand's story:

"Approximately 10 a.m. one morning in March 1953, I was scrambled on a routine patrol mission. We were expecting the Navy to try and penetrate our air defenses in the local area for practice purposes. After about 20 minutes of flight, the radar site controlling our flight gave us a target to our left at about eight o'clock position. Upon visual checking, my airborne radar operator and I could see tiny specks in the sky which appeared as a ragged formation of aircraft. Our position at the time was approximately three miles northwest of downtown Detroit. The targets appeared to be over the city's central section.

"The objects were a little lower than our aircraft so we were in a slight downhill run at full military power, without afterburners, on the intercept. I can recall thinking more than once I should be able to start identifying the aircraft any second—but couldn't. Their tails, wings and aircraft features just didn't seem to 'pop out' as they normally do when you close in on an aircraft to identify its type.

"All the while we were on a quartering head-on intercept, my radar operator in the back seat was trying to pick up the targets on our airborne radar. The ground radar had both our aircraft and the unknowns painted as good, strong targets, but we were still unable to get any positive identification, and the objects seemed to be getting a little larger all the time.

"About this time, the radar operator in the back seat started receiving some returns on his scope and thought he was picking up the targets. I was watching the objects until I looked in the cockpit, trying to inch out a little more speed without going into afterburner. When I looked up again—after no more than two to four seconds—*the objects were gone.*

"I had estimated the number of UFOs to be between 12 and 16. We had been expecting to find and identify Navy fightertype aircraft. But now, nothing. *Every one of the objects had disappeared from sight.*

"Immediately I asked the ground radar controller where they were and he told us the targets were still there—loud and clear. We continued to fly the headings given by the controller, right into the center of the targets. We flew and turned in every direction, but there was still nothing in sight. Gradually the targets disappeared from ground radar after we had been amongst them for three or four minutes, as close as 2,000 feet according to radar. Our airborne radar had picked up nothing after the initial fleeting contact before the objects had disappeared from visual sight.

"No UFO report was submitted by the aircrew for one reason. This was the era when it seemed the Air Force was denying even the possibility of UFOs and was attempting to make everyone who thought that they had seen such objects look silly or stupid. In retrospect, I have personally come to two conclusions about my sighting.

"Number one: that I could not identify the objects as aircraft, because they weren't—there were no wings or tails to pop into sight for identification as aircraft. At that time I had no thoughts of flying saucers; therefore, I made no efforts to identify them as such. If I had even so much as thought of it at the time, I never would have taken my eyes off them.

"I can say definitely that the objects were not conventional or jet aircraft, due to the fact that no aircraft could have turned around or 'gotten away' so to speak, in the two to four seconds I was looking in the aircraft cockpit. Remember, all the while we were bearing down on the objects at approximately 500 mph in a quartering head-on pass.

"Number two: that the objects went straight up, out of sight to me and my airborne radar operator, but still visible as targets on the ground radar. Other sightings have been made where UFOs have gone straight up for tens or hundreds of thousands of feet in one or two seconds, then hovered or moved slowly at the new altitude. At the time of the sighting, I had 1700 hours flying time, accrued in nine years. Today I still feel the sighting on that perfectly clear day in 1953 was valid, that it was no figment of the imagination or trick of the eyesight. I have had no other sightings since that time."

Serious thinkers on the UFO subject owe Lt. Col. Strand a debt for making his experience available for study. The case is one that is worth ten million words of evasive jabberwocky from official science or twenty-five years of intellectualizing. Let us hope that the high priests do not force Colonel Strand to recant facts stated so cleanly and clearly. Those facts must be faced.

Man requires for the period of human development now opening a mode of mentation that will permit his thinking to follow, with ease and in total harmony with his cosmic origin, transitions of substance from one level of tangibility to another. Such a transition obviously occurred in this Detroit case. Why cannot man follow such transitions?

Man lives inside an *armored organism*, a rigid structure intended to protect him from painful stimuli and also from his own natural organ sensations. In his musculature this armor is a demonstrable biophysical reality. Half-strangled by his armor, the somatic anchorings of his neurosis, armored man is much less than half a man. He is 90 percent inefficient. How then does this situation determine his relationship to UFOs, or rather, his inability to relate to UFOs?

Mankind in the mass, and the overwhelming majority of individuals, cannot at the present time tolerate the organismic movement of biological energy, because of the armoring process. The armor stops bioenergetic movement, and throttles or diverts organ sensations, including those connected with vision. UFOs have their technical roots in the same biological energy—orgone energy or chemical ether—that is trying to move in man and move man forward. UFOs as a cosmic manifestation are already cracking through the armor, and with it the mass neurosis and all its retardative social manifestations.

Man's intolerance of bioenergetic movement is exemplified in his intolerance of *firm UFO evidence that is biological and bioenergetic*. What is alive in UFOs stimulates through resonance and correspondence that which is alive in man. He cannot stand this movement. His own biological energy, with its prime bodily expression in the sexual processes, is inhibited by socio-religious taboos imposed for thousands of years. These taboos have hamstrung and blocked man at a time when his technical, electrical cleverness has led him to cut across—and to become aware of—a heretofore unsuspected biological realm. Man must therefore face *himself* before he can face intelligently the cosmic beings upon whose dimensions he has begun to impinge technically with ever-increasing intensity.

Man's armoring against life, and against the movement of his own life energy, is responsible for the process of evasion that has characterized official science in its attitudes toward UFOs. Evasion of facts intimately interwoven with the cosmic life energy—the energy that is central to the UFO mystery—is neurotic and anti-Life. The unscientific evasion of radar evidence demonstrates its irrationality. The UFO "problem" lies not in a lack of hard data, but in the orientation of armored man away from anything that leads to the roots of his own existence. That is where UFOs, understood and approached aright, inevitably lead.

To Wilhelm Reich we owe the elucidation of all this irrationalism. Understanding his discovery of the orgone energy brings automatic understanding that one may follow UFOs only functionally, that is, in accordance with living processes and their continual dynamic transitions. Fixed, anchored, rigidized conceptions are worthless in any attempt to understand the dynamic diversity of UFOs. These strange flying things have to be permitted to speak to us in their own way, and their language for now is the mode of their manifestation.

   Functional thinking and functional methodology are required before we can begin to understand UFOs and all that they portend for mankind. A comprehensive UFO theory stands outside the boundaries of what is attainable by mechanistic science, no matter what resources are employed. Cosmic dynamics and mechanistic rigidity are incompatible, and always will be.

   Mechanistic science has produced incredible wonders and benefits for the human race, but it is an inadequate tool for cosmic investigation. So is the mind sterilized by its lifeless conceptions. There is no way that such minds can comprehend or cope with the cosmic dynamics that UFOs now thrust at us from space. The task is for a new generation of free, functionally-minded, vitally alive young men and women who will be pro-Life when they approach UFOs. They have the task, as this book will show in due course, of snatching Earth evolution literally from the clutches of the Devil, of surmounting the Great Impasse.

NOTES TO CHAPTER ONE

1.Orgone energy is known to modern etheric science as "chemical ether".
2. The U. S. Joint Technical Advisory Committee 63.l.4 provided data for the Frequency Spectrum Chart published by North American Rockwell in 1968 that delineates some of these interfaces.
3. For an authoritative account of these happenings, read the interview with Canadian scientist Wilbert Smith reported in Frank Edwards's *Flying Saucers: Serious Business,* pages 48-50 in the Bantam Books edition. Head of the Canadian Government's "Project Magnet" to study UFOs, Mr. Smith was loaned a fragment shot from a UFO by an American aircraft, and the loan was from an agency of the U.S. Government. Mr. Smith so states in the interview. Washington, named for him who could not tell a lie, has become one vast lie factory, created by public default.

Chapter Two

DIMENSIONS UNMEASURABLE

A man who has bought a theory will fight a vigorous rearguard action against the facts.
— JOSEPH ALSOP

Mechanistically-minded humans have accepted uncritically the theory of interplanetary spaceships as the fundamental explanation of UFOs. This theory has dominated the subject from the modern advent of the phenomena down to the present day. Involving only a linear projection of extant earthly technology, and doing no violence to the mechanistic cosmo-conception, this simplistic theory has paralyzed the thought processes of several generations of human beings interested in UFOs.

Acceptable as one theory relevant to *certain types* of UFOs, the ships-from-other-planets approach was elevated irrationally to the status of a foregone conclusion for explaining *every* UFO sighting. The bankruptcy of official science in the empirical phase of the UFO field is due to its having sought to prove this foregone conclusion. The time has come to establish a more rational perspective, from which we will allow the phenomena to tell us about themselves in their own way. Manifestation is a language of its own, and one we must learn. Compulsively demanding that phenomena respond to mechanistic criteria has been barren of results.

The ships-from-other-planets concept was the bedrock upon which an "establishment" in ufology was erected. Immature notions of cosmic workings characterize this approach, together with overconfidence in the current crop of mechanistic scientists. Habile and pitilessly efficient at perfecting engines of destruction, these men have drawn blanks on UFOs. The charm of the ships-from-other-planets notion lay for years in the expectation that UFOs could be understood with existing scientific knowledge, or with linear extensions of such knowledge that were deemed imminent. The approach least likely to disturb the neurotic *weltanschauung,* ships-from-other-planets therefore became automatically the most popular, despite its irreconcilability with a large corpus of observations.

Subscribers to this theory as the primary explanation for UFOs exhibit a marked blindness in connection with UFO propulsion. Ships-from-other-planets as they have impinged on earth life, command a power source impenetrable to official science. Energy in some arcane form is being used for propulsion in a way that earthmen do not yet understand. Since extensions of existing technology in no way approach UFO propulsion capabilities as observed, native common sense suggests, with much evidential support, that progress may be made by taking a *wholly new* approach.

Ships-from-other-planets devotees usually recoil from this idea. Common sense suggests a possible beginning in offbeat, borderland areas of investigation and thought, where human beings of novel bent have always labored outside and usually beyond official science. The mavericks of this borderland include many qualified scientists who explore the field avocationally. These scientists tackle phenomena that do not square with mechanistic concepts

14

and methods, or which seemingly spill over into the methodologically forbidden realm of faith. Such individuals are increasing in number, and they are true to the ideals of science.

Since UFOs stand outside mechanistic concepts, and have also evoked a powerful mystico-religious response among humans, this borderland area of original, untrammeled work and thought might be expected to yield valuable indications to any UFO project mounted by official science. No such approach has been made. On the contrary, scientists who had studied UFOs avocationally—sometimes investing thousands of hours of their leisure in this way—were ruled out of the Colorado University project for fear that their objectivity would be adversely influenced. Organized ufology has also been unable to extend itself even to the fringe of the borderland.

Evidence that punches holes in mechanistic conceptions—such as the multitudinous examples of materialization of UFOs—is papered over or shunted aside in favor of evidence considered "harder", that is, more accessible to mechanistic method. When the UFO subject began to take on an inevitable mystical and occult aspect, establishment-type UFO organizations responded by elevating ministers and rabbis to their boards and committees. They have proved themselves as helpless as the official scientists. The main idea in all these machinations was to maintain comfort and avoid tackling the invisible.

Contact stories were dismissed as unworthy of scientific consideration, and every establishment-type UFO organization has a list of such "cranks." The irrationality of these organizations seems incredible, since they obviously conclude that UFO intelligences should communicate only with two groups of earthmen:

1. Scientists already beaten by the phenomena, which had torn the fabric of Mechanism to rags.

2. Political leaders who were using the full machinery of government to suppress UFO evidence and discourage its discussion.

This irrational bias toward an orthodox, thoroughly safe approach to the subject has barred the way to the comprehensive theory that the facts—be they welcome or unwelcome—demand out of their own stuff and substance. A comprehensive theory of UFOs cannot evade what has been observed, experienced and recorded by human beings in connection with UFOs, nor may it exclude the mass-psychological factors that militate against free discussion of the subject. Many dimensions of the UFO problem are indeed immeasurable.

The UFO mystery is intimately involved with the whole question of how human beings perceive phenomena, and how their perceptions are bioenergetically and biopsychiatrically distorted. In short, we confront the inherent errors of man's natural philosophy. The idea that UFOs and their technical principles can be adequately dealt with by unaided physicists, engineers and aerospace specialists, supported by ministers of religion—is recklessly naive. The days when we can delude ourselves that we are investigating this subject when we run deferentially to panels of physicists for approval of our findings, or when we try and get Congress to act, are gone forever for all realists. The experts of this earth are experts in what this subject—these phenomena—are *not*.

UFO phenomena are all around us, unseen. This basic fact has been established by radar and further demonstrated by my pioneering photography. Accurate understanding of such phenomena requires the New Knowledge, in which formal qualifications may well run second, third, or worse to actual participation. Cosmic tides wash strongly against the ivory towers of mechanistic science, and will tumble those ivory towers just as soon as old ways and methods are transmuted by the brilliant young men and women already entering upon careers in science. They are a new and different breed of human, and formal education lags far behind the exceptional powers and capacities that they have brought with them into the world.

Native common sense, activism, unblocked perceptions and a free human being's understanding and acceptance of his own basic life processes, are the primary qualifications for facing UFOs on their own ground. The new young scientists have these faculties and capacities. Their diligent scientific labors in years to come will codify what is merely broached and indicated in this book—in short, they will make it into the science of tomorrow. By summarizing here the tradition-wrecking aspects of UFOs, we can illustrate strikingly the need for a fundamentally new thinking—strong and vital enough to break man out of the straitjacket of the past.

The role of radar in demonstrating the invisibility of many UFOs has already been described, and its role in *provoking* the phenomena broached for later elaboration. Subsequent to the large-scale development of radar, UFOs also appeared as visible, physical objects, and they have been with modern man ever since. Atomic explosives and atomic power are also interwoven with the nexus of electrical events already cited. The mode of this atomic implication will be clarified when the discovery of orgone energy is elaborated in later chapters.

A comprehensive theoretical approach to UFOs must provide an acceptable explanation, preferably with experimental support, for the visible manifestations as well as for electronic sightings of invisible UFOs. Scientific honesty requires that nothing observed should be evaded. Since UFOs have been seen to appear and disappear on numerous occasions, we must be prepared to deal with transitions of substance from physical to invisible-physical. In my personal experience, I have never once seen a UFO in the normal physical state that did not vanish while I had it under observation.

Since I could not follow these objects with my sense apparatus, my adventures stem largely from following them with my thinking. External apparatus was then used to objectify and verify what my thinking commended to me as being truthful and lawful about their disappearance. By their mode of manifestation, UFOs are inviting us to follow them. That is what they are "saying" when they disappear before our eyes.

The facts as they have unfolded in the UFO field force us to go much farther than understanding even such seemingly incomprehensible happenings as materialization and dematerialization of flying discs. *Ignoring the evidence of his scientific instruments has brought into question the methodological and epistemological basis of man's modern science.* We have been forced into mass-characterological considerations of a dimension and complexity almost as staggering as the UFO phenomena that have thrown these considerations into relief. Errors and inadequacies to date lie less in scientific instruments and materials than in the character structure of those using these tools of investigation. The instrument is worthless in the hands of the man who cannot tolerate biophysically and biopsychically what it records.

Railing against a dying but still powerful order in science is of no value. Mechanism is there. We are dealing with a definite psychophysical *structure* in Mechanism that can be understood, but not changed in the mass sense, other than prophylactically. Children raised in accord with New Knowledge principles and emerging into healthy, nonneurotic adulthood, will not easily accept the mechanico-mystical splitting of the human psyche to which prior generations have been forced to succumb. The youth revolution has its roots in the new, healthy wholeness of the coming humanity. This is another of the numerous immeasurable dimensions to the UFO problem.

The mechanistic scientific mind has been oriented for generations to denial of the invisible. This mind has accordingly been unable to approach UFO phenomena that beckon human attention to invisible strata of energy and substance, wherein lie the roots of life. That invisible UFOs suddenly appear to our gaze, and that conversely, visible UFOs suddenly disappear to our gaze, simply tells us that the *invisible and the visible are functionally unified.* The structural tendency of the mechanistic mind to split apart functional wholes—to shatter to fragments and then bewail the complexity of nature—is well illustrated in this aspect of ufology.

As this book proceeds, the reader will be made aware of what has already been achieved, although not officially accepted, in exploring our invisible-physical borderland and its denizens. My experience has taught me the futility of seeking formal recognition for any of these findings, because such a venture reduces itself always to a hopeless battle against character structure—against modes of reaction and behavior inculcated since infancy. One is forced to choose between continued quiet work and exhaustion in the labyrinths of scientific bureaucracy.

My choice was the former, seeking to demonstrate as conclusively as possible that the invisible is upon us, and to illustrating and illuminating the cognitional impasse in which mechanistic science has landed itself at the time of its greatest triumphs and highest influence. The immeasurable dimensions of the UFO mystery include this constant pressure for change on the old, classical scientific order. Youthful attention turns inevitably toward functionalism—the ability to follow with the mind the perpetual dynamic changes of the living. A cosmo-conception backed by nearly two centuries of continuous academic and cultural support is now under fire through UFOs. We are already a long way from mere ships-from-other-planets.

Tied in with the chronological breakthrough of UFOs came the "foo fighters" of World War II. Scientific non-results in dealing with this phenomenon demonstrate many of the inadequacies of existing cognition and method. To this day the foo fighters have not been satisfactorily explained.

These elusive, seemingly intelligent small objects played luminously around warplanes toward the end of World War II in Europe. Allied reaction was to classify them as a new German weapon under tests, since no casualties resulted from their presence. The Germans thought they were a new

Allied invention. After the war, both sides found their evaluations incorrect. Foo fighters belong to no one on earth. During the Korean War, they were seen again.

Harvard University's Dr. Donald Menzel opined that they were reflective eddies created around battle damage to Allied aircraft, but Dr. Menzel didn't carry out his usual comprehensive research. His assertion in his book *Flying Saucers*,[1] that battle damage to Allied bombers was greater in the final stages of the war is historically insupportable. In the Korean War furthermore, U.S. aircraft suffered little battle damage, due to the lack of consistent enemy air strength.

Foo fighters should be amenable to explanation by the kind of comprehensive UFO theory of which mankind stands in need. *Ad hoc* theories to cover specific, isolated instances can usually be formulated by specialists when the phenomena appear within the scope of their disciplines, but rarely are such theories useful beyond the specific instances. An approach is needed that can help break down the compartmentalization of knowledge—in itself a consequence of man's propensity to split apart and artificialize phenomena that are functionally unified.

The UFO theory needed should permit an understanding not only of the determinism of foo fighters, but of the functional relationship they bear to all the other invisible objects encountered in ufology and to the more conventional flying discs. Now if we take a highly sensitive Super 8mm camera to 30,000 feet in broad daylight in an airliner, load that camera with standard Ektacolor 160 film and cover the lens with an 18A filter, we have an experimental arrangement that can record evidence. An 18A filter is scientifically designed to absorb all visible light and color. Don't let that worry you. Shoot a few rolls of such color film, with a filter designed to block all color. You'll find the whole gamut of UFOs are out there, and you'll record them *in color.* Impossible? Don't even discuss the matter until you have acted, until you have been there and done it.

A vista of staggering portent beckons with the use of digital camcorders, similarly fitted with an 18A filter. This equipment was not available when this book was first written over thirty years ago, but now offers those willing to use it a number of amazing possibilities. From a technological standpoint there is no better time in history, for looking straight into this invisible world, than right now.

This is cited to illustrate how, by sometimes going 180 degrees against conventional theories and ideas, objective evidence of UFOs may be acquired. The man to beware of in this field is the narrow scientific specialist. He will be the first man stretched thin by the sheer width of the ufological spectrum. Only activism—work and participation—count in this field.

Solidly verified concomitants to UFOs include interference with electrical systems. Such UFO interaction with man's electrical works date from World War II. The late Harold T. Wilkins reported an early incident involving suspension of an aircraft's electrical system in his *Flying Saucers Uncensored,* a 1955 opus from Citadel Press of New York. Writes Wilkins on page 209:

"In 1944 an American pilot, flying over the Burma Road, said his plane was held motionless and propellers stopped, while far aloft a mysterious disc appeared to be putting a sort of immobilizing ray on his plane. After this seeming 'inspection' his power came on again, his propellers resumed turning and the mysterious object disappeared into the far blue."

Since that time, the world has seen numerous instances of commercial power failure, suspension of auto and aircraft ignitions, and a variety of magnetic and electrical interference indisputably connected with UFOs. Such happenings must fit into a comprehensive UFO theory, and not be split off for study as discrete phenomena.

Caution towards the narrow specialist is enjoined by the occurrence of such phenomena within the much wider body of UFO phenomena in their totality. In earth science we cannot yet duplicate this ability to paralyze electrical activity. Ford Motor Company tried it unsuccessfully on automobile engines. The theorist concerning such paralytic activity should be asked to account also for kindred and connected UFO phenomena. If we apply this principle practically, we soon learn, practically, that much of our so-called "hard" knowledge is illusion. Here again, it is the orgone energy—the ether—with its demonstrable antagonism to electromagnetic energy, that provides the technical break-in.

Every scientific specialist faces a stupefying spectrum of phenomena connected with UFOs that lies outside his discipline, in areas in which he is not qualified technically even to hold an opinion. That is why Dr. Wilhelm Reich was right to say that there are no "authorities" and no "experts" now that cosmic science—the New Knowledge—is being literally forced upon us. The rational approach is a sharpened awareness of the inadequacies of mechanistic science in dealing with Cosmic phenomena. We are all brethren in ignorance, facing immeasurable new dimensions.

Hostility on the part of certain UFOs is another factor that must find its place in a comprehensive UFO theory. Establishment ufology has a blind spot here. This aspect of the UFO problem has been steadily resisted, despite the evidence that aircraft have been destroyed in the air—and sometimes kidnapped complete with crew. In my 1958 book *They Live in the Sky,* the affidavit of a veteran French pilot, M. Pierre Perry,[2] was presented. Perry recounted a shocking incident he had observed from the ground in the wilds of Arizona in 1943.

A USAF aircraft with two occupants was deliberately destroyed by balloon-shaped UFOs. The bailed-out crewmen had their parachutes set on fire by heat rays from the UFOs. The unfortunate victims of these weird entities from space were crushed to death by their fall to the ground. Few indeed are the ufologists who will look directly at such happenings and see them for what they are. So-called "objective" investigators have preferred to disappear into the mist of wishful thinking.

In the same book, I presented another sworn case from Paris, Illinois, wherein a USAF jet fighter was observed from the ground by Mr. Eugene Metcalfe to be abducted in flight by a large, bell-shaped craft of unknown origin. The jet was simply swallowed into the underside of the hostile vehicle. This was one more instance of seriously unethical acts by entities from space, but the establishment in ufology declines to see such acts in all their clean clarity.

There was also the abduction over Lake Superior on 23 November 1953 of a USAF F-89 jet fighter piloted by Lt. Felix Moncla Jr., whose aircraft was vectored to a UFO by ground control intercept at Kinross AFB, near Sault Sainte Marie, Michigan. Moncla's fighter merged on the radar screen with the UFO, the two objects becoming one large blip 70 miles from Keweenaw Point. Moncla and his radar observer, Lt. R. R. Wilson, were never seen again, their aircraft was never found, no wreckage was recovered, and the pursued UFO also disappeared.

Major Donald Keyhoe gave a full account of this baleful incident on pages 13-23 of his *Flying Saucer Conspiracy*. Despite this and many other similar incidents, until the end of his tenure as Director of the National Investigations Committee on Aerial Phenomena (NICAP), Major Keyhoe believed there was no convincing evidence of UFO hostility. Lts. Moncla and Wilson had vanished from human ken, complete with plane, but even this is deemed unconvincing evidence of hostility. Again we find, in a new way and in another facet of the UFO subject, that same *evasion of the essential* that has kept ufology tied to the skirts of the mechanistic world conception.

There have been numerous cases involving hostility on the ground in encounters between humans and a variety of queer entities who have dismounted from spacecraft of various kinds. Humans have been attacked and clawed, their abduction attempted, and others have been knocked senseless by various ray weapons possessed by the intruders. These incidents have occurred year in and year out, in areas as widely separated as South America and Scandinavia, and have been verified by responsible investigators.

Some of these incidents will arise later in this book, in context with new findings, but the serious student of ufology desiring a steady flow of such information can do no better than subscribe to the *Flying Saucer Review* from England.[3] FSR presents the stories of these vital incidents after investigation by its qualified representatives in foreign countries. The publication is produced avocationally by a group of scientists, engineers and physicians whose qualifications are beyond reproach.

Self-styled skeptics avoid looking at these unsavory, unwelcome and disturbing facts. Weak jokes about "little green men" constitute the maximum effort mounted by the media of the western world with regard to these epochal encounters. Any intelligent, alert, unblocked and discriminating individual can satisfy himself quickly concerning the validity and the increasing incidence of these landings and encounters. They are no joking matter.

The hostility of certain visitants must be woven into a comprehensive UFO theory. Since the incidents continue with the years, and apparently began with the war period, we should expect a functional connection to exist with the other complex and seemingly impenetrable aspects of UFOs. Once more we may note that the narrow scientific specialist in a technical discipline can bring little to bear on this serious and far-reaching problem of ethics and behavior. Conventional psychiatrists and psychologists who might assist here cannot deal with radar or electromagnetic interference. The discovery of the orgone energy—alone among all events of the past several centuries—gives functional access to all these riddles. A new epoch has opened.

The obverse aspect of hostility is the reluctance of most UFOs to make contact with humans. Most UFOs are elusive. In most cases, the objects make off at high speed from the vicinity of human observers, from aircraft or from happenstance encounters with humans. This instant readiness to conceal themselves from human beings is well established, and must take its place in any comprehensive UFO theory. There must be a reasonable basis for saying why these things happen so frequently as to be among the basic characteristics of most UFOs.

The immeasurable dimensions of the UFO problem may be seen also in a worldwide outbreak of psychism. Associations between psychic phenomena and UFO phenomena are intimate, broad and profound. Dozens of books have already been written about UFOs on a psychic or quasi-psychic basis, and there is a potent mystico-religious overtone to all UFO affairs that may not be ignored—any more than we may ignore the evidence of radar.

The coming of the flying saucers has been the subject of numerous sermons by ordained ministers, and in the more modern churches unballasted by tradition, UFO lecturers are sometimes invited to speak at Sunday services. These speakers are free to ventilate the metaphysical aspects of the subject as they see fit. The heads of large aerospace concerns and distinguished men of science may be observed often in the congregations. Some of the braver ones have even come to my talks.

A sociological fact of life in today's world is that millions of people have been led to absorb themselves in the metaphysical aspects of UFOs, whether or not they are possessors of formal scientific training. There is no manifestation in human record that comparably straddles science and religion, straining each division of thought at its foundations. The purportedly objective approaches to UFOs that ignore these staggering sociological facts must be characterized as fraudulent.

Our comprehensive UFO theory must look at this situation squarely, and establish how it has arisen, and why. Ignoring the irruption of psychism that is coincident with the modern advent of UFOs is tantamount to ignoring the radar evidence, the photographic evidence or any other kind of evidence. There must be a place in our theoretical mosaic for these undeniable psychic, mystic, occult, spiritualistic and religious aspects of the UFO subject because they not only exist, but also are extending their influence continually into the vacuum left by scientific abdication.

The socially-endorsed posture of the mechanistic scientist, that these manifestations are "mysticism" beyond ken and therefore excluded from scientific attention, is not only cowardly, but mindlessly evasive. By giving these manifestations a name—whatever that name may happen to be—the mechanistic investigator deludes himself that he has explained what is happening. The deep stirrings evoked in millions of humans by UFOs have taken place at the dawn of mankind's Cosmic Age. Conceiving of this additional mighty coincidence as an accident is the hallmark of a simpleton.

Our comprehensive theory of UFOs must face these socio-religious phenomena, and the undeniable impact of UFOs on the inner life of Man. Honest investigation of UFOs and corollary observations of 20th century life, will convince us that spiritual forces are at work on, in and through the human being in a decisive fashion, and with definite and comprehensible earthly goals. Right now, this is yet another immeasurable dimension to the UFO problem, but it is there and must be faced.

Contact and communication with UFOs have been derided not only by establishment-type ufology, but also by a lamentable coterie of qualified scientists. A comprehensive UFO theory must provide an understanding of the psychic encounters with UFOs that far outnumber physical encounters. Processes beyond the reach of official science have obviously been brought to bear on human beings by the aliens. These encounters are as indubitable as the physical landings and incidents, even if inaccessible to mechanistic method. New ways must be found and opened. UFOs are too important to be left to mechanists.

A viable UFO theory must account for the lack of communication through orthodox, electrical communication methods. There must be sound reasons why advanced UFO intelligences do not utilize our communications systems to signal us in some way, and we must have at least a theoretical guideline to the means of communication they do employ. Psychic communication has been extensively employed in many variants, and the only rational attitude is again to face these facts and let them lead us where they will. Our existing radio receivers have got us nowhere. Our psychic receivers have brought in plenty.

Man is irrational when he makes Herculean efforts to become a cosmic voyager on the one hand, and denigrates communication with alien intelligences on the other. This is the situation today in the official attitudes toward communication with space. Extraterrestrial life remains the largest challenge to mankind growing out of UFO phenomena, and mechanistic biology, rooted in sterile chemistry and physics, has practically no chance of cosmic survival. UFO data already on record sharply illustrate its inadequacies.

The fundamental characteristic of everything that is alive, as opposed to what is inert, is *pulsation*. To official biology, pulsation remains inaccessible, inscrutable—a mystery as vast as the UFOs themselves. Prompted by observed pulsations of UFOs sighted at high altitudes by its pilots, the USAF once broached the idea of "space animals" in a public release dated 27 April 1949, stating that the objects appeared to behave more like animals than anything else.

The airmen who observed these aeroforms, and expressed the feeling that they were living organisms, were closer to the truth than scientists in the grandiose discipline of exobiology have yet come. A comprehensive UFO theory must incorporate living aeroforms within its structure, for such living creatures do exist, have been extensively photographed by myself and others and are now being recorded inadvertently on NASA videotapes in the space environment. NASA's exobiologists literally do not know what they are looking at, so intent have they been on establishing the sterility of the moon. Some of these creatures are monstrous in size.

Many persons have theorized that UFOs are alive, including the late Kenneth Arnold, who put the term "flying saucer" into the language. I happen to be the earthman who first photographed these life forms extensively, proving their existence and simultaneously penetrating their bedrock involvement in the UFO mystery. A comprehensive UFO theory must include them, with all their revolutionary portent for the new life sciences, and their power to shatter some of mechanistic man's cardinal illusions about the origins of the earth.

UFO propulsion should have received instant and unrelenting scrutiny by world science, for it is beyond doubt that a new mode of propulsion is involved. Existing instrumentation and knowledge admit us only to fringe physical effects of this power. *A different kind of power, a heretofore unknown or unsuspected mutation of energy is involved.* UFOs exhibit mastery of gravity, and can attain velocities and execute maneuvers far beyond the farthest reach of mechanistic science or its most optimistic projections. The UFOs are in the here-now with all this. Air-supported craft are destined for the dustbin.

Man has nevertheless been content to waste his scientific substance in chemical power plants of monstrous inefficiency, with precarious control of space ventures exercised only through the precise cooperation of battalions of highly skilled technicians and specialists, who crouch convulsed with fear at launch time lest their plaything fall over and blow up on the pad. All this tragicomedy goes on while space vehicles are already present in the atmosphere, and beyond, that eclipse our ashcan spacecraft as a Cadillac surpasses Ben Hur's chariot.

Brave astronauts riding these clumsy, manmade contrivances into space have sighted and photographed vehicles otherwise propelled and controlled, yet the focus of aerospace activity stays on rockets and kindred devices *as though there were something to be feared in the energy system that is propelling the UFOs.* This fear and avoidance of the new power by human beings in the most relevant of all fields, has to be explained by our comprehensive UFO theory. Again, we will find as we assemble this theory—with both thought and experiment—that it is the orgone energy that makes such explanations possible.

Our theory must concern itself not alone with UFOs and their scientific determinism, but with the thorny, prickly question of irrational human attitudes toward the phenomena. Our concern here cannot be confined to the layman or to the nut. A solid theoretical approach should allow us to understand why scientifically trained individuals have not only avoided this subject, but in many cases have become active and even ardent agents of the new obscurantism.

There are the visible manifestations, stunning in themselves, with their irrefutable invisible aspects; there is the connection between the electromagnetic nexus of the Second World War and the modern, worldwide advent of the phenomena; there are objects that are seemingly spacecraft and objects that are obviously biological; there is the paradox of hostile visitants with heat rays, murdering humans in remote areas, and the general avoidance of contact with humans; there are the psychic, metaphysical and occult aspects of the mystery; there are the great enigmas of communication with the piloting intelligences; there is the riddle of pathological evasion of the UFO problem by human beings deemed responsible in all other normal ways; there is the new mutation of energy that is the key to unraveling the tangle—the swaddling cry of a new Life-positive science that the doomed priests of the old order seek to strangle at birth. There is the conceptual dead end at which mechanistic science finds itself at its moment of greatest glory and influence; and there is the glittering promise of the New Knowledge arising from the ruins of the old.

When our comprehensive theory is assembled, we should be able to understand why men, enlightened and educated to the best formal standards, turn away in the clutch from science in favor of scientism. We should be able to understand why those scientists who accept that UFOs are ships from outer space, appear structurally incapable of dealing with the inevitable, consequent question of contact with the piloting intelligences. A comprehensive UFO theory must elucidate, and thereby prepare for the eradication of, these obstructive human dilemmas that affect all people to some degree.

These are just the primary aspects of the greatest mystery in human history—a mystery that requires a changed human being for its eventual penetration. Consider the range, scope and depth of the UFO problem without prejudice. Ask yourself if the so-called extraterrestrial hypothesis, or ETH, is adequate. Ask yourself if your own present thinking and manner of viewing creation is adequate to cope with these presently immeasurable dimensions.

The towering theoretical problem involving UFOs is a challenge to the best that is in man. The Little Man that is in all of us, usurper and suppressor of all that is great and godly in every one of us, wants us to hang on like drowning people to that seedy ETH. If we catch the sweeping magnitude of the mystery, if we let its grandeur and cosmic power live in us, then there will be room no more for our Little Man. He is the one who whispers to every one of us that we can do nothing great—that all people are Little.

The Little Man wants you to recoil from the vastness of this subject. When the problems are presented as they have been here, already interwoven with each other so that we can see the folly of segmental, indecisive approaches, we can already see why there have been no government announcements about UFOs that have any meaning whatever. Only an idiot—a Little Man— can expect or feel the need for such announcements. The subject is too vast, too far-reaching, and too radical in the primary meaning of that term, "root"—to permit quick assurances to the Little Man. This subject is *big*.

There are functional connections between all the ramified and seemingly irreconcilable factors thus far outlined. Functionalism gives us a new beginning and a new way to proceed in the future. UFOs are manifold and multiform phenomena that are at once ancient and ultramodern, containing within themselves a range of physical, biological, biosocial, biopsychic and bioeconomic principles new to mankind. UFOs are the space age bearers of the New Knowledge, knowledge that has the power to renew human life and culture.

Man reached this cosmic rendezvous with a fragmented and essentially contrived mode of scientific cognition, and with his intuitive intellect beaten back into a corner like a whipped dog. Man must make himself whole and healthy if he is to plunge farther into the cosmos, and have commerce with the beings who are even now all around his planet Earth. He will need to keep a firm hold of the best in what he has learned, but his greatest need is to get a firm grasp of the new. What is worthless and devoid of value in his Cosmic Age, no matter how old or honored by tradition, can then be allowed to slide into the limbo.

For the individual, a prime task is to push his usurping Little Man off his own inner throne and put the King in his place. Dr. Franklin Thomas taught me this, and thereby opened all that followed. Every human being can do likewise.

By turning a substantial portion of my life energy and my earnings for many years into the pursuit of the New Knowledge; by taking advantage of opportunities that came to me to learn from several magnificent human beings; and above all by going out and participating as an innovating experimenter—daring to do—I believe that I have reached the end of the beginning of the UFO mystery. My Little Man would have convinced me, years ago, that an ordinary man could never do such a thing alone and unaided.

I believe I can formulate for you the comprehensive theory that seemed so frighteningly complex only a few pages ago. What it took me painstaking years, grinding struggle and immeasurable sadness to find out, you can learn in the short space of this book. Stand by my elbow and follow my story as it happened. Grapple with the UFOs at first hand as I did, and share my great adventure.

NOTES TO CHAPTER TWO

1. Published by Harvard University Press, Cambridge, Mass., 1953.
2. The author met M. Perry personally, and questioned him in detail.
3. Ufology's best buy. *Flying Saucer Review,* FSR Publications, Ltd., PO Box 585, Rickmansworth, WD3 1YJ, England. Website: www.fsr.org.uk.

Chapter Three

FINDING A NEW PATHWAY

...the scientific community has been largely corrupted or silenced by military domination.
— DR. RALPH E. LAPP

My introduction to UFOs came through Major Donald Keyhoe's books *Flying Saucers Are Real* (1950, reissued by Book Tree 2007), *Flying Saucers From Outer Space* (1953) and *Flying Saucer Conspiracy* (1955), which rightly rank as UFO classics. He also wrote *Flying Saucers: Top Secret* (1960). Major Keyhoe laid the basis for establishment-type ufology by collecting verified data, recounting dependable sighting stories and endlessly chasing down evidence. This approach essentially regards UFOs as extraterrestrial spacecraft, and has high confidence in existing scientific theories and those who adhere to such theories. Purely mechanistic speculations are made from this seemingly safe base in reality. Personal experience forced me to diverge from this approach, but I pay unreserved tribute to the integrity, courage and perseverance of Donald Keyhoe.

A professional U.S. Marine Corps officer and pioneer aviation writer, Major Keyhoe risked a sound reputation and his literary career when he took up flying saucers. Through his books and labors with NICAP he has made large personal sacrifices to make the world aware of alien visitation. He did this on the basis that was most likely to find the widest acceptance, working diligently to keep the subject respectable. Like all pioneers, he had plenty of mud thrown in his face.

Keyhoe has been treated shabbily by official Washington and unchivalrously on many occasions by the broadcasting industry and the press. I once saw the sound deliberately shut off while he was speaking on TV. For years he urged the creation of a formal UFO project by top scientists with Federal backing. The Colorado University project under Dr. Edward Condon nominally provided the elements for which Keyhoe had long asked. Within a short time, he found that NICAP data supplied to the Colorado project were not being properly utilized, and he withdrew NICAP support. Keyhoe has always deserved better than he has received from official sources. He was an idealist in realist's clothing, and this field would not have been as far along as it is without him.

On the basis of the evidence Keyhoe presents in his book, a rational person is compelled to accept the presence of UFOs, although not necessarily to accept that they are spaceships and nothing else. Keyhoe's books and other classic UFO literature left me puzzled by one element that I found sharply anomalous. There has been no communication between these alleged spacecraft and human beings. Such a silence did not jibe with my own personal experience of the way our own world is run. Practically all my life up to the time I learned about UFOs, I had been a professional in communications, both in a technical sense and in the realm of ideas. As a staff writer for radio programs in New Zealand in my teens, and also as a radio actor, I eagerly learned and applied the art of communicating ideas from the written word via the human voice. In a technical sense, I qualified myself as a broadcasting engineer and as a shipboard radio officer.

In this latter capacity, under the flags of both New Zealand and Great Britain, I traveled all over the world before I was twenty-five. This experience made me aware of the role played in civilization by communications, because of my daily involvement in this field. The living standard of any country could be measured by its communications development—a criterion valid to this day. The more refined the communications of a given country, the greater its technical advancement and the higher its living standards.

Communications are an index of technological achievement and material progress. The American people were able to see their astronauts aloft and walking on the moon, and to broadcast the spectacle simultaneously to the world. American communications are unsurpassed, and there was no insuperable difficulty to communication with earth from the moon.

To me it was logically untenable that a civilization capable of building a spacecraft would not be capable of communicating adequately. While I tended to accept Keyhoe's general view that there had been no orthodox radio communication with humans by UFOs, it seemed obvious that UFO designers and operators would be using a different system of communication to earthman's radio, as far beyond our methods as ours are beyond the tree trunk drum or Indian smoke signals.

There seemed to me to be no rational dissent from such a view. The propulsion methods employed by UFOs were generations in advance of anything planned by our brightest minds. The vehicles traveled in multiples of the best speeds attainable by mechanical aircraft. Communications development must have proceeded synchronously with these technical advances. Earth life has taught us this lesson in our own experience.

Our communications have progressed from the invention of wireless telegraph to modern television within the lifetimes of many human beings. By the middle 1950s, under the stimulus of the space program, scientific attention was already turning to so-called extrasensory perception, or ESP, as a possible means of space communications. By November 1958, the august Westinghouse Corporation, citadel of corporate and scientific orthodoxy, had set up studies into mental telepathy as a means of long distance communication, under the direction of Dr. Peter A. Castruccio. Other corporations with pipelines to the Federal treasury have since followed this lead.

Even a ten watt mentality could project the likely results of 50 years of such work, since science has almost routinely achieved any goal to which sufficient resources are applied. Blowing up the entire planet, or extinguishing its life, is achievable today because men wanted this power. College-educated killers by the thousands have devoted their lives to explosives and the slaughter of children, and it seemed inconceivable that Life-positive scientific work aimed at the mastery of telepathy would not be just as successful.

Scientific mastery of telepathy in 50 years seems reasonable, if it is pursued. Similarly reasonable is the proposition that entities utilizing propulsion methods 50 years or more ahead of ours would have communications 50 or more years ahead of ours. On the basis of what we on earth have been driven to explore therefore, we might expect the aliens to be using telepathy. This expectation is reinforced by their non-use of radio in any mode we are capable of detecting, although UFO entities have at times communicated with pursuing USAF pilots via their VHF radiotelephone receivers.

In an ever-broadening investigation of the UFO subject, with my attention drawn to the communications anomaly, I became aware of the contact stories. Involving encounters of various kinds between alleged extraterrestrials and human beings, these stories already made up a bulky literature by the middle 1950's. I tackled them to see what I could find out about UFO communications. I began to hear about *psychic communications,* with which I entered a new and wholly alien world.

Breaking down my structural resistance to such things was a primary problem. Such resistance is present in all persons who have been given either religious or scientific training. Religionists consider such communications to be "of the devil". Scientists consider the same area to be one of fantasy and illusion. I had been raised as a Christian Scientist. In that benign and tolerant religious system, the occult and psychic were deemed to be things better left alone.

The discs continued in the skies, and the communications problem could not be ignored. Already I detected an unhealthy and irrational fear, outwardly manifesting as ridicule, among the many trained persons of my acquaintance to whom I broached my general theories about communication with space entities. Friends in the aircraft industry passed on rumors of a man living in the high desert of Southern California who claimed to have telepathic contact with the entities piloting the flying discs. Thus I came to meet Mr. George Van Tassel, a pioneer in such communications, who later died in 1978. His distinction in this world was to obtain from his other-world communicants, advanced technical information.

This technical information is impenetrable to conventional scientific knowledge and to the mechanistic mind. Armed with the revolutionary discovery of the orgone energy—to which UFO intelligences have extensively referred as "Primary Energy"—Van Tassel's early communications become a technical bonanza. Any trained scientist whose thought processes have been functionalized will be able to penetrate most of the early Van Tassel communications, and conceivably could create instruments and devices of a wholly new character.

At the time I went to see George Van Tassel at Giant Rock, he was regarded by scientists as a crackpot. Most laymen who came in contact tangentially with his writings[1] undoubtedly regarded him as just another flying saucer nut. History will probably have a different verdict.

Investigators in this field must learn to *participate*—to bore in with body, mind and heart as a total human being—if they are to find their way to the truth. The old, onlooking, withdrawn and purportedly objective approaches have failed in this field. The case of George Van Tassel, as I experienced his doings and as they influenced me, is an example of how participation and the use of all one's personal resources of experience can open doors in UFO research. Sometimes the consequent problem of how to get some doors closed again becomes critical, but one must *dare to do in order to know.*

Giant Rock derives its name from a colossal boulder that rests near the end of a remote, dry-lake airstrip in the high desert of Southern California. Around the Giant Rock, Van Tassel had created, through the years, a small living complex including a restaurant and his own home. He had also created the physical facilities for his annual spacecraft convention. In the days when I first visited Van Tassel's place, he used to conduct communications with UFO intelligences from a chamber hewn out of the ground beneath the Giant Rock. There was no technical apparatus for this communication. Stumbling down the rough stairs in semi-darkness, one was seized by the horrendous possibility that the giant boulder might roll and crush the chamber, complete with occupants. There were usually about 60 or 70 persons present at these seance-type gatherings, varying from college professors to ordinary working people, and always including a sprinkling of reputable and successful businessmen.

Van Tassel's preparations for contact consisted mainly of focusing the power all present into an energetic unity. Songs were sung in unison. Prayers and chants followed. Then came silence, while Van Tassel waited for contact.

Singing, chanting and prayers constituted the "call" phase of communications, differentiating the group of people with their common desire for contact with UFO intelligences, from the random psychic racket of the ordinary world. The call phase is intended to stand out above the psychic background noise as a strong radio signal does above radio noise and weaker stations. Such a call in radio receives the immediate attention of any skilled operator monitoring the particular frequency in question. He reads the signal if it is in code, and if the person signaling is calling him, *he answers.*

Van Tassel in effect functioned as a psychic radio operator, focusing the power of his transmitter—*the biological energies of his gathering*—to attract the attention of someone with whom he wished to communicate.

His calls were answered. Out of the darkness from Van Tassel's direction would come booming the most dynamic and powerful voices I have ever heard. For years as a youngster I studied voice and speech, and as a young radio actor put these studies to practical use. I was always around accomplished radio actors and announcers, and on this account was extremely voice and speech-conscious. I understand fully the art and technique of changing one's voice so as to act out an entire radio play doing two or three, or sometimes more, separate voices.

The voices at Van Tassel's were like no others I have ever heard.

George Van Tassel was an ordinary man, and not highly educated. He spoke well, but did not have an unusual vocabulary. He made occasional grammatical errors like most people. The intelligences who *spoke through* Van Tassel, utilizing his vocal apparatus as though it were a physical amplifier or transducer for their own thoughts, made no grammatical errors. The intelligence flowed in strong, forceful language without a split second of hesitancy. The effect was like having Orson Welles present, orating at the height of his powers.

Several of these intelligences would speak, one after the other. There was in each case a distinct change of the speech pattern, pace, voice timbre, accent and subject matter. No radio actor could have done it. Van Tassel's vocal mechanism provided a sound carrier, as it were, upon which all these other voices were impressed. There was no doubt that they were speaking through Van Tassel, not only because of these objective facts, but also because of something that I experienced, and therefore learned, by actually being there.

I found that I could hear, in a way new to me, the thoughts of these beings *before* they were transduced by Van Tassel's vocal apparatus. The effect was akin to monitoring a tape recording, the broadcast of which is briefly delayed. I knew what Van Tassel was going to say before he got it out as audible sound. Prior to this time I had no psychic abilities, or experience with any kind of psychic phenomena. All of it had been to me a tightly closed book. Now it was opening.

Blundering in where angels might be terrorized, I asked Van Tassel how he was able to develop his receptive ability. He gave me certain routines to follow. Not aware at this time of any element other than idealism in connection with UFOs, I followed Van Tassel's indications with diligence and persistence. Once, at a later time, I became for a brief minute, the "loudspeaker" used by these strange intelligences.

There was no doubt that communication was being carried on by these means, no matter what official science might think of it all. On one occasion in my presence, and subsequent to their promise to do so, luminous UFOs manifested in profusion above the nearby desert until dispersed by Marine aircraft from the 29 Palms base that were scrambled on intercept. The troglodytic professor on the USAF payroll may smile, but he does not *know*. The New Knowledge is the property of those who participate, and I was unequivocally convinced that the entities piloting the discs were quite capable of communicating with human beings, without any need for radio apparatus. My empirical work had begun with a little research involving my own psychophysical person.

With consuming interest I read the technical matter that these intelligences had communicated to Van Tassel. My electronics background was insufficient to penetrate this material, but everything about it seemed to throb with life, even though it was beyond my comprehension. Reading the communications was an adventure in itself. Worlds within worlds, densities within densities, energies, polarities, ethers—material with no seeming connection to earthly concepts—but all of it discussed with an overlay of good humor and love. From this time on, I began to think that just being an ordinary man might be an advantage in investigating this subject. I had only a little junk to jettison before tackling this New Knowledge.

The communications referred often to the *primary energy* in connection with spacecraft propulsion, and I knew that there was at this juncture no readily intelligible connection to earth science. There was something missing— something crucial—from the ever-multiplying trivia of modern knowledge at human disposal. Intuitively, I got on the trail of that missing something right then. These intelligences were doing things differently to us, and our men of science weren't trying to find out the answers. They were trying to kill UFOs, and pouring their energies instead into the rackety and uncertain fireworks at Cape Canaveral.

In *They Live in the Sky* I detailed my first psychic experiences and will not therefore repeat them here. Suffice it to say that by persisting with the techniques learned at Giant Rock, I set the stage for an irruption of the unseen worlds into a consciousness—*mine*—not prepared for such an impact. Becoming sensitive suddenly to spectra of vibration with which one is totally unfamiliar can be an unhinging experience. In recent years, the so-called "psychedelic revolution" has exposed untold thousands of persons to the consequences of chemical tampering with consciousness. All such forcing open of doors is destructive of orderly inner development, no matter what the academic qualifications of its advocates.

There was in my case no visions of the unseen worlds or astral phantasmagoria, but I did develop extreme sensitivity to telepathic impulses. I found that I could barely control the situation. In daily business life in the aviation industry, I would hear a sentence psychically before a client ever spoke the words physically. When the telephone rang, I knew who was calling before I picked up the instrument.

A constant struggle soon ensued for control of my physical vehicle—myself against unseen interlopers. I was fighting continually against various forms of automatism. Anyone who doubts the reality of occult things would have no doubt whatever concerning them were they to endure an experience of this order. I emerged from it all with a solid respect for the reality of the occult that I have never subsequently lost.

My difficulties were extreme, and I felt that I was slowly losing my battle to retain the mastery of myself. I bitterly regretted ever having meddled in UFO communications. The "intelligences" into whose realm I had broken poured confusing rubbish into me. There did not seem to be anyone to whom I could talk about my difficulties without already seeming to be "around the bend". All was not lost, however, for a great man was at hand.

Dr. Franklin Thomas was a publisher of many small UFO books, as the owner of the New Age Publishing Company in Los Angeles. Many of these books dealt with the things I had stumbled upon. He used to give lectures in his Glendale Boulevard bookshop on Friday evenings, and I resolved to attend.

Franklin Thomas was a diminutive, slight, sharp-featured man, wrinkled of face and generally shabby. He seemed indifferent concerning his physical person. His lectures were delivered in a soft, low-pitched voice, but with a lucidity that I found enthralling. He could cover the esoteric aspects of widely divergent subjects in such a way that he constantly wrought seemingly unconnected things into a higher synthesis. I knew that he could assist with my problem.

He was heavily burdened in his struggling business, and his health was failing. He knew he was dying, and at a period when he needed all his reserves of strength, he spent as little time as possible after his lectures talking to those who pressed him with questions. He listened carefully to me. There was an understanding nod and a knowing glint came into his eye. I asked him to help me. He agreed.

He was the most accomplished occultist I have ever known, an adept and master teacher functioning as one of us in the workaday world, but otherwise something much more. At a time of loose talk about the occult and occultism, to have known and been taught by Franklin Thomas was an occult experience in itself.

This shabby little man had conscious control of the hidden forces of nature, and he wielded his powers in setting me free. I became in a short time, complete master in my own house again, and the ability I had forced on myself to function telepathically was brought under control. This was the first step toward suppressing the faculty entirely.

The essence of regaining control was to confine the receptions to a given time of day, and never to depart from this regimen. Then it became every other day, every third day, and finally the spurious door to the unseen was closed and sealed. Every student of esoteric matters who wishes to make genuine progress in the development of his inner life, will find the avoidance of psychism and psychic phenomena—as an involved party—absolutely essential.

During this period when I had these daily contacts and before all such activity was halted, some information was passed to me concerning the UFO mystery in many of its aspects. I had learned, through Franklin Thomas, how to sort out the other-world telepaths. Those intelligences which would seek to communicate *without controlling* had what seemed to be the valid information.

Through Franklin Thomas and his gentle guidance, I began to discern that the mystery required a sound working knowledge of occult science for its overall comprehension. My knowledge of these things was so limited that it was absurd to think that anything of great value could be passed through me. I was an inadequate instrument for any such purpose. Recognition of this fact by me accelerated my phased withdrawal from psychic activity.

During these limited contacts, information was nevertheless passed from the "other side" that went against many of the commonly accepted conceptions of UFOs. Later publication of this material was to make me a sort of pariah even in the way-out UFO field. The truth is unwelcome in this world, and it hurts because of its innate power to disturb. The entities I contacted conveyed to me, in essence, the following basic information:

1. UFOs are spaceships, but their vibratory makeup is not fixed in the physical-material density. They are *mutants*.

2. UFOs have their main existence in a density that is invisible to human beings of normal vision.

3. The intelligences behind the spaceships are various orders of etheric beings, that is, beings differently constituted to man and normally invisible to him, yet capable of materializing at will when necessary or required.

4. A war is in progress for control of the mind of man, a veritable battle of the earth that will determine the future course of evolution.

5. There are negative forces from beneath man seeking to drag him down and positive forces assisting him to fulfill his destiny in freedom.

6. There are normally-invisible living things in space that are *not* spaceships.

7. Space is filled with primary energy currents of which existing earth science knows nothing.

8. Infrared film, exposed between dawn and sunrise in high, dry locales will frequently objectify invisible objects of various kinds living in and passing through the atmosphere.

This information permitted the formulation of a skeleton UFO theory, and also opened a pathway to obtain objective proof of the presence of UFOs. The latter would come to hand *only if the communicating intelligences had told me the truth.* I had work to do, of both a theoretical and empirical nature.

At this point, Franklin Thomas had a serious talk with me. I pass on now his fundamental wisdom.

"You have reached a point where you have a sort of assignment—a task to discharge. I cannot tell you where it will lead—to discovery or to disappointment—but I can tell you with certainty that success, if it is to come to you, demands that you cut off all psychic activity and abandon it entirely. Everything for you, henceforth must be in the *full light of consciousness,* with no communication of any kind with any unseen forces, no matter what their purported wisdom. Contact must be with your own High Self—your Atma—and with nothing inferior to that."

This advice was followed. All psychic activity was terminated. My experiences had convinced me that the disc occupants communicated by means of thought transference—telepathy in its many variants—and that communication with earthlings would be unlikely via any other method. Communication seemed to require either that the human being surrender in some way the function of his organism to beings that he could not see, or that new devices for manipulating the primary energy of which these entities spoke, would have to be designed.

These new cosmic electronics would use biological energy. Experience had already shown that such devices were quite different to and enormously advanced over anything possessed by earthmen. Later on, my adventures would bring me to a remarkable personality who had made revolutionary strides into cosmic electronics, but for the moment, I did not know where to begin.

I did know, in a fashion that permitted of no contradiction, that if ever UFOs were to be understood and comprehensive theoretical approaches made to the subject, I would have to resign myself to *years* of study and research. This endeavor would, by its very nature, lie outside the boundaries of official science. Despite the resources and facilities that they commanded, the official scientists were getting nowhere in penetrating the UFO mystery, and my brushing personal involvement with the technology behind UFOs taught me why they had failed and would continue to fail: *humans were looking in the wrong places for their answers, and they were looking in the wrong way.*

Man, to their mode of cognition, was $2.98 worth of chemicals organized in a complex way. Complexity of organization was held to be the only difference between man and the chair on which he sat. I was already interested in the energy that propelled me across the room. When I asked learned physicians about this energy and what it was, I found that they became angry.

Already I was beginning to think that there might be a functional identity between my personal power source—the biological energy that drove me across the room—and energy that drove the discs across our skies. Learned, able men with the highest academic qualifications were evading confrontation with both.

Franklin Thomas continued to tutor me in his quiet way. Everything he taught me concerning occult knowledge was given to me in such a way that I could dovetail it with something I could personally observe in the outer world. When I reported having noticed something different about a common feature of our environment—thereby illustrating that I had dynamically understood and applied his teachings—he would give me something more.

He steered me to the work of Dr. Rudolf Steiner, the Austrian philosopher, scientist and founder of the Anthroposophical Society. The scope and sweep of this universal genius was staggering. As I plunged into Anthroposophy, I realized that I was in contact with a body of New Knowledge, and a methodology for penetrating the unseen worlds, that was *wholly healthy* and demanded the best that was in me for its pursuit and application. My whole life began to change.

Gone forever was the craven search for "messages" from spacemen. Dr. Steiner had presented in modern scientific form, for modern consciousness, the necessary cosmo-conception and valid indications for the renewal of human life and culture along functional lines. His work was not confined to things normally deemed of the spirit. He went deep into such seemingly compartmentalized realms of thought as medicine and economics, pedagogy and biology, pharmacology and zoology, drama, philosophy and agriculture. Everything he touched he renewed and revivified.

Jules Saurwein, the Paris savant who had known all the great thinkers of the times, deemed Steiner the most impressive figure in European cultural life in the 20th century. With my mounting suspicion of orthodox conceptions, I contrasted the two or three lines devoted to Steiner in the encyclopedias with the overwhelming, massive literature he left his fellow humans. I could not square the incomprehension and the sequestration of Steiner's genius with the inflated biographies in the same encyclopedias limning the lives of life-killing developers of nuclear weaponry.

Franklin Thomas presented me with a monumental book entitled *Man or Matter* by Ernst Lehrs, Ph.D.,[2] one of Steiner's own students. An electrical engineer and mathematician, Dr. Lehrs presented in this book not only the history of man's cognitional impasse—out of which has sprung the Great Impasse of ufology—but also the method of surmounting these difficulties. All of it was presented in a healthy and modern way, on the bedrock laid by Goethe and later modernized and expanded by Steiner.

Lehrs spread himself out over such seemingly unrelated disciplines as optics, biology, meteorology, geology, botany and mechanics, unifying these diverse fields in a functional approach so that one could see the inner connections they all bore to each other. This book had a decisive influence on me. I had already begun field research in UFO photography. I was putting to the test the fundamental suggestion for photographing invisible presences in the atmosphere of the earth that had been given to me during the final phases of my experiments with psychic communication.

Man or Matter acquainted me with the Goethean approach to optics, light and color, and the more I experimented, the more Goethean I became. Lehrs also dealt in detail with the visual ray or eye beam—a ray of biological energy by which consciousness is carried outside the body to objects in space. Armed with this knowledge, some simple occult exercises and conventional cameras with conventional infrared film, I began the practical field work. The knowledge involved and methods employed will now be described.

NOTES TO CHAPTER THREE

1. Considerable technical data is presented in early issues of *Proceedings of* the *College of Universal Wisdom,* and in *Religion and Science Merged*—the latter a retitling of Mr. Van Tassel's *Council of the Seven Lights.*
2. Published by Rudolf Steiner Press, London, 1958.

Chapter Four

PROBING THE NEW REALITY

Sight is energy leaving the person.
— LEONARDO DA VINCI

The UFO intelligences with whom I had dealt in my brief experiments stated clearly that they were etheric beings, *invisible to our normal eyesight and optics.* They were in no sense "spirits", but inhabitants of another density and as "real" on their plane as we are on ours. In the work of Dr. Rudolf Steiner I found that such conceptions were by no means strange to intelligent, learned humans who had given these matters attention. Dr. Alexander Cannon, Max Heindel, Alice Bailey and Manly Palmer Hall further expanded my horizons.

The various esoteric investigators emphasized different aspects of their studies and investigations, but were unanimous in their cardinal descriptions of the unseen constitution of the human being. The human etheric double, called in various systems of thought the functional body, the vital body, formative-force body, ether body and orgone body, is actually a double of the physical body and duplicates the physical body in all its parts and organs. The etheric double surrounds and interpenetrates the physical body, and is perceivable by any person of normal vision who uses the correct techniques.

A readily practicable method is to obtain a set of dicyanin screens or goggles,[1] and systematically sensitize the eyes so that the etheric double and its associated radiation become visible. Goggles are the most convenient arrangement. Through the goggles, gaze at a frosted, incandescent light bulb of 75 watts, as used in normal domestic lighting. Arouse the function of vision fully by trying to see the filament of the bulb through the frosting. Invest at least 60 seconds in this effort, and preferably twice that time.

Have available nearby a shaded, fluorescent light fixture. Arrange this fixture so that you may look at your hands beneath it, without having the fluorescent tubes shining directly into your eyes. After spending at least a minute attempting to penetrate the frosting of the incandescent bulb through the goggles, move over to the fluorescent fixture and look at your hands beneath it against a white background.

The most important technique required in looking at your hands under the fluorescent light is to set aside, for a few brief seconds and as an act of will, the lifelong habit of looking *directly* at your fingers. Look *past* them to the white paper beyond. The radiation of your etheric double will appear as a soft luminescence standing out around your fingers. Move your hands together and separate them again. Note how this strangely fluidic, sticky energy trails behind your hand when you move it, seemingly violating the laws of optics.

Once you become aware of this strange radiation, you will also note its presence on certain theatrical motion picture films made in color. The effect is prone to appear when the bare head of an actor appears in dim or diffuse lighting against a sky background. The effect is also unequivocal on mag-

36

netic videotape recordings used in television, particularly when individuals are videotaped in conditions of diffuse light. In diffuse light and against a snow background, the effect is strongly evident.

If you can master the temptation to look directly at your fingers, you will see this radiation quite readily. Approximately 95 percent of all persons invited at random by me to see this phenomenon for themselves, under my direction, have been successful at first attempt. Any person wishing to know what is behind UFOs, and wanting to know how to get at them experimentally, will be well advised to not just follow this experiment with his mind, but to actually do it.

This energy is not a spiritual energy in any sense of the term, but a demonstrable physical presence that only requires the proper technique to make it perceptible. Once you have seen the energy for yourself, changes have already commenced in your cognitional powers. Persist with the goggles and screens. Learn to see the differences between the energy around the living organism and the energy around inert matter. Inert matter also has this luminous margin around it, but it is not as wide around most substances as it is around the hand or the head of a healthy human.

A human being extends farther than the immediately visible limits of his flesh.

Persistent use of the goggles will eventually result in your being able to see a certain amount of this radiation around everything, all the time. Soon you will not need the screens to see the emanation around pillars, posts, pictures and people, particularly in areas illuminated by fluorescent tubes, or where lighting and background are otherwise favorable. If your bank, for example, has fluorescent lighting, and peach-blossom or light green walls, you will be able to see that your teller, your bank manager and your fellow depositors all walk around in a bath of this radiation.

If you persist with the new technique of looking past objects rather than directly at them—looking *beyond* the object whose radiation you wish to see—you will soon find that you will be able to see this luminescence at will. You will learn to switch your focus. Bear in mind that this is nothing spiritual or mystical. The energy would not be discernible at all if it were not physical. The same radiation around living things is also detectable with a sensitive electrometer whose output is coupled into an audio amplifier.

In the Soviet Union, Semyon and Valentina Kirlian developed a method of photographing this radiation by exciting various parts of the human body, and also plants, with high frequency electromagnetic radiation. Living tissue was found, under this stimulus, to emanate definite patterns of this heretofore unsuspected energy, characteristic seemingly of vital activity in the various tissues. Here we may hark our minds back to the chronologically sudden, luminescent appearance of UFOs coincident with the injection of high frequency, pulsed electromagnetic radiation into the ether around the earth.

The Kirlian work has been essentially duplicated in the U.S.A. by such workers as Stanley Krippner, Ph.D., William Tiller, Ph.D., Thelma Moss, Ph.D. in association with Kendall Johnson, and also by numerous others. There is by no means full agreement as to what is being recorded, but the implication that the radiation patterns around living organisms are intimately connected with vital processes is inescapable. The pioneering Soviet investigation of this "bioplasmic" energy is detailed in *Psychic Discoveries Behind the Iron Curtain*, by Lynn Schroeder and Sheila Ostrander. The book was published by Prentice-Hall, Englewood Cliffs, N.J., in 1970.

Using an old color photo technique in 1969, Dr. E. Lewis of London reported having demonstrated a kindred phenomenon. Dr. Lewis stated that man is surrounded by a halo of warm air, charged with particles and with a high content of microorganisms. According to the British physician, the layer moves slowly in an upward direction, with circular currents as it goes around corners and into crevices.

All these manifestations will be shown in due course to be fully understandable in light of the work and discoveries of Dr. Wilhelm Reich and Dr. Ruth B. Drown, two pioneers who were persecuted in the United States for working with this form of energy, while many scientists at our major universities devoted themselves to drugs, napalm, explosives, weaponry and similar craven support of the military-industrial complex. If you will persevere with the simple techniques herein described you will be able to make a giant step into the new Cosmic Age of mankind. Unlike the mechanistic scientist in the employ of NASA, you will be able to look at the colored films of U.S. astronauts walking on the moon's surface—in the middle of clouds of bluish energy—and see something familiar.

You can see this radiation for yourself in your own living room, and get a fair idea of what it is by *experiencing* it ocularly. That is preferable to squandering 24 billion dollars of public money to stumble on the effect in an embarrassing scientific accident. This happened when Apollo 12 objectified the phenomenon accidentally, as the astronauts stumbled around *in vacuo* on the moon, exciting their orgone energy fields with the electromagnetic radiation from their portable radio transmitters.

Consider at this point, a germane statement of the immortal Goethe, who did more for mankind than just write poetry and immortalize Dr. Faustus:

"Every process of nature, rightly understood, awakens in us a new organ of cognition."

You have with your own eyes, and in an extremely simple way, seen that there is more to man than 165 pounds of flesh. There is associated with man—and with everything living—a field of radiation which official science is at present ignorant. If you still don't accept this after experimenting, then by all means accept the Apollo 12 color movies. All the time you have lived on this earth you have most likely been unaware of this energetic presence in your environment.

Many human beings, however, see this radiation from infancy and never lose this perception. They go through life seeing it all the time, and because their own perceptions are their only standard of objectivity, they assume naturally that everyone else sees this radiation around people, animals and things. This was brought home to me when I gave demonstrations of the perception of this phenomenon. Many times I have had people say to me: "Oh, I see *that* all the time. Always have. Doesn't everyone see that?" In such simple things lie the grossest errors of human philosophy.

Here is something that is *invisible but physical*. You are on to something that in its basic description connects functionally to those invisible yet physical objects detected by radar during the Second World War and since. Already you are on the trail of the UFOs. Already you begin to realize that perhaps with your new organ of cognition you might begin to win access to more of the amazing reality that exists unseen around you. In our introduction to the problems of a comprehensive UFO theory, we saw how official science had no method for approaching invisible-physical UFOs. You have begun to build that method for yourself.

As your perception of these things improves, you will be able to see, in normal room lighting and unequivocally, that there is a double body around your physical body. Grayish in color, consisting of this softly luminescent energy, this double stands out about 1/32 of an inch from the physical periphery. The cloud of radiation you have been seeing emanates from this double. The human etheric double is therefore indeed as Steiner and numerous other outstanding teachers and investigators have stated.

All of these great personalities also state that man boasts a still higher vehicle, which may be said to *irradiate* the etheric body and thereby give it the luminous quality you have observed. This higher, even more subtle body—the seat of consciousness and of the emotions—is termed the astral body. Man has his etheric double in common with plants and animals, and an astral body in common with animals only. The astral body is of emotional substance, and not accessible to physical detection like the etheric body and its radiation. We therefore leave the astral body with the status of a postulate. We already have sufficient topics to deal with for the moment.

Return now to your shielded fluorescent light and this time, without any goggles but using a black velvet background, hold up your hands with fingers spread slightly apart so that the shielded fluorescent light is between your eyes and your fingers. Avoid any direct sight of the fluorescent tubes themselves. Bring the fingertips of each hand close together, and moving the fingers randomly, use the same visual technique that you used to perceive the etheric radiation. Look *past* your fingers toward the black velvet. In between your fingertips you will see subtle, but quite distinct rays of energy connecting the various fingertips.

The effect is like a beam of energy escaping from the fingertip in the direction in which the finger is pointed. Move your fingers and hands around slightly and watch the way the rays intersect and shift from one connection to another. The emanations from the four fingers on each hand form a sort of gossamer web of radiation as you move them about. There is no question of their reality. These rays of energy have been escaping from your fingertips all your life.

The great universities have just begun to tackle this evident physical presence so intimately connected with life. Therefore you will find that most qualified biologists and medical doctors cannot tell you anything about this radiation. If you broach the subject too insistently, you will see them, on many occasions, become angry and irrational. You are on your own.

If you take away one hand and substitute a group of four pencils for its fingers, you will find that the pencils do not have this ray of energy escaping from their tips. This energy is therefore what the great esoteric investigators and teachers of the ages have termed it: life energy, vital energy or functional energy. In the modern scientific approach, as represented by the science of orgonomy, this energy is orgone energy—so named to permanently identify it with life.

You have now seen your own orgone energy field around your fingers and hands, and the rays that are escaping continually from your fingertips. Some additional experiments that you may simply perform will further verify the presence of this energy. You will find that irrational people will try, irrationally and without empirical backing, to reverse your conviction that these things exist. The position of the mechanistic scientist, often stated to me with sublime illogic is: "If such energy existed, *we would* know about it." Logicians may shudder, but thinking of this stripe still paralyzes progress.

Put your fingers together, with your thumb nestled comfortably on top of them. Stand face to face with a cooperative companion and have him stand facing you with his hand—right or left—similarly arranged and standing out at right angles from his body. Bring your fingertips to within an inch of his. Slowly oscillate your hand past his. Each of you will feel, as the rays escaping from your bunched fingers meet, a slight but definite impact.

This impact may travel far up the arm of a sensitive person, but virtually anyone who attempts this experiment in a serious manner will feel this energetic response for himself. A thin sheet of cardboard will overcome the objection that one is feeling draughts of air. The rays penetrate cardboard.

As a final experiment to convince yourself that you are surrounded by a physical energy that is new to scientific cognition, and to your own, take a standard 20 watt fluorescent tube that is used for desk lamps, and go into a room that is totally dark. Allow five minutes for full eye accommodation. Hold the tube firmly between your two hands, grasping the glass as though you were going to break it across your knee. Your hands should be separated by about six inches. Hold the tube firmly for a few moments, then quickly whip one hand clear of the glass while holding on firmly to the tube with the other.

A flash of light will appear in the tube. Continue grasping and releasing. The flashes will continue. If you persist, you will see that the tube sometimes flashes as you return your hand quickly to the glass. Awakening once on shipboard, I struck the bunk light with my right hand. The tube burned dully for more than five minutes, without any application of power from the 115v supply. We are not far from "free" energy.

In Rudolf Steiner's writings, the role of the ether body in all the processes of perception is described. The rays that you have seen escaping from your own fingertips, if you have followed the experiments outlined earlier, actually escape similarly from all body orifices and pointed areas. Departure of a ray of energy from the human eye is an integral part of the process of vision.

In *Man or Matter*, Dr. Ernst Lehrs calls this eye beam the "visual ray". This term is appropriate because the visual ray acts similarly to a radar beam, going out into space from the human being and extending consciousness out into space. Each eye emits energy, but one eye is the "master eye" and controls and dominates the whole process. Through the visual ray, the living energy in you makes contact with external objects, reaching out and grasping those objects, as it were.

The visual ray is of the same quality as the rays leaving your fingertips, but is much more sensitive and more functional because of its direct connection to consciousness. The visual ray is also under the direction, in a mechanical sense, of the exquisite human ocular apparatus. Brief reflection reveals why old conceptions of vision are no longer adequate.

Standard notions of human vision are mechanistic, and in their descriptions of the anatomy and physiology of the eye, perfectly valid as far as they go. Decades of dedicated investigation were necessary to assemble these mechanistic interpretations of vision, but in a fashion typical of such notions they fail at a crucial point. The unsolved enigma is: how do the pictures get into consciousness? A dead man may have had 20/20 vision on his deathbed, but sees nothing after the cessation of his heartbeat. His visual ray has withdrawn. His fingertips no longer emit rays. He no longer has an etheric double.

Seeing for the living human is a function of his etheric forces. The mechanical conception of sight is not sufficient, now that we know about this radiation. We are forced, by what we have observed and experienced of this energy surrounding and leaving the person, to entertain the conception of the visual ray. We do this now in highly practical fashion.

Lie quietly in a darkened room, and allow yourself to be come accustomed to the environment. Shut off all radio, TV and distracting sounds from any source. Remember that you are trying to bring into function and into your own awareness, faculties and endowments of your organism that have been beaten down by the daily assault of civilized living. You have also been brainwashed by mechanistic thought that such things do not exist. Give the grander realities a chance to live in you and to show you their wonders.

Hold your hands up in front of you in the dark, palms facing you. Remember that the visual ray is a subtle beam of energy that leaves your eyes like a radar beam—the same quality of subtlety as the rays you have seen and felt from your fingertips. Attempt now to feel your own visual ray.

Swing your eyes back and forth across the spot in the dark where your hands are. Do it slowly, carefully, persistently. Extend your touch out through the visual ray. Unless you are an especially insensitive individual, or one determined not to succeed, you will, after a time, feel the gentle passage of the visual ray across your hands. You will find its gossamer subtlety a new experience in feeling.

Persist and your sensitivity will improve. Focus all your faculties of touch into the visual ray and send it out into the darkened room to feel all kinds of protuberances. If there is a venetian shade in the room, run the visual ray up and down over the slats. Once you have developed a little sensitivity, the effect is akin to running the teeth of a comb over your fingertips, but immeasurably more subtle.

Paper forms that are directly responsive to the visual ray, complete with cutouts for immediate assembly, are contained in John P. Boyle's *The Psionic Generator Pattern Book*. Published by Prentice-Hall of Englewood Cliffs, New Jersey, U.S.A., this document will significantly assist private researchers in this field.

You may demonstrate the presence and the *power* of your visual ray to yourself in the meteorological realm. This direct proving is overwhelming.

Once again, you do your proving by participating. You may dissipate clouds with your visual ray by following a simple procedure. Do not attempt short cuts, variants or modifications until you have performed exactly in accordance with these requirements:

1. Choose a day when you have numbers of *small*, discrete clouds that are approximately the same size.

2. Select one cloud to be dissipated. Have a friend take a Polaroid photograph of the scene, and mark your selected cloud before you begin dissipating it.

3. Resolve inwardly that you are going to dissipate that cloud, drawing the energy down your visual ray into yourself.

4. Concentrate on the cloud and slice your visual ray back and forth across it. Then bore into it. Then slice back and forth. Then bore some more.

5. Within five minutes the cloud will have begun to disperse after clearly having lost cohesion. Surrounding clouds of comparable size to the selected cloud remain essentially unchanged. A friend with a Polaroid will objectify it all for you.

6. Concentration upon fragmenting portions of the cloud will secure their final dispersal.

Arguments and intellectualizing about this procedure are no substitutes for participation. Go out and *do it,* and leave the mechanistic skeptics to verbalize themselves to death. The proof and the changed cognition that comes with it will belong to you thereafter, and no one can ever take these things from you.

The most important thing to bear in mind when attempting this for the first time is that there is always a *delay* between the application of the bioenergetic stimulus of the visual ray and the manifestation of gross physical efforts, i.e., the visible dissolution of the cloud. Allow for this inherent bioenergetic delay. You are no longer dealing with mechanical things in a mechanical fashion. You are entering the realm of the living.

You will find that your whole ocular apparatus becomes strongly stimulated by visual ray cloudbusting. This stimulation is due to the orgone energy that binds the moisture in the cloud being drawn down your visual ray into your organism in accordance with the Law of Orgonotic Potential. This law will be elaborated fully later as we deal with the discovery of the orgone energy. Suffice it to say for now that orgone energy is negatively entropic and sweeps away the Second Law of Thermodynamics as a universal law. All you need to know right now is that if you are aware of your visual ray and use it as described, you can produce *tremendous physical effects at a distance.*

With the unaided human organism, one may not bust clouds indefinitely. Once the organism is charged bioenergetically to capacity, it cannot absorb more orgone energy from the clouds without discharge of that energy, either into water or through the sex organs. Cloudbusting is repeatable within these limits, with the visual ray.

Construction of a multilayered orgone accumulator at home, which will be described later, will enable you to further verify the presence of the orgone energy to yourself. When you stare at the *outside* of the accumulator from the other side of the room, you will find your eyes will start to sting gently as the energy travels back down your eye beam and fuses with the fluid on your eyeball. There is no commercial or battery power connected with such an accumulator. Here again, to participate is to know.

Another convincing proof of the visual ray, as a bioenergetic emanation from the human being, is to observe carefully your reactions and sensations when you see an image in a fun house distorting mirror, or look at the world through glass that has been physically bent at a sharp angle. Wrap around windshields on 1955-58 vintage cars provide such surfaces. A sharp sensation of *strain* communicates itself to the organism as a result of the *bending* of the visual ray.

This inner sensation of discomfort, and sometimes of pain, has nothing whatever to do with viewing a distorted image as such. Study a photograph of an image from a distorted mirror. This does not produce the sensation of inner discomfort that arises from being present physically, and therefore involved bioenergetically, in the actual distortion process. Careful self-observation is an important part of this new kind of work.

You have in your possession now most of the fundamentals essential to the serious understanding of UFOs. Scientists who frowningly kid themselves that only their work is serious have been unable to do anything with this ubiquitous mystery. Much of what you have learned will already have brought to you a lightness of heart, a feeling of being slowly reborn and renewed. Serious work on UFOs can only be undertaken with the investigator already willing to accept change in himself as a part of a vast learning process. You have begun.

Pursuit of this subject will occupy generations yet unborn, but the place to start is where the reality of things begins. That reality lies in your own orgone energy field, and with a new, basic understanding of human vision — an active, dynamic understanding that supplements and vitalizes the purely passive, mechanical concepts of human sight that are now current. The mechanist always likens the human eye to a fine camera, and he is right. To this we now add that from what we have experienced, we know that the eye with its visual ray also *closely resembles a radar set*. That is how we intend to use our eyes empirically. We are going to remember as we go that it was radar that first uncovered the UFOs in the modern period.

If your knees quake a bit because the USAF and its satellite scientists seem so awe-inspiring in their negation of UFOs, remember that they have officially thrown it all out as junk. They don't want any part of it. In December 1969, the USAF abandoned its 21-year investigation of UFOs because it "no longer can be justified either on the ground of national security or in the interest of science."

That is the official position of the USAF and of official science. They have publicly abdicated from the subject. That means that the field is all yours. Take courage from the scientist who did more to put humanity into space than anyone else, the distinguished Dr. Hermann Oberth, the "father of astronautics". Writing in *Fate* magazine in May of 1962, Dr. Oberth had this to say concerning the role of the intelligent layman in UFO research:

"If the universities already had institutes for ufology then I would say: 'Leave the research to them;' but there are no such institutes today.

Furthermore, this is an area which (a) embraces several special fields of knowledge; (b) can be handled at the moment quite well by intelligent laymen because all there is to do at present is to collect and screen observations, and often a layman is better at this than a specialist in another field, because he has a more general education and a better knowledge of people."

There is therefore no reason for the intelligent person to feel disqualified or unqualified in ufology, or that he cannot contribute to knowledge in this field. We do well to bear this in mind as we take unto ourselves the knowledge of the visual ray and the human orgone energy field and apply this knowledge to phenomena with which they have an obvious kinship. These simple, yet radical steps move us closer to the inner reality of UFOs than any conventional investigator has yet come.

Before we step off into the field, let us summarize what we have learned.

1. We have seen our own invisible double body, and the field of fluidic energy that surrounds us. Conventional investigation of this same energy is proceeding apace, sparked by the Kirlian work in Russia and employing the name "bioplasmic energy." This scientific work confirms, from a different standpoint, what you are able to perceive with the right techniques.

2. What we have seen by these techniques is borne out by astronauts having unwittingly photographed their own bioplasmic energy fields in the vacuum of the moon's surface. These fields are the characteristic blue of Wilhelm Reich's orgone energy excited in a vacuum.

3. We have introduced ourselves to the visual ray, and its role in vision and perception. We note that it is an invisible yet active part of ourselves that we can extend into space, and that every human being has a personal radar set in this bioenergetic guise. We have found that we can perform a quite prodigious task with this ray by dissipating small clouds.

The relevant and kindred UFO phenomena are:

1. Abundant evidence dating from World War II that their (the UFOs) main existence is an invisible-physical existence.

2. Detections by radar, often mass detections, without visual corroboration by pilots vectored directly to their location in space, have confirmed this invisible yet physical UFO presence.

We are bringing into consciousness, in our work with ourselves and upon ourselves, a realm that is similarly invisible yet physical. We note that we are not normally aware of this level of form and substance, but must bring it into consciousness gradually. Largely, it is a question of opening ourselves to something that has always been present, but from which our attention has been diverted. By consistent use of sensitizing methods we will become able to extend our seeing faculty until we see a portion of this invisible-physical realm around us all the time. We become aware of something new and consciousness follows the mental focus.

The time has come for you to stand by my side with this basic New Knowledge you have won, and let me show you how I took my first UFO photographs. Let me explain how I captured on film some of the unseen intelligences present in the atmosphere of the earth. You will be surprised, and perhaps even staggered, by what was possible to two ordinary men who wouldn't allow themselves to be put off or put down by self-styled experts— in and out of science—who have misled you, defrauded you of tax money, and openly lied to you concerning this greatest of all mysteries.

NOTES TO CHAPTER FOUR

1. Dicyanin goggles are available in the U.S. from Mankind Research Unlimited, 1315 Apple Avenue, Silver Spring, MD 20910.

Chapter Five

EYE INTO THE ETHERS

A bonafide example of extraterrestrial life, even in a very simple form would revolutionize biology. ...it would be truly immense.
                                                    —DR. CARL SAGAN

Cosmic workings leave nothing to chance, and rightly understood, they eliminate the concept of accident. Such workings were in process when a casual acquaintance of mine began in 1956 and 1957 to manifest a strong interest in my early dabblings in ufology. At this time I needed help, morally, spiritually, financially and physically. James O. Woods filled the void to perfection. We became good friends, then close co-workers, and finally brothers living together the greatest adventure of our lives. His premature passing in 1984, after a long illness, was a heartbreaking blow. He was an outstanding human being and a magnificent friend.

Without Jim Woods, none of the fieldwork could have come to fruition. My debt to him can never be repaid in kind. He stepped into my life at precisely the right time, and with a perfect background and temperament to aid me with my photographic work. Some of the most amazing photographs ever made were shot by him, and verify for all who view them the endowments and sensitivity that he brought to this work.

An account of some of our early adventures with UFOs in the high desert of California has already been given in *They Live in the Sky.* Space limitations make it necessary in this present book to skip over the early gropings. Our goal here is to turn those elements of the New Knowledge already conveyed to you in previous chapters, into field photographic methods—and results that you may pursue on your own.

Any application of photography as wild and revolutionary as the objectification of invisible aeroforms[1] in our atmosphere, should by its nature prepare the participant for a departure from old norms and old forms. The majority of people who have passed the age of twenty-eight find even the conception of space animals and craft hard to tolerate and as participants they are largely disqualified. One must be prepared to relinquish, when one undertakes the fieldwork, the comfortable, accepted and "safe" conceptions of reality with all their technical and academic underpinning. A new reality is upon us, forced into human ken by the UFOs.

There are no textbooks on this infant art, and it remains an art—with a scientific basis—up to the time of the writing of this book. Lack of an authoritative, traditional text disqualifies automatically the unenterprising and the timid, but everyone who is willing to risk time, film, money and contemporary ridicule is contributing to the New Knowledge and serving the New Age. The more evidence that is acquired of these new things, the shakier becomes the intellectual dictatorship of the old order.

There are no authorities in this subject, and no authorities on the objects that are being recorded. The mechanistic mentality wishes at the earliest possible moment to begin putting the various manifestations into boxes, but classification efforts are premature and so is classification thinking. The phenomena need to be allowed to flow freely into human ken, and in due time we will gain, from what we are able to objectify, the essential keys to their determinism.

Everything that a person raised in the mechanico-mystical mode of cognition has learned is violated when he goes forth to photograph things in the air that he cannot see with the naked eye. He is outraging all that he has been taught. When you stand for the first time with a camera around your neck, facing the empty heavens at dawn in some remote locale, the whole process can seem oppressive and even monstrous.

Space ventures are usually attended by the most complex and expensive equipment that the technical expertise of man can devise. In the new photographic art under consideration, this approach must be reversed. You must be willing to gamble that you can do, on your own, more than the mechanico-mystics have been able to do in great, multi-billion dollar projects. Your guiding maxim is to be: "Simplicity is the keynote of efficiency." The Little Man who has usurped your inner throne tells you that you cannot do anything great or revolutionary. Shove the little bastard aside. He has no right to obstruct the God-power in you.

If you surround yourself with ultrasensitive, highly complex equipment at this break-in stage to a new and dynamic realm, you will destroy your innate clarity of mind. You will become a slave to your instruments instead of master of the situation. When the arcing spheroids appeared on videotape, exposed on and near the lunar surface by our astronauts, the blocking of free perception by complex programming and instrumentation was adequately demonstrated. As an individual, participant-investigator, you must let through what programmed human beings shut out.

The mechanistic mode of scientific cognition is dependent upon instrumentation, all of which is valid in its own way and for its own purposes. In this new field, these things become an impediment. You must make contact with an energy that is neither electrical nor electromagnetic, but something quite different. Because this energy produces magnetic effects when it is manipulated, huge errors can appear in the instruments upon which the mechanists depend, most of these instruments being dependent upon the magnetically actuated D'Arsonval movement.

Fear not, therefore, over the paucity of your equipment. The human body is the most marvelous laboratory in physical creation. Begin on a minimum basis as I did, and you will find that simplicity and directness are the handmaidens of success.

Obtain a simple 35 mm camera, the simpler the better, provided that it is fitted with a good lens. For your own convenience and to maintain a lightproof control of your photographic materials, select a camera that you can unload and load easily in total darkness. This is the only satisfactory way to handle infrared film in a project of this kind. Be sure also that your camera is adapted to quick and easy loading and unloading within the confines of a daylight loading bag. You will need a loading bag on your field trips. Numerous camera bodies, including many famous makes, are absurdly overdesigned and cannot meet this primary requirement.

You will be working with ranges of light to which normal light meters and measuring devices do not respond. Therefore your camera does not need to be equipped with a light meter. The direction of the work goes once again toward simplicity of external apparatus, and heavier involvement of the investigator. Young people who see the modern world as being steadily dehumanized should leap with delight at the new realm we are penetrating, for here, the individual is all-important and not a punched-card cipher.

You, yourself, are going to measure the light with your own faculties. You are going to develop the ability to judge exposures. Therefore, you can do your work with a simpler, older type of camera to which no light meter is fitted. If your present camera has a light meter, you can easily learn to ignore this instrument, the functions of which are not specifically designed for your present purposes. Bear in mind as you begin to measure light by "eyeballing" it, that some of the top commercial aerial photographers in the world use this "eyeballing" method.

Your camera must have a sports finder that will mechanically frame for you the area into which you are aiming. In this type of photography you are going to use the full faculties of your vision, and you need to have both eyes open for unimpeded contact with the atmosphere and with the atmospheric orgone energy. If your camera requires that you look through any kind of prism or telescopic arrangement in order to aim it, and does not have a completely open sports finder, make a simple sports finder out of wire. A paper clip is satisfactory. Make it into a frame and slide it into the accessory shoe on your camera. Adjust the size of your finder so that it frames approximately the area covered by the lens at infinity.

You will need no other setting than infinity in this work. In some applications you will be using opaque, black filters, so if your camera is a reflex type, you will not be able to see through the lens anyway. A sports finder is essential.

You now have roughly the same type of camera that I had when I started photographing UFOs. Suitable cameras can be obtained in the used camera market for $50 or less. Most of my work was done with Praktica FX2, Praktiflex, Leica G and Bolsey cameras, all essentially minimum-type units with good lenses, but nothing extraordinary. The cameras are less important than the photographer, and what he brings to the total process.

Obtain an 87 filter to fit your camera. This filter is black in color, and if you are unfamiliar with such things, you will find it incredible at first that any kind of photograph could be taken through such a black glass. Hundreds of photographs are made every week in scientific and medical photography in this way. You will make unorthodox use of a well-established process. You have your camera, and a suitable filter, and now you need only your film.

You are going to use a material called *infrared* film. All this means is that the film has been sensitized to the electromagnetic radiation that occupies the spectrum before the color red. *Infra* is a Latin term meaning "before", and infrared simply means "before the red". Since the color red is the lowest vibratory activity perceived by the normal human eye, you are going to objectify on your film anything that happens to have form in the range before the red.

The film has been sensitized to this invisible range by treatment with special dyes. At the opposite end of the light spectrum as seen by the human eye lies the color violet. Beyond the violet lies the *ultraviolet, ultra* being another Latin term for beyond. Later on, we are going to probe this realm in a new way also, but for now it is enough that you should find nothing in these concepts to overwhelm or disqualify you.

When you place your 87 filter over your camera lens, only the invisible infrared radiation passes to your film through the lens. If you use this film without a filter, you will widen your window on reality by almost two-thirds of its normal width. Remember too, in an echo of the fundamental mysteries surrounding UFOs, that the infrared portion of the spectrum adjoins the microwave spectrum—the portion of the spectrum in which radar operates and in which radar first detected the invisible UFOs.

You are not reaching into the microwave portion of the spectrum, but you are getting nearer to it than you can with your naked eye. You can see now that it is not so very crazy after all to take this photographic trip into the new reality. You are taking a novel step, but facts, logic and experience justify that step.

There are two types of infrared film on the U.S. market that are suitable for ordinary 35 mm cameras. There is Kodak IR-135, a low speed film that is sold in standard 20 exposure cassettes ready for you to use. No special handling is required for this product, although it is wise to load and unload your camera in total darkness for full protection against fogging. There is also Kodak Highspeed Infrared Film HIE 421, sold in 100-foot rolls, and also in 20 exposure cassettes.

If you spool the HIE-421 into cassettes yourself, this must be done in total darkness. Your camera must be loaded and unloaded in total darkness. Cassettes of this film must be placed in cans or other light-tight containers and transferred to and from your camera only under conditions of total darkness. This requires either darkroom loading, or the use of a loading bag. These precautions preclude fogging. Because the film is so superbly sensitive, the tightest controls on handling and processing are essential. Onerous at first, the "darkness all the way" procedures soon become routine.

When we discussed earlier the seemingly irreconcilable observed fundamentals of UFO manifestations, you will recall our mentioning heat and heat rays in connection with UFOs. Only one case is necessary from classical ufology to verify this and there is none better than the 1 July 1954 Walesville, N.Y. case. Two USAF airmen were nearly overcome by appalling heat that filled the cockpits of their jet when they pursued a UFO. The heat was so intense that the flyers were forced to bail out, leaving their jet to crash in a populated area. In case you are not aware of it, infrared radiation is simply a scientific term for *heat.*

You see that every step you take along this direct line of investigation is widening your ordinary sense of reality to meet the reality of the UFOs. Reality is reality, of course. The point here is that ordinarily we tune in to only a small part of the spectrum. Most of reality is ultra- or infra- to our tuning, as in infrared radiation. With your camera sensitive to heat you are perhaps chasing this aspect of UFOs more closely than your realize, perhaps more closely than you can tolerate. If your solar plexus is starting to tighten, you will know from experience how we felt at the beginning of all this, because no one had ever tackled these phenomena in this head-on way before.

Early in this book I stated that certain UFOs are biological, i.e. living organisms. There is a basic principle connected with mammals—body-warmth—that is worth noting here. All the formal biological wisdom of this world has been unable to determine where the warmth comes from, or where it goes at death, but only to say with certainty that it is a condition of life in the mammal. Your camera and its special heat-sensitive material may well be expected to react to anything hot, or even moderately warm, in the atmosphere.

You are in a fair way ready, with your simple equipment and dauntless heart, to cut right across anything that is living and has form in the invisible, infrared range. Something is there, invisible, because radar has detected its presence. Forget what man's natural philosophy has conditioned you to accept about form, substance and life. Be curious enough to look, to raise evidentially-backed questions, and to accept what you find in freedom.

In the telepathic encounters previously described, the following basic advice was received concerning photography of UFOs:

1. Remember that UFOs belong to an invisible realm, and if you want to see beyond the physical, try and *look beyond* the physical. (This tallies with the descriptions already given for perceiving the radiation around your own body.)

2. Use infrared film, and find out for yourself how to adapt this material to a new application.

3. Operate in the period between dawn and sunrise. This will allow you to make best use of the balance between light and dark, and to catch elemental forces that are in full function at that time.

4. Choose a site where you can count on clear skies, low humidity and freedom from disturbance. The high desert was suggested, near but not at Giant Rock.

5. Cooperation would be forthcoming but there would be *no physical materialization*.

Dr. Franklin Thomas added two valuable suggestions. First, expect the unexpected. Second, be prepared to persist by going back to the same site on a regular basis, preferably every week at the same time. We chose a site between Yucca Valley and Old Woman Springs that was accessible yet isolated. We now had what we believed were the right materials and equipment, a good place to work and an approach that we believed would expose the UFOs *in situ*. There was a missing element. How to attract UFOs to our particular location.

We have already encountered one aspect of this problem in describing Van Tassel's telepathic communications. He developed a method by which he would make the presence and purpose of his group stand out from the random psychic noise of humanity. Unless we were to be entirely passive in our activities, depending upon the UFOs to manifest, we would have to have a method of attracting attention to our selves. We would have to reach out to the cosmos and not be onlookers. The method we developed involved the invisible life energy of my body, the same energy that you have been able to see around your own hands with the aid of the dicyanin goggles or screens.

Once again, it was the knowledge and insight of Dr. Franklin Thomas that made our experimental beginnings possible. He suggested cyclic repetitions of the "Star Exercise", an esoteric procedure by which the human body force field may be strongly energized. Since those early days, the Star Exercise has been openly published elsewhere, and I am not violating any confidences or secrets by describing it here.

Face the quadrant (north, south, east or west) to which the sun is closest at the time you are commencing the exercise. In the morning, before dawn, you would naturally begin facing the east. At noon, you would face south, and so on. With legs spread apart easily, and with arms parallel to the ground, gently oscillate the body a few degrees across the north-south or east-west line of the earth.

If you are sensitive, you will feel small warm spots in the downturned palms of your hands. In the region of the thymus you will also feel a slight impact at resonance. These indications enable you to "tune" yourself to the magnetic and diamagnetic fields of the earth.

For a substantial part of my life I have been professionally connected with the art and acts of tuning electronic apparatus, so this new kind of tuning came easily to me. Practice will bring it within reach of virtually anyone. The main thing is to do it—to participate. You will inevitably become conscious of these subtler natural forces.

Later in our work, we captured the radiation jumping out from my fingertips at resonance, recording this on high speed infrared motion picture film. Therefore, it is not theoretical, imaginary or merely a concept that there is such radiation. Securing proof on film of this resonance point as my body was rotated across the west-to-east lines of the earth served to verify the assertion of electrical pioneer Michael Faraday that man, in his totality, is *diamagnetic.*

Our operations began in this general way in the summer of 1957. I acted as the "bioenergetic beacon" and Dr. Jim Woods as the photographer. I would stand on top of a large rock, silhouetted against the western heavens, and Jim would site himself to the east of me with his back to the dam, ready to shoot his camera into the sky behind me. This procedure obviated spurious reflections, and unless the sky were cloudless, the operation would be abandoned. Under these conditions, we could be reasonably sure that anything supersensible yet within range of our film—would be at least partially objectified for further study.

Performing the Star Exercise involved finding the resonance point while facing east, then turning 180° around to the right and finding it again. Then 90° to the right and finding it while facing north, then 180° to the right and finding it again while facing south. Thereafter, I would turn 270° to the right and return to face the east and begin again. I would perform from sixty to one hundred repetitions of this procedure in the dawn period.

Performance of the Star Exercise in this fashion results in a regular pattern of bioenergetic pulsations in the ether. Dr. Wilhelm Reich's work in weather control, which we have carried forward in recent times, has verified the extreme sensitivity of the earth's orgone energy envelope to any kind of stimulus or disturbance capable of affecting this continuum directly. My purpose was to get myself noticed by unseen intelligences not of our own physical density and polarity, but having their existence in an unseen borderland. This was the whole purpose of these long repetitions of the Star Exercise.

Draw a parallel from everyday human life. Traffic passes your home day and night, something you accept as routine. The occasional blowing horn, cries of children or revving automobile engines do not attract your attention. If someone should turn a strong spotlight on your living room window however, and flash it fifteen times a minute every night at 9 p.m., what then? You are going to go to the window to see who is trying to attract your attention and why. This is what we did with the borderland to our own world.

Under extremely dry conditions, and with a stiff breeze blowing, Jim could "see" by looking past me into the western sky—just as you look past your hands to see the rays of energy from your fingers—that my body was surrounded by a huge, luminous sphere that pulsed regularly with my movements. As the exercise was continued, the sphere would grow slowly larger, extending at times to a perceivable diameter of well over one hundred feet.

This highly unusual yet intelligent activity by a human being in a remote place—a locale where elemental forces normally held sway—quickly precipitated a response. Blindingly fast flashes of movement began appearing around me, and Jim sought to capture these on film. They were at no time tangible in the regular ocular range, but required the same subtle shift in the focus of vision as is necessary to see the radiation of your own etheric double.

From inside the luminous sphere that Jim reported seeing around me, I could also see luminous things darting at incredible speed. They moved so rapidly that it was not possible to react fast enough with the camera shutter. *Something* was there. No spaceships showed on our films in an unequivocal fashion, but we found that against the black sky background produced by infrared film, there were present in ethereal, barely tangible form, the shapes of many of the standard bell-shaped flying discs.[2]

The kindly injunction of Dr. Franklin Thomas to expect the unexpected was not in mind when we completed our operations on the morning of 25 August 1957. We felt inadequate. Our groping approach left much to be desired. Yet we had seen enough to know that we were stirring up something and that we had to keep going until there was a breakthrough. No one had told us it couldn't be done.

Consciousness follows the mental focus, and with the pursuit of UFOs dominating my waking thoughts, combined with persistent repetitions of the Star Exercise, I was becoming extremely sensitive. Impacts that had never reached me in the past struck me like a punch in the nose. When I passed people in the street, I could feel the intersection of their body force field with my own, just as though they had touched me. Sometimes I could feel this impact thirty feet from another person. This expanded awareness and superfine tactile sense worked for me on the morning of 25 August 1957.

I suddenly became aware of a presence above me as I sat at breakfast with Jim. The impression was overpowering. Springing to my feet, I looked up overhead. "There's something right above us," I shouted, snatching up the loaded Leica. Against the clear blue sky I could see a strong pulsation, and a shimmering variation of the otherwise smooth blue background. Jim said he couldn't see anything.

From the inception of our probings into the unseen we had an agreement concerning any perceptions we had of objects not directly visible. If I were to report to Jim a certain perception, and he had not observed the same thing, he would so state unequivocally, without regard to my sensibilities. I would similarly confirm or deny his perceptions. From this procedure we never departed. As a result of this strict agreement, the law of probabilities has always been on our side.

On this occasion, Jim did not "see" the pulsation that I saw. At this time, I believe I was more sensitive than he was. The Star Exercise is a powerful sensitizing procedure. Acting on my own, I began making a series of exposures, describing to Jim how the pulsations moved above us. First I photographed the object directly overhead. As it moved over toward the south and seemed to lose altitude, I became able to frame portions of the desert terrain in the picture with the pulsation. Securing relationship with known physical topography or objects was a primary requirement from the inception of our work, so that with the UFOs, wherever possible, there would appear a known terrestrial reference.

Six successive exposures of the object(s) were made by me, and four of these are published in the Alpha Series in this book—in the sequence in which they were made. Similar creatures have since been captured on Apollo mission video tapes above the lunar surface and against the black cosmic sky. This NASA accident vindicated my original claims that these supersensible organisms exist and are an intimate part of the UFO scene in its totality. What NASA stumbled over—and still does not recognize—I obtained by deliberate pursuit of UFOs into the invisible.

Certain observations are permitted by these photographs concerning the nature of these creatures and the mode of their involvement with UFO phenomena:

1. In the first exposure, the object has recorded on the film with great intensity and considerable detail. At the time of writing this book, there is no comparable photograph available to the public, from any source, showing such close-up, in-focus detail of a biological UFO.

2. An observer whose perceptions are undistorted by any kind of neurotic overlay or mechanistic bias, will see in this photograph not a spaceship from another planet, but some kind of living organism.

3. Proceeding through the series of pictures, the dark, T-shaped mark in the body of this organism may be seen in every photograph.

4. From its tightly contracted and radiant state in Alpha #1, the object grows larger and less dense, as well as more transparent, as the series progresses. The impression of something swelling, *i.e.*, passing into the "charge" phase of a bioenergetic pulsation, is virtually undeniable.

5. This series of photographs verifies the subjective perception of an object pulsating in the air above my head, the manifestation that first drew my attention to this particular type of phenomena.

When I first outlined this story in *They Live in the Sky*, little credence was given to "sensing" UFOs in this way. Official science has in the meantime verified that certain persons can, as it were, "hear" electromagnetic radiation. Professor C. E. Ingals of Cornell University performed test experiments of this kind using a low-power radar beam, and proved that the radiation was detected in the brain itself, independently of hearing processes.

Further tests by Dr. Ingals established that the ability to sense the presence of pulsed microwave radiation existed above the forehead. With UFOs, one is not dealing with pulsed microwave radiation *per se*, but with cyclic disturbances of the ether itself, most accurately described as orgonotic pulsation. Since my report of this particular detection incident predates the work at Cornell, there is some satisfaction in learning that independent, conventional investigation has verified what I long ago set down in print.

By attempting in my work to give objective evidence of the truth of my findings, I have sought to discharge, to the best of my solitary ability, a prime scientific obligation. As the distinguished French mathematician Aime Michel has pointed out,[3] theories are the cheapest part of our research in ufology. Although my work is evidentially backed, it has nevertheless been almost totally sequestered. Later on, we will understand better this quirk of the mass neurosis that sends men swerving away from the living element.

There was a thorough understanding on the part of Dr. Woods and myself of the *unwelcome* nature of these photographs and the dozens of similar, supporting photographs subsequently made by the same methods. These living creatures, these bioforms, were neither what we wanted nor what we expected. We wanted spacecraft. At that time, we stood in ignorance of any biological element in UFOs, and to begin with we hardly appreciated seeing what looked like unicellular organisms when we were looking for spaceships.

While these photographs will probably at a future time be recognized as the break-in to genuine exobiology, they were for us at that time, a definite emotional letdown. In the intervening years I have observed with interest and fascination the disquieting, disturbing effect that they have on all persons whose approach to UFOs is mechanistic. This experience has taught me that purported scientific "objectivity" is a fiction—an ideal to reach reality among men eons hence.

Returning to the first series of photographs, we should be mindful that they objectify a seemingly "solid" organism that was not directly visible despite its immediate proximity, and obviously large size. I wanted a spaceship photograph, and went after such a photograph in a way that seemed to me to multiply my chances of success—all of it based on the radar evidence that some UFOs hail from an invisible yet adjacent realm. I got the pulsation not from a spaceship engine, but from something alive, precipitating its form repeatedly into a film emulsion.

Carefully observe a salient technical factor concerning the invisibility of these objects that is obvious from the pictures: The organism recorded on the negative solely because an 87 filter over the lens absorbed the blue radiation from the sky above the desert. As the bottom area of the exposures, immediately above the desert floor reveals, a normal sky return causes the organism to be absorbed into the background, *becoming invisible even to infrared.*

Study of thousands of exposures through the years leaves me in no doubt that the sky background of infrared photographs contains all kinds of artifacts, organisms and forms that are new to mankind. Regular film exposed in an orthodox way absorbs all this into its regular sky background. Any person who carefully studies infrared sky photographs will find the truth of this for himself.

The existence of these organisms seems to be essentially *plasmatic*, i.e., having their form expressed in heat substance. They travel in pulsatory fashion, swelling and shrinking cyclically as they move through the air, much as we pulsate with our heartbeat and swell and shrink with our lung movements. They are spheroidal, and therefore often look like discs.

Being plasmatic, they are capable of returning radar echoes, and often do. These things may be read more or less directly from dozens of photographs of these critters that we have made. Motion pictures we made verify that these plasmatic organisms are capable of velocities and changes in shape and size that render worthless any attempt to handle them with the mechanistic mode of mentation.

By following a direct theoretical line into practical experiment, keeping our approach as simple as possible, being willing to use our own organisms as instruments of investigation, and insisting on objectification of our findings, a breakthrough was made into the invisible whose ultimate consequences cannot be foreseen. That there are vehicles of intelligent design involved in the UFO phenomena seems indubitable. So also is it indubitable that there are invisible living organisms in the atmosphere around us. These organisms share a common functioning principle (CFP) with the spacecraft.

Franklin Thomas had warned me to expect the unexpected. He was plainly delighted when he saw this first set of photographs. "So you found them" was his comment. Like many accomplished occultists, he was able to recognize our achievement because with their more highly developed sense centers—dormant in the untrained person—such occultists perceive these organisms directly. Young people who have precipitately blasted open such perceptions with chemicals have also seen these queer atmospheric fauna. They are no longer any kind of secret from mankind, but in 1958 they were an overdose, and aroused much irrational fury.

Dr. Thomas had been right in his guidance. His expectations had been borne out. The intelligences who had made the basic empirical suggestions had also stated quite clearly that there was more to UFOs than spacecraft. Their basic suggestions had been followed and had produced certain clear results. There was indeed more to UFOs than spacecraft.

Further pursuit of the subject along the same lines seemed justified. In a short time, we had broken into a new and dynamic physical borderland of which there was no knowledge in official science, and to which, as I was to discover, there existed an almost psychopathic resistance. What mattered, in our eyes, was not who we were or who we thought we were, or whether we were right or wrong, but what we had *done*.

We had brought to ufology its first qualitative element. Rational persons no longer doubted that some kind of intelligently controlled objects had been appearing persistently to mankind since the Second World War. Our biological UFOs were a much-needed addition to the basic spacecraft concept.

In the mode of their manifestation to mankind, spacecraft and biological UFOs had been mutually confused, all through the modern period and down to this day.

Verifying this was not particularly difficult if one went at it directly. Two amateurs had done it with the simplest equipment. The phenomena were accessible. They could be understood if only we would let them manifest in their own way and "read" that language. The blockage lay in man, who had already made up his mind that UFOs were ships from other planets—and nothing more.

No comprehensive theory can ever be formulated concerning UFOs that does not include biological UFOs along with the concept of intelligently controlled craft of some kind. Establishment ufology has clung tightly to its two-plus-two assertion that UFOs "must be" ships from other planets or outer space, and all purportedly "objective" investigation is biased toward proving this foregone conclusion.

A biological revolution such as Dr. Carl Sagan envisions from the discovery of extraterrestrial life, "even in a very simple form," has already occurred in truth. Staved off for a few more desperate years by life-haters, this biological revolution has our strange atmospheric amoebae at its core. Their discovery has doomed the crude and lifeless cosmo-conception of the mechanists, and with it, their illusion of neurotic security.

We called these creatures "amoebae" because of their resemblance to unicellular lifeforms. We chased them hard into the borderland we had opened. Soon we were running across their functional companions in the invisible—constructs propelled by an unknown mutation of energy. In the coming months and years, the crude eye into the ethers that we had fashioned was to give us an accurate perception of our own ignorance, and fuzzy but unequivocal flashes of an interpenetrating world.

NOTES TO CHAPTER FIVE

1. The term "aeroform" was contributed to ufology by Mr. John Bessor of Pittsburgh, Pa., veteran UFO researcher and theorist.

2. These forms were not reproducible by standard half-tone printing processes and have therefore never been published. They have been shown in exhibits on prints directly from the negative. We obviously lacked a technical step that would change the polarity of these objects. They were *positive to light.*

3. *Flying Saucer Review,* "*In Defense of the E.T.H.,*" Nov./Dec., 1969.

Chapter Six

ETHER SHIPS

The occultism of today is the science of tomorrow.
— MEADE LAYNE

They were adventurous, hard-driving days for us in the summer of 1957, on through the fall and winter and up until the spring of 1958, when we took our first break in our photographic pursuit of UFOs. All else in our lives shrank to nothing by comparison. The challenge consumed our energies and resources like an ever-present dragon. Our women folk became virtual widows—abandoned on weekends and left at home weeknights while we labored in the darkroom. Ordinary business activities, essential for income, were begrudged the time they demanded for the pursuit of UFOs. All spare money was poured into material and equipment costs, laboratory bills, gasoline and traveling expenses.

Every Friday night, after gulping a meal, we would pile into my Hudson Hornet convertible and drive to our high desert site. Being alert and fully functional for the vital period between dawn and sunrise required that we roll out of our sleeping bags in the pre-dawn darkness. This routine became grueling in the perishing chill of the desert in winter. Had we not had results with this work, we never would have been able to keep going in the face of these rigors.

The inordinate disproportion of our two-man efforts, plugging away in an immense, all-new field, was a source of concern to us in the beginning. Surely it must seem inane to many people, even now, that two ordinary men without outside assistance, were tackling any kind of research program. Ours was all original work, including the methods themselves, in a field that had defied penetration by this planet's best-trained scientists. Many today will find it strains credulity that useful results could flow from such a wild project conducted by amateurs.

The leading minds of academia have, up to the time of this writing, declined to tackle UFOs. Their belief is that the *government* should do something about it, as exemplified in the NICAP publication, *UFOs—A New Look.* In this booklet, some fifty qualified men of science signed a statement that they doubted the objectivity of Air Force investigation, pledged their support to any thorough scientific program on the subject, and urged Congress to initiate an investigation. The universities and organizations employing these fifty gentlemen control billions of dollars worth of facilities and resources—but the government should do something!

Our view in those early days of our UFO work—and it is one we have never seen reason to change—is that *we are* the United States government. By exploring the borderland of nature in the 20th century, and saying our farewells to the comforts of the neurotic plane, we became the spiritual brothers of the mountain men. They had left the civilization of their time behind, and faced another kind of wilderness with primitive tools. Our wish, from the beginning, that Americans should face and penetrate this mystery never left us. Often we kidded each other that an immigrant and a native son tackling this new wilderness of the mind was a modern echo of the American past.

Numerous nights on the high desert of California enabled us to perfect a basic observational technique that is passed on here. Anyone who wishes to see UFOs for himself, and thus take the subject out of the realm of verbiage and evasion, can master this technique. I developed it by synthesizing my knowledge of the human visual ray with an observational technique I learned during my early seafaring years.

In 1947, I sailed in the British Prince Line with a chief mate named Stevens. He seemed to have a fantastic pair of eyes for detecting distant objects. My eyes had always been sharp, but I was no match for him. I once asked him how he was able to pick up a ship coming over the horizon minutes before I could see the vessel, especially since in normal applications his eyes seemed poorer than mine. He wore glasses for reading. He was 38 against my 22, and therefore by the standards of youth, very nearly burnt out. Meeting Stevens was another of those brief encounters that paid unexpected long-term dividends.

"Your trouble, young fella," he said, "is that you stand up here looking for ships. You glare at the horizon. That's why you don't ever see them until the stack or something massive appears. The art of seeing small things at a great distance is not to look for them, but to *let your eyes find them.*"

He went on to explain the simple technique of relaxing one's eyes, and just allowing them to roll loosely around the horizon, easily and naturally. This was the polar opposite to driving the will into the ocular apparatus in an effort to pierce the distance. Applying his suggestions, I found he had spoken the truth. When the eyes were relaxed and allowed to seek in their own way, I was astonished at their ability to detect instantly the bare tip of a mast or the spider-thin upper rigging. The relaxed, gently rolling eye hooked on to the object and brought it into consciousness unerringly.

These experiences returned to me with force when I learned years later of the visual ray and its role in human vision. This subtle, energetic extension of the human being will detect objects at great distances. The technique was applied to observation of the night sky for UFOs, and anyone who spends a little time developing this ability will acquire quickly all the evidence he needs of the presence of strange aeroforms in our atmosphere.

Here is the procedure to follow. Choose a site where you have clear skies, away from population centers and as high up as possible. The high desert of Southern California, where this technique was evolved, is around 4,000 feet above sea level, and most nights are clear. At 5,000-6,000 feet in the mountains of the southwestern U.S.A., one approaches ideal conditions, but there are thousands of sites throughout the world that meet the basic requirements.

Lie down comfortably, warmly clad or covered, so that you have the celestial sphere above you. Have a companion with you who will also observe. You strengthen your confidence and your accuracy if someone shares and confirms your experiences. Agree to be completely honest with each other in confirming what is observed. This way, the law of probabilities becomes your ally, and you have a counter balance to misobservation. Avoid idle conversation, concentrating entirely on the celestial expanse above you.

Allow your eyes to relax, and do not focus on any one point in the heavens. Look instead at the whole inverted bowl of the night. Avoid conversation so that your attention is not diverted by other sensory stimuli. Once you have overcome the lifelong habit of focusing on, or staring at, one particular spot in the heavens—transferring your gaze only from spot to spot—you will find that any movement in the cosmic expanse above you will attract your attention instantly. You will find that the night is full of movement, and that it is barely possible to spend even ten minutes without catching sight of something moving in the sky.

You will catch glimpses of meteors, which enter the earth's atmosphere thousands of times daily. You will sight man-made satellites easily. Time and again however, your vision will be drawn instantly to something that is not a meteor or satellite. You will become aware of the passage across our skies of whitish, translucent objects traveling at enormous velocities and becoming visible only for instants.

Your vision will adjust to catching these objects, if you persist in practicing the simple observational technique described. Occasionally, in the quiet of the night, you will hear a distant crackling like that of static electricity, as the objects zip across the heavens. Often they will leave a luminous trail etched against the stars for several seconds. An example of this trail appears in the Alpha photograph made above George Van Tassel's home at Giant Rock. The "trail" appears like a pair of electrostatic rails along which these strange vehicles slide at thousands of miles per hour.

One hour of patient work is usually sufficient to permit sighting of a UFO using these techniques. Persistence over a period of months will verify for you that the time between midnight and 4 a.m. is the most fruitful—a time that etherian physics assigns to the maximum contraction into the earth of the chemical ether, which can be identified with the orgone energy discovered by Dr. Reich. During this significant time period you will be astonished by what it is possible to see.

You will certainly see the well-known bell-shaped craft materialize occasionally during maneuvers, and make the right-angled and U-turns that you have heretofore only read about in books. You will observe for yourself that not all the objects you see glittering in the heavens are stars or planets, and that, occasionally, things that look like stars to the casual or impersistent observer move in the sky relative to nearby stars and planets. Sometimes you will see several of these phony "stars" subtly come together for a few minutes, then separate and return to their previous positions.

Anyone who thinks that all seemingly fixed objects in the sky move steadily, should make a simple time exposure of the heavens, as did Dr. Reich. He found star-like objects notably diverging from the courses traced on his film by the rest of the heavenly bodies. Tackle your viewing venture seriously, and you will find I speak the truth, regardless of any scoffing that may emanate from stiff-necked astrophysicists or others. Where I have been asked to take serious groups on such viewing ventures, I have never had a failure. People can indeed see UFOs when they go about it the right way.

Two important things establish themselves from protracted experience with these techniques. First, extended sightings of UFOs are highly unusual.

Never once, in thirty years of observations, have my sightings been protracted enough to raise and aim and set a camera. Furthermore, the objects are usually at tremendous altitudes, something borne out by friends of mine who have chased them as USAF pilots—only to find the UFOs far above them at the ceiling of our hottest fighters.

Second, the objects will be *observed to come from and disappear into an invisible state.* This firsthand experience reinforces the radar evidence gathered from World War II onward, that UFOs are essentially invisible, and their appearance to the eyes of earthmen essentially sporadic. Our experimental work continued side by side with constant observations of the heavens and of the atmosphere around us, in tune with the new realities that had been opened.

We soon became aware of a basic problem in *polarity* that is central to the photographic recording of UFOs in their native, invisible state. What the visual ray would detect against the clear post-dawn, pre-sunrise heavens as a bright pinpoint of bluish/silvery light would print out on the finished photographs as areas of total blackness. We did not know it at the time, but we had collided with the basic determinism of primary energy or orgone energy, and were trying to solve our problems on the basis of secondary, electromagnetic energy and the electro-mechanistic theories attached to this energy.

Let us retrace our steps to George Van Tassel, and his early booklet, *I Rode a Flying Saucer,* for some hints and clues as to what was now turning up on our negatives as we groped in the invisible. We were not afraid, then or now, to test unorthodox or unusual information in our search for a break-in. In his first booklet, Van Tassel reproduces a communication received telepathically from his otherworld contacts regarding spacecraft propulsion.

The space-being Ashtar, well-known to those who have perused contactee literature, is the personality from whom this technical data is purported to emanate:

"For the information of your scientific minds throughout the planet, Shan, our ventlas do not spin. The emanation of spiral radiation from our ships gives the illusion of spinning. The upper or positive polarity of a ventla radiates emanations outwardly from the center. Due to the collection and concentration of light particles through a vortice funnel in the center unseen, these light emanations radiating outward appear as grooves on one of your phonograph records. The lower negative polarity operates in a reverse manner. This light substance emanation is contained within a field of zero circumference which is void, giving the impression of an edge. Your spectroscopic camera will reveal us only as light in the spectrum, plus elements in your atmosphere."[1]

Later, Van Tassel published a cross-sectional drawing of a ventla-type vehicle as operated by the entities with whom he had contact. This drawing is reproduced in the photo section of this book. Since my own involvement had arisen out of my visits to Van Tassel's place, there is reason to expect that something would arise in my photographs that would verify in some way what Van Tassel had published.

Beside Van Tassel's drawing appears Alpha #5, also called "UFO Ahoy!," a photograph made in April 1958 by James O. Woods. This photograph captured this ventla-type UFO *in situ*, invisible to the naked eye. The dense black dot atop the two-decked and partially materialized body of the vehicle corresponds to the "lens" or multifaceted arrangement provided for energy ingress to the vehicle. The whitish lobe of radiation around the object is obviously a plasma created in the atmosphere by the spinning, orgonotic field of the vehicle. The heat of this plasma has been detected by the infrared film.

This is one of the few photographs made of a person in deliberate proximity to a UFO, and the only photograph published at the time of writing showing a human being and ventla vehicle *in situ* in the same photograph. The propulsion clues it yields are worth a hundred thousand hours of neurotic hassling and evasion. No would-be "expert" will ever knock this photograph over or discredit it. Those who thought they could have, returned it sheepishly through intermediaries, and there have been no blazing exposes of fakery. Compare Alpha #5 with Ashtar's description of ventla propulsion given above, and you will realize that Jim Woods pulled off one of the greatest photographs ever made.

This particular photograph, and Van Tassel's drawing, duplicate in essential detail the form of the disc-shaped UFO photographed by a French military pilot near Rouen, France on 5 March 1954, the well-known McMinnville, Oregon photographs of later date, and the enigmatic series of photographs made at Cluj, Rumania, on 13 August 1968 by a construction technician and former military officer. Chance? Accident? Beloved as chance and accident are by neurotic mechanists as a means of evading the essential, they do not apply here. Nor does collusion or fakery. The existence of an advanced technology, with its base in bioenergetic phenomena and etherian physics, is the explanation that makes sense.

Early in our field work, we became aware that the proximity of these invisible vehicles produced in my own body force field, or orgone energy field, striations of light that were visible to me at the center of that field. Occasionally, Jim would catch glimpses of this peculiar orgonotic heterodyne, but inside my own sphere, as it were, they often achieved a remarkable brilliance. That it is an orgone energy effect hardly seems dubitable today, for its functional counterpart may be produced simply by moving a household fluorescent lamp rapidly past a piece of suspended polyethylene sheeting in darkness.

The first time I saw the latter manifestation, my mind returned instantly to that April morning on the Mojave Desert. When "UFO Ahoy" was made, I had myself run out of film. The orgonotic heterodyne of the vehicle's propulsion field with my personal field appeared brilliantly to my left. I at once called to Jim to shoot accordingly. The photograph itself verifies the essentials of this account.

The visual ray, once its presence and nature are understood, will unerringly detect the lens area that appears on top of these ventla-type craft, and also the high orgonotic potential areas underneath the vehicles. The potentials in this area are high enough to nullify film emulsions, and thus show as disproportionately dark areas, devoid of all detail. The visual ray detects areas of high potential, after practice and application, as brilliant bluish-white points. On a photographic *print,* as distinct from the negative, these areas appear as completely black. The photographic print is thus an orgonotic *negative*—the reverse of your sensory detection of the object. This is a classic instance of the way in which the judgment, and not the senses, can mislead the unwary investigator.

Van Tassel's drawing shows the lens as a multifaceted structure through which primary energy is absorbed into the propulsion system of the vehicle. What the spacemen refer to in their technical communications with Van Tassel as "positive primary energy" from the sun, best equates with the orgone energy discovered for earthmen in the 20th century by Dr. Wilhelm Reich. Dr. Reich established that high concentrations of orgone energy desensitize film emulsions. In the light of this knowledge, we may see that our infrared photograph does not lie, but truthfully reproduces the energetic manipulations involved.

The black spot indicates high absorption of light—from a conventional viewpoint. This black spot, viewed on the original negative, is an area where no emulsion reaction took place. The black spot may also be interpreted as a concentration of an energy form known from earthly experiment to desensitize film emulsions. The visual ray supports the latter by conveying to consciousness an impression of intense, bluish light with concomitant *suctional* effect upon the visual ray.

The only portions of a ventla-type vehicle that are likely to be detected by an investigator—when the ventla is invisible—are the whirling energy field around the disc, the brilliant point of energy formed by the lens, and very occasionally, the high-potential underside of the vehicle. The energy field appears as a pulsation of varying frequency, probably dependent upon the proximity of the vehicle, and whether or not the vehicle is moving or hovering. This field shows in "UFO Ahoy!" as a white, quasi-spheroidal and discrete envelope of radiation, plasmatic and probably created by atmospheric resistance to the energy field.

The spherical shape of the field, as we shall see when we go deeper into etherian physics, is wholly consistent with the formative signature in nature of the warmth ether; and the "breaking free" of this ether is also a central factor in the functioning of Wilhelm Reich's orgone energy accumulator—an invention that torpedoes the Second Law of Thermodynamics,[2] and should preclude this falsehood from being compulsively crammed into any more young minds in the universities. The flying discs are thus telling us where to look for the answers for which our technologists lust in secret.

Two other examples of flying discs captured *in situ* are offered in Alpha #6 and Alpha #7. In both cases, the central, dark area is probably a disc on edge. The surrounding plasmatic force field corroborates "UFO Ahoy!" with its obviously whirling field. These objects do not appear to be biological, but rather, appear to be *constructs* of intelligent design. That they are using bioenergy or orgone energy in their propulsion will become increasingly clear as we proceed.

In capturing these constructs on film we usually "saw", with our special techniques, bright streaks with the fleeting impression of a disc on edge. What we saw as bright streaks to our extended vision, turned out to be *black* on the finished prints as herein reproduced. Our observations and verifying photographs again and again brought home to us that the energy involved with these discs was desensitizing and sometimes nullifying film emulsions. Would-be experts in the science of photography as it is presently understood should beware of condemning *any* photographs from *any* source as fakes just because shadows do not appear on discs consistently with the sun's position. If the camera "stops" the whirling high orgonotic potential field on the camera side of the disc, there is likely to be a "shadow" where it has "stopped" the whirling field.

The high orgonotic potential of the "stopped" field will desensitize the film emulsion and an entirely anomalous shadow will appear on the materialized structure. This happened with the Cluj, Romania photographs, and deeply puzzled the engineers and scientists who analyzed this remarkable series. A full account of this sighting appears in the *Flying Saucer Review* for November/December, 1969. Many photographs have been discarded by "experts" in the past because of anomalous shadows, and it may be here stated that any purported photographic expert who assigns authority to himself in these things, or who allows it to be assigned to him by others, is nothing more or less than an apocryphist. A *working knowledge of orgone energy functions is a minimum condition even for limited investigation of UFO photographs.*

The orgone energy is positive to light, and therefore is *not* perceived by an individual who sees a materialized disc in the atmosphere. Nor is the energy field detectable photographically other than by the desensitizing or nullifying process. Infrared film permits direct detection by reacting to the plasma created in the atmosphere by the whirling propulsion field, when the blue sky is absorbed by filters or otherwise in the exposure. The spinning plasma then reverses itself out of the black background of the finished print.

These facts, which appropriate work by others will confirm and expand, enjoin a healthy skepticism toward the claims of self-styled experts. What earthmen *don't* know about light, color, energy, substance and tangibility would fill a bloody great hole in the ground. Ignorance in such magnitude rules out, for decades to come, claims of expertise in the technical realities behind UFOs. My work is no more than a break-in, the cry of a newborn baby.

Groping along in the early days of this work, I began working on UFO photographs away from the desert site. I had an office in North Hollywood, California. By convenient bankruptcy, the other tenants in the building were forced to evacuate, and I had full use of the premises for the necessary attracting process prior to taking the pictures. Many photographs were thus obtained of invisible critters and craft right in the North Hollywood industrial area. The proximity of hundreds of workers within a few blocks fairly well shatters the notion that the objects were visible.

Alpha #6 was made with employees of an adjoining manufacturer passing by the telephone pole to the left of the picture. They saw nothing. The infrared film, with its reach into the invisible, saw a disc on its side with associated propelling fields. Bravo #4 was photographed above the top of Mt. Wilson, little more than a mile from the Mt. Wilson Observatory with its staff of astronomers—trained observers all.

In my solo work, in North Hollywood and on Mt. Wilson, as well as in all the work done with Jim Woods assisting me, I continued to get both basic types of UFOs—craft propelled by an intelligent application of an energy form unknown to official science, and biological UFOs of often horrible configuration and monstrous size. On one occasion, I made two successive photographs of a large, serpentine creature, essentially transparent even to infrared, curling around near the telephone pole seen in other photographs made at this site. This etheric monster had his head curled around toward me, and in one exposure I was photographing through perhaps six or eight feet of his body lying directly behind his head. This made him dense enough to register on the highspeed infrared film.

These photographs are unequivocal, but like many in our massive collection, do not print out suitably for book reproduction because of their subtlety. They have been displayed many times in public at lectures and exhibitions, and careful study convinces even the skeptics of their authenticity. Great, spherical critters that looked like atmospheric jellyfish, fish-like forms evading USAF jet fighters, serpents, elementals of various kinds, soon made up a cosmic rogues' gallery of borderland residents.

Knowing that things like this were in the atmosphere, even if of a different order of tangibility to ourselves, did not make my personal task any easier when I stood up on that desert rock at dawn. Knowing that the simple ritual I was performing *attracted* such denizens of the borderland reinforced the apprehensions that often arose. Many times my thoughts turned to the "amoeba", with whom I had been cheek-by-jowl on that unforgettable morning. I frequently had to fight down my fears, and I could hear my teacher, Franklin Thomas, within my mind's ear: "You will need nerves of steel for all this."

Most of the objects that appeared to be biological were fusiform or discoidal, and thus set up a ready confusion with man's preconceived notion that all UFOs were ships from other planets. Pulsation of these biological aeroforms—the natural and inevitable concomitant of life is *pulsation*—provided further opportunity for confusion with mechanical notions. The pulsation was interpreted wherever seen or otherwise encountered as due to the "power source" or "power system" of the UFO. The concept that such pulsation was equivalent to man's own heartbeat never found a home with the engineers, chemists and astrophysicists who sought to unravel the mystery on the basis of their own sterile world conception. The "kill-it-then-study-it" biologists did no better.

Immense and endless confusion arose from the two-fold yet basically biological nature of the UFO phenomena in all their diversity, and the mechanical preconceptions and foregone conclusions about UFOs that formal experts were attempting to prove. This confusion has extended down to this day. The two main manifestations are *mutually confused,* and it is neurotic man's block against the living element—the bioenergy as the common functioning principle—that has brought ufology to stagnancy.

Experience and practical provings, first-hand experience over months that rapidly grew into years, and the precipitation of these heretofore unknown forms into photo emulsions, delivered us from most of the common illusions about UFOs. In the spring of 1958, several hundred feet of motion picture film were made of UFOs spinning and bouncing around me as I stood on my rock before sunrise. Six frames from one of these lengths of film are reproduced with the Alpha series of illustrations to show what was obtained.

On these frames, six in all, shot at twenty-four frames per second and therefore showing what happened in the space of a quarter second, one may see the futility of mechanists' concepts of UFOs. The objects are seen to come and go, change their shape, number and position, all in this brief period. Ships from other planets? The scene is more like the action perceivable through a microscope!

No one who has ever seen these films conceives of them in the spaceship context. The biologist is much more at home with them than the aerospace technician, for they are obviously some ultradynamic lifeform existing beyond the reach of our unaided sense apparatus. Their plasmatic nature is evidenced by the peculiar emulsion reaction they produced, as though blistering the chemical coating. They also appeared able, when focused by the lens, to splash right through the closed shutter between frames and manifest with diminished intensity on the film. The commercial laboratory in Hollywood that processed the two successive rolls of film on which these things were recorded said that nothing like this emulsion reaction had ever been seen by their technicians.

This work took place between dawn and sunrise. Our operations and recording of these critters far preceded the first orbital flight of Colonel John Glenn. He was the first U.S. astronaut to report seeing "fireflies" following the dawn around the earth and at times floating outside his capsule. One cannot tell what the relationship is between the activity we recorded and the fireflies reported by Glenn and at least two Soviet cosmonauts. Suffice it to say that we had something as real as the fireflies—and probably of a kindred order—literally at our fingertips. These plasmatic forms were as new to man as Colonel Glenn's fireflies. The difference in the two activities was that billions had been spent to boost Colonel Glenn into orbit in his elaborate ashcan, and we were just two ordinary men with our feet on the ground.

We had broken into the biological reality of an invisible, interpenetrating, yet objectifiable realm, while NASA sought comfortable, compatible evidence of "extraterrestrial life". Every mechanistic space venture brought endless applause and blasts of windy rhetoric, but nowhere was there *LIFE*—a situation for which this planet's population of neurotics seemed grateful. The astronauts and cosmonauts offended no one, and could be admired by all. Public reaction to space exploration of this formal kind etched it all indelibly into history.

Reactions to my personal exploration of the physical space borderland around us were something different. I found not fairness and scientific objectivity, but irrational evasions of essential findings, and malignant stupidity that forced upon me not only a new view of my fellow men, but the need to understand why they were so mortally terrified of Life.

NOTES TO CHAPTER SIX

1. *I Rode a Flying Saucer,* by George W. Van Tassel, New Age Publishing Co., Los Angeles, Calif., 1952.
2. The orgone accumulator proves that the heretofore assumed universality of the 2nd Law of Thermodynamics is inapplicable in certain specific arrangements of materials.

Chapter Seven

THE CASE FOR THE CRITTERS

*...the mistakes and errors are the price for the great romance of doing
something for the first time.*
— SIR FRANCIS CHICHESTER

The concept that certain UFOs could be living organisms did not originate with me. Sir Arthur Conan Doyle's story *Horror of the Heights* gives a startlingly prescient description of the "air jungle" and its denizens. Doyle makes other references to borderland residents in *The Edge of the Unknown,* published by Putnams in 1930. Charles Fort also speculated about critters in his *Book of the Damned,* reprinted by The Book Tree in 2006. The first modern theory encompassing this idea with specific reference to UFOs was probably the original "etheric interpretation" of the flying discs, formulated in 1946-47 by the late N. Meade Layne, M.A.

Mr. Layne published his theory in a penetrating 1950 monograph *The Ether Ship Mystery and Its Solution.* He was founder and first director of the Borderland Sciences Research Foundation, Inc., in San Diego, California.[1]A former university lecturer and an outstanding writer, Meade Layne produced one of the classics of ufology in his monograph. His achievement was to produce the only theoretical treatment of UFOs to survive the entire modern period. This document is still available today with added comments by Riley Crabb, second director of BSRF.

Mr. Layne's "etheric interpretation" is also readily extensible to cover new UFO facts, and is required reading for any person with a genuine desire to break out of the bondage of mechanistic inking. There can be no doubt that Meade Layne was a generation ahead of his time with his mobile thinking. He also emerges as easily the greatest writer of ufology's break-in period.

The methods of obtaining advanced information that were employed by this pioneer, at a time when UFO data were sparse, will be dealt with later in proper context. A sample of his writings will illustrate that his work disqualifies with equal alacrity both the faint-hearted and the weak-minded:

"The aeroforms are thought-constructs, mind constructs. As such, they are, in effect, the vehicle of the actual entity who creates them. Just as our own terrestrial minds rule and become identified with our bodies, so does the entity of the Etheric world make for himself a body or vehicle out of etheric substance.

"This body may be of any shape or size, any one of a hundred *mutants* — such as the indefinite and changing shapes reported by observers of flying saucers throughout the world. The shapes may be a wheel, a globe, a fusiform or cigar shape, a fireball, vapor or gases. It may have any density, any rate of vibration desired. The impenetrable steel of landed discs, is, as it were, a sort of etheric isotope of our terrestrial steel, or we may call it 'etheric steel'. The shapes and vehicles and the entity operating them form one being, just as a human being is a psychophysical mind-body unity. The body of this Etherian entity is a thoughtform which can go anywhere, and penetrates our earth and sea as easily as our air."

Etherian physics in all its significance for true ufology—the ufology of the Future—will be enlarged upon as this book proceeds. The important point for now is that I was not the originator of the theory that some UFOs could be living organisms, and that Meade Layne—and also John Bessor—preceded me by about ten years. When Meade Layne saw my collection of more than a hundred photographs of these mutants not long before his death, he was exuberant. He termed their capture on film "the death knell of the old order."

Meade Layne's daring hypothesis—a brilliant synthesis of physics and metaphysics—was shown to be valid when I literally stumbled across these organisms in the summer of 1957. This experimental accident, and my subsequent photography of dozens of these critters *in situ*, merely confirmed the extant theories of far brighter minds than mine. This borderland breakthrough was crude, awkward and somewhat uncomprehending, but it nevertheless threw light immediately into the darker, previously impenetrable corners of history's greatest mystery.

Endless implications ensued. Technical, theoretical, cognitional, philosophical, psychological, methodological and scientific questions arose of staggering magnitude. Unraveling all this will take the best efforts of far better and *younger*, more functional minds than those that currently dominate world science. The response we found, outside the small circle of friends and associates who understood our work, was essentially confined to fear and anxiety. We understood some of this only too well.

When these strange living forms burst into our ken, we found it essential to be light-hearted about our work—if only to diminish the psychological impact of unknown terrors. The sight of these queer, plasmatic fauna in photographs was sufficient to repel sensitive people otherwise interested in UFOs. Motion pictures of them sometimes caused psychically sensitive persons to bolt from the room during screening. Numerous persons today who have tampered chemically with their perceptions by ingesting LSD have become aware of these strange organisms. Back in 1957-58, however, few people were willing to attempt chemical extensions of perception,[2] and prior knowledge of the critters was confined to learned and accomplished occultists, who knew only too well how real the things were that we had beenable to capture on film.

We conceived of ourselves as children, struggling to stay upright by clinging to the side of a crib. Our falls and spills and mistakes were almost perpetual and often comical. Everything we touched was new, and there were no technical texts to consult. Our ability to see the funny side of our adventures never left us, and was always a counterbalance to what might otherwise have become overwhelmingly serious.

Kidding in no way diminished our curiosity. We kept on after the critters. No one had ever done anything like this before, and it was a tremendous thrill. The laws of these etherian realms are essentially functional and biological, and they open themselves only to the researcher who, in his experimental work, *has not lost his ability to play*—to play like a child. The searching organism is almost always playful. A sense of humor is essential, for experience quickly teaches the field worker that the stupidity and cupidity of contemporary man must surely be the comedy of the gods.

The critters had even aroused the USAF to some public theorizing as early as 27 April 1949. On that date, the USAF stated in an official release:

"The possible existence of some sort of strange extraterrestrial animals has also been remotely considered, as many of the objects acted more like animals than anything else. However, there are few reliable reports on extraterrestrial animals."

This USAF release, which may have been originally intended as the opening wedge to further revelations of observations made by USAF aircrews, was unwelcome in ufology. Spaceships had become *idée fixe,* and nothing biological was wanted to disturb or modify the basic extraterrestrial spaceship hypothesis. At the other extreme stood the original skeptic, Dr. Donald Menzel of Harvard University, eternally ready to discharge his self-appointed duty to debunk UFOs. He didn't like the USAF "space animals" release either.

In his book *Flying Saucers,* Dr. Menzel took exception to this release. He pointed out that even *one* reliable report of an extraterrestrial animal would be sufficient, let alone a "few", as the USAF had implied were in existence. What the USAF has in its secret files in this connection will not soon be revealed. Suffice it to say that the USAF uses infrared photographic apparatus and detectors extensively, and penetrates round the clock into the stratosphere with piloted aircraft. USAF radar blankets the United States.

My opinion is that they have objectified a great many things aloft that science does not presently understand, including animal forms. I know that USAF aircraft have chased my critters because I have personally photographed USAF fighters carrying infrared homing rockets chasing them above the Mojave Desert. The USAF, like numerous agencies of the U.S. Government, lies to the American people, and only the gullible believe otherwise. At the time of the USAF "space animals" release, there was in print at least one reliable report of an atmospheric animal form, *available since 1934.*

This report was brought to my attention by Mr. Adrian Cox of London, England, who saw its significance after reading an illustrated article of mine in the *Flying Saucer Review* for July/August of 1960. Mr. Cox connected this article, entitled "Space Animals—A Fact of Life", with an incident recounted in the book *Everest 1933* by Hugh Ruttledge.

The Ruttledge book was published in Britain by Hodder and Stoughton, and in the United States in 1935 by the National Travel Club under the title *The Attack on Everest*. On page 228 in the American version, climber Frank Smythe writes of the second assault on Mt. Everest as follows:

"The second phenomenon may or may not have been an optical illusion. Personally, I am convinced it was not. I was still some two hundred feet above Camp 6 and a considerable distance from it when, chancing to glance in the direction of the north ridge, I saw two curious-looking objects floating in the sky. They strongly resembled kite balloons in shape, but one possessed what appeared to be squat, underdeveloped wings, and the other a protuberance suggestive of a beak. They hovered motionless, but seemed slowly to pulsate, a pulsation incidentally much slower than my own heartbeats, which is of interest supposing it was an optical illusion.

"The two objects were very dark in color, and were silhouetted sharply against the sky or possibly a background of cloud. So interested was I that I stopped to observe them. My brain appeared to be working normally, and I deliberately put myself through a series of tests. First of all I glanced away. The objects did not follow my vision, but they were still there when I looked back again. Then I looked away again, and this time identified by name a number of peaks, valleys and glaciers by way of a mental test. But when I looked back again, the objects still confronted me. At this, I gave them up as a bad job, but just as I was starting to move again, a mist suddenly drifted across. Gradually they disappeared behind it, and when a minute or two later it had drifted clear, exposing the whole north ridge once more, they had vanished as mysteriously as they had come. It may be of interest to state that their position was roughly midway between the position of the 1924 camp 6 and the northeast shoulder. Thus they were at a height of about 27,200 feet, and as I was at about 27,600 feet when I saw them, a line connecting their approximate position with my position would not bring them against a background of sky, but against lower and distant mountains. It is conceivable therefore, that it was some strange effect of mist and mountain magnified by imagination...."

Mr. Smythe prefaces his account by saying:

"Men under physical and mental stress have experienced curious things on mountains, and instances are described in the *Alpine Journal*. Furthermore, the effects of oxygen lack on the brain are complex and but little understood."

Frank Smythe was a member of several famous expeditions of this type, and from the point of view of character and background, he is a witness of integrity. A graduate electrical engineer, he was educated at Faraday House Engineering College, was a former Royal Air Force officer and a Lieutenant Colonel in the British Army. He was in three Everest attempts, and was a member of the International Kanchenjunga Expedition of 1930. An accomplished author, he wrote several books and many articles on mountaineering. Smythe died in 1949, but will undoubtedly be accounted among the earliest modern observers of biological UFOs—when in due course the determinism of those UFOs is established.

Frank Smythe's observation must be accounted a good one by any sound standard. He observed with extreme care, did everything possible to eliminate the possibility of hallucination—or to so identify it if it were a hallucination—and verified that the pulsation of the objects differed from his own heartbeat. He did all these things systematically, after the fashion of a man trained in engineering. He also established the approximate altitude and location of the objects. Most important of all, he was extremely wary at the time of the changes in perception and consciousness that can take place at high altitudes.

Skeptics eager to torpedo the obvious biological implications of this sighting might well seize on the abnormal location of the observer, and the unusual physical conditions, as a means of calling it all a hallucination. The surmise here would be that what Smythe "saw" was due to altitude induced sensory changes in his organism. Now that creatures answering Smythe's general description have been photographed with materials sensitive beyond the range of the human eye, such skeptics are forced more on the defensive. Furthermore, the objectification of these extra-ocular regions of the spectrum is constantly progressing. The discovery of more and different critters is inevitable; scientific development will not cease.

The probability is high that Smythe perceived something objective and real, and if altitude, oxygen deficiency and exertion happened to extend his vision just a few millimicrons beyond the normal range, then he would perceive directly this adjacent range of physical form that has since been photographed—complete with its strange critters. The role of altitude in the perception and objectification of these critters seems significant. While this role is poorly understood at present, subsequent work will elucidate much that is now murky.

The imponderables involved will probably include the diurnal "breathing" of the planet Earth. This will be dealt with in due course. From personal experience in field work, I can report that observation of the human orgone energy field at an altitude of approximately 5,000 feet on Mt. Wilson, in California, shows that this field is vastly extended over its sea-level size. Perception of the borderland critters we have photographed is also much more difficult at sea level than at 4,000 feet altitude and above.

Perhaps the extension of the orgone energy field and its lumination that we achieved via the Star Exercise happened spuriously to Frank Smythe through his being in rarefied air near the summit of Everest and breathing oxygen.[3] Since the vacuum of the moon's surface also revealed objectively the blue orgone energy fields of the astronauts—extending far beyond their bodies and external to their space suits—there is independent evidence that these things happen approximately as I have described them.

The functional connections between my methods and findings, the "mysterious" blue emanation around the U.S. astronauts on the moon and Colonel Smythe's high-altitude observation of biological UFOs, are self-evident. The Smythe sighting took place when I was six years old. I was in no position to implant my ideas "by suggestion"—a tired old bromide routinely supplied by the anxious.

Colonel Smythe was not the first man to see the critters of our atmosphere in modern times. A number of years prior to the Smythe sighting on Everest, an American named Don Wood, Jr., got a closer and more terrifying look at a couple of these critters. He saw them in full physical density and in light-reflecting negative polarity as a result of his hobby and interest in flying.

The experience shook him sufficiently to make him keep his peace for several decades. Mr. Wood's story was originally published in Ray Palmer's publication, *Flying Saucers*, in October of 1959. Few people have done more to advance ufology through the years than the dogged and able Ray Palmer, who died in 1977. He sacrificed much in his lifetime in order to keep publishing material from which orthodox publishers recoiled. Mr. Wood's experience, as recounted here, found the light of day thanks to Palmer's open-mindedness.

Mr. Wood states:

"I must write you of what happened to me in 1925, which I think solves most UFO reports. I have never told this to anyone, but can get a signed affidavit if needed. Four of us were flying old 'Jennies' (OX5 motors)[4] over the Nevada desert. One plane was a two-seater, the one I was in. We landed on Flat Mesa, near Battle Mountain, Nevada. The mesa is about 5,000 square feet and the walls are too steep to climb unless a lot of work is done.

"We wanted to see what was on top of this flat place. We landed at 1 p.m. While walking about the top of this place we noticed something coming in for a landing. It was about 8 feet across and was round and flat like a saucer. The undersides were a reddish color. It skidded to a stop about 30 feet away. This next thing you won't believe, and I don't care but it's the truth. We walked up to the thing and it was some *animal* like we never saw before. It was hurt, and as it breathed the top would rise and fall, making a half-foot hole all around it like a clam opening and closing.

"Quite a hunk had been chewed out of one side of this rim and a sort of metal-looking froth issued. When it saw us, it breathed frantically and rose up only a few inches, only to fall back to earth again. It was moist and glistened on the top side. We could see no eyes or legs.

"After a 20 minute rest, it started pulsating once more. (We stayed 10 feet away.) And so help me the thing grew as bright as all get out, except for where it was hurt. It had a mica-like shell body. It tried to rise up again, but sank back again. Then we saw a large, round shadow fall on us. We looked up and ran. Coming in was a much larger animal 30 feet across.

"It paid no attention to us, but settled itself over the small one. Four sucker-like tongues settled on the little one and the big one got so dazzling bright you couldn't look at it. Both rose straight up and were out of sight in a second. They must have been traveling a thousand miles an hour to get so high so fast. When we walked over, there was an awful stench, and the frothy stuff the little one had bled looked like fine aluminum wire. There was more frothy, wiry stuff in a 30-foot circle where the big one had breathed.

"This stuff finally melted in the sun, and we took off. So help me, *this was an animal.* I have never told this before as we knew no one would believe us. I only write now because this animal would be one big 30-foot light if seen at night. I don't expect belief, but I simply had to write."

History should have a place for Don Wood, Jr., for making what might well be characterized as the most important observation thus far in the history of this confused UFO subject. I wrote him at his home in one of the southern states of the U. S., and am satisfied that he reported his experience accurately. Truly it is that one example worth a thousand, containing all within itself — what Goethean scientific thinkers call the *ur* example.

Only a compulsive-neurotic doubter can make a spaceship out of what landed next to Mr. Wood. His story is strong and full of *life.* Since this experience runs counter to the cherished, mechanistic conceptions of flying saucers, little attention has been paid to it, despite its significance for a genuine exobiology. Sequestration of such experiences by establishment-type ufology is their typical fate, lest disturbance be caused to cosmic conceptions rooted in sterility.

Here is an example of a man viewing something close up—a full twenty-two years before the term flying saucer was coined—that was discoidal, pulsating, and had its own light source. The object was obviously *alive* and also injured, as though it had been attacked. Metallic froth issued from the wound, not blood. The object had a mica-like shell, such as today's would-be exobiologists speculate might be possessed by Martian lifeforms. Such a shell might be expected to return radar echoes, especially if it happened to be thirty feet in diameter like the rescuing critter.

There was no radar then. Radar came later. The critters were there *before* radar, and if ten years before radar, then why not millennia? A question surely of far more weighty import for current notions of both life and reality than any thus far raised by the ships-from-other-planets advocates, resting their case on coincidence and accident, and minus any coherent concept of the relationship between life and pulsation.

The functional connections between the Wood sighting of 1925 and a huge corpus of evidence gathered since are everywhere apparent. Ufology is very largely made up of reports of pulsating, glowing, super-performing discoidal objects, precipitately conceived of as ships. Since we added to Don Wood's observation, photographs of glowing, discoidal objects that appear like giant amoebae, there is clearly a new and *wholly biological dimension to ufology.*

Anyone who labors under the illusion that modern scientists do not contribute to and support obscurantism in anything pertaining to genuine findings on extraterrestrial life should peruse biologist Ivan Sanderson's book *Uninvited Visitors.* Mr. Sanderson was a scientist who ransacked the world for evidence concerning visitation from space, and his biggest problem always appeared to be getting the facts from fellow scientists. He relates an instance in Chapter VI of *Uninvited Visitors* where an enormous object similar to that described by Don Wood was washed ashore in southwestern Tasmania in 1962.

A wealthy amateur naturalist who heard about this critter went after it when it washed ashore on the south side of Sandy Cape. He was accompanied by a government zoologist. The intention of the naturalist—a museum backer—was to obtain a sample of the flesh. He was unable to do so because *even an ax could not cut into the thing.* Let us pause and refer back to the sample statement from Meade Layne's writings quoted earlier in this chapter, dealing with the hardness and density of etheric matter. Reference is also justified to the "hard, mica-like shell" reported by Don Wood to encase the critter that landed beside him in Nevada.

Ivan Sanderson recounts the pathetic, deplorable measures set in motion by the Australian government and its satellite scientists to suppress the discovery of this critter. Sandy Cape was placed off limits to everyone, including the Australian press. One enterprising reporter nevertheless got to the critter, and after a husky cop failed again to cut it with an ax, touched a cigarette lighter to its edge. He noted that the thing "withdrew" from the flame, later returning to its original contour.

The case for the critters was advanced significantly by this incident, and reinforced by subsequent revelation in Australia that seven other, similar critters had been washed up on that country's southeast coast during the previous twenty years. The critter in the incident described had been washing in and out on the tide for over two years, but had not decayed. Touching it with flame indicated some kind or residual life function, reminiscent of Galvani's basic experiment with a frog leg. On the basis of the facts presented by Mr. Sanderson, the conclusion seems inescapable that the critter was not a previously known species of terrestrial life.

Secrecy and febrile obscurantism surrounded this incident, and will be all-too-familiar to the student of ufology as typical of official response to UFOs all over the world. While the handling of the Australian case may be at odds with the spirit and ideals of science and scientific inquiry, realists are aware that such machinations are the norm nowadays. Spirit and ideals were trampled under long ago in the economic stampede. Science is a means of making a living, and it is the socioeconomic blackmail thereby opened that is responsible for scientific knuckling under to government edicts that suppress scientific information.

Australians should not feel that they are being put in a bad light. In the U.S.A., the courts have actually ordered—and carried out—the burning of scientific literature and bulletins that would help us understand these denizens of space. American courts and jurisprudence have been corrupted and misused to silence scientific pioneers of the new cosmic biology. This book will in due course describe two such individuals—and their fates—as verification that these evil things occur right now in modern times.

Fear of Life is the power source for this irrationalism. Life from space is feared most of all. The fear climate has worsened as the Space Age has advanced, as though man has come so to love the darkness that it is the light that terrifies him. Understanding these queer peregrinations of a frightened humanity, standing as it does at the portals of a new epoch, has become an essential element in ufology. Evasive human attitudes are more important at present as Space Age phenomena than the investigation of the arcane propulsion plants being used by the aliens. Man can have no access to things that his neurotic world-conception compels him to evade.

Ufology stands to receive enlightenment and extension from Don Wood's sighting, as well as from the corroborative Australian critters. Tolerating the critters is a crucial psychological problem, for as Ivan Sanderson so aptly remarks of ufologists in his *Uninvited Visitors*, "The mere suggestion that there could also be a biological aspect to their subject invariably seems to upset them." Mr. Wood's silence in 1925 is understandable. There was no scientific or quasi-scientific frame of reference into which his experience could be fitted. Even the way-out wilderness of ufology did not then exist. He would probably have found himself fitted into a straitjacket by the same kind of myopic mentalities who, at the same period in history, felt that a straitjacket was the correct garb for American rocket pioneer Robert Goddard.

Straddling the Wood sighting and that of Colonel Smythe is the 1926 sighting by the Roerich Expedition of the American Museum of Natural History. Expedition members saw a shining disc high over the Altai-Himalaya. Alert and diligent research will uncover more of these early-modern encounters, forcing acceptance of the presence of the critters in our atmosphere—with all their revolutionary consequences for science.

The helplessness, and sometimes the irrational opposition of official science in the face of these pressing questions should not deter any unblocked, free person from facing up to their further investigation. Progress depends on the raising of questions. The answers will almost certainly prove disquieting to the old order in science, and utterly destructive of its central dogmas. That is why it will take *young* people—free of neurotic dependence on the mechanistic world-conception—to press these matters forward.

An example of this is contained in Appendix IV, included with the translations into European languages that were undertaken in 1987-88. Mr. Larry Arnold, a young friend of mine, obtained from historical records additional solid evidence of critters in the skies of this planet, long before either aircraft or radar. Appendix II recounts the work of the GRCU research group in Genoa, Italy, under their late leader, Luciano Boccone. These people extensively photographed the critters in Italy, without even knowing about my work twenty years previously. Similar photographs have been made in Rumania by a group under engineer Florin Gheorghitza.

Critter reports such as those described, enable us to accept rationally that biological forms, *capable of easy confusion with the idealized flying saucer,* have landed on the earth many times in the past. They landed at least once in close proximity to humans, twenty-two years before flying saucers were ever mentioned. Acceptance of the critters does not invalidate the spaceship hypothesis. On the contrary, we strengthen and broaden our investigation. Furthermore, we are put on our guard against two kindred sets of phenomena—sharing a common functioning principle—that are prone to mutual confusion.

By identifying his two UFOs as living organisms, Mr. Wood has done ufology a powerful service. Actual observation and obvious inference tie these creatures to a *natural power of flight* using their own life energy. These critters flew naturally at 1,000 mph on this power, just as we have the power to walk and run on the earth using our own life energy. That conventional science has evaded the investigation of biological energy does not alter the facts. Man is functionally connected to UFOs of this critter variety by the animating energy. In due course, we will show that it is the orgone energy discovered by the late Dr. Wilhelm Reich.

The leading brains in official science, all the Nobel Prize winners of the world combined, all the resources of all the great universities combined, cannot explain to you how you are capable of walking from your chair to the door. The energy involved is *not electrical,* and all the laboratory facilities and sterile conceptions of mechanistic biology do not give us access to this energy. Clearly, if we pursue the source and nature of our own *biological power,* we will be cheek-by-jowl with a power source for space travel—such as reposes naturally with the critters.

NOTES TO CHAPTER SEVEN

1. Borderland Sciences Research Foundation, Inc. (BSRF), now in Eureka, California, was known as Borderland Sciences Research Association (BSRA) while under the directorship of Meade Layne.

2. No inference is intended here that the author supports in any way, drug-induced "extensions" of perception. The author is opposed to all such tampering with consciousness, whether attempted by the uneducated, or by academicians.

3. There is an intimate relationship between the element oxygen and the human astral body, or seat of consciousness, according to Rudolf Steiner.

4. The "Jenny" was a World War I-vintage aircraft, mainly used for training and popular in the postwar years for sport flying.

Chapter Eight

EXPANDING THE CASE
FOR THE CRITTERS

proof: the cogency of evidence that compels acceptance by the mind of a truth or a fact.
— WEBSTER

There is an incident involving a persistent UFO recounted in *Flying Saucers: Serious Business* by the late Frank Edwards that supports the conception of some UFOs as metallic life-forms. The UFO in question began appearing frequently in the middle of 1964 in Rio Vista, California, sixty miles northeast of San Francisco. The relevant passage in the Edwards book appears on pages 14-15 of the Bantam Books edition.

'The reports which kept coming in to the sheriff of Solano County described the thing as torpedo or dirigible-shaped[1] about five feet in diameter and twelve to fifteen feet long. It glowed a warm red and it moved silently, witnesses told the sheriff. One housewife brought in some color transparencies of just such a thing, which she said she had taken in 1964 as the object hovered near her home.

"Deputy Sheriff John Cruz of Fairfield told the *San Francisco Chronicle* that he finally became interested in the case on September 22,1965, and went to have a look for himself. Residents told him that the thing generally appeared near a water tower about five miles from town. That night an estimated 300 to 400 persons standing patiently in the dark on the hilltop around the tower were rewarded by the appearance of the strange object: glowing softly red, cigarshaped, hovering or moving slowly about only a couple of hundred feet above the treetops or the top of the water tower.

"What makes this incident noteworthy is not the repeated appearances of the same object in the same area... but the fact that some of the witnesses told the deputy that boys with .22 rifles had shot at the object one night—and the bullets had made a metallic 'twang'—and caused the object to flare up bright red for a second."

One must disagree with the late Mr. Edwards as to what is noteworthy in this incident, since his fundamental bias was always toward the extraterrestrial intelligence hypothesis. Probably because of this bias, he has overlooked the *biological* significance of repeated appearances of the object near to a source of water. Such a cyclic return to a water source—or to an otherwise favored locale—is a fundamental trait of animal behavior.

Waterholes are recognized by hunters as ideal places to trap and shoot animals. Any pet owner will verify that dogs and cats cyclically favor certain places for sleeping, dozing and hunting. There is therefore a basically biological backdrop to the appearances of this UFO, which lends a completely different character to the portion of the sighting that Mr. Edwards found significant, namely, the firing of a .22 bullet into the visitor. The thing "flared up bright red for a second," exactly as you would flare up in fright, anger and pain if struck by an unexpected BB pellet. Your whole organism—in its seen and unseen totality—would flare up. The point of impact would most assuredly be *red*.

Machines as they are known to earthmen do not flare up in this plainly biological fashion when struck with .22 slugs. We have already seen that we are dealing in some phases of ufology with queer creatures that "bleed" something akin to steel wool when injured. We have also seen that when they are in our density and polarity, these critters are extremely hard, and capable of glowing and pulsating with blinding intensity. They are probably of metallic or metal-like constitution. One should therefore not be carried away by the metallic "twang" of the impinging bullet to the extent that the biological flare-up is ignored.

Investigators have serious need to accustom themselves to a significant point arising from these encounters, which so often involve metallic objects. The significant point is that metallic appearance, metallic effects and metallic properties *do not necessarily, and in all instances, indicate a machine.* Humanoid entities dismounted from landed UFOs have sometimes been described as metallic-looking. Some of these humanoids have appeared as resembling metallic asparagus. A perusal of the 1966 *Flying Saucer Review* issue entitled *The Humanoids* well verifies this.

In the Rio Vista case, as in many kindred incidents throughout the world, UFOs speak the language of the living—when they are allowed "free speech" and not overridden by mechanistic preconceptions and bias. UFOs are frequently all-too-obviously biological. Earthmen nevertheless compulsively impose upon these phenomena the language of machines. Biological implications and inferences of the most obvious, glaring kind are evaded, ignored or uncomprehended. Mechanistic thinkers subject to these pervasive, culturally-induced compulsions should properly be disqualified from authority in UFO investigations. Instead, they are placed in charge, barging into realms where their parameters and modes of thought are inapplicable. Obsolescent troglodytes of this stripe largely run science today. Their end is in sight, but they will die hard.

Frank Edwards recounted also in *Fate* magazine a number of years ago how two naval officers reported to him that they had watched a couple of UFOs retreat hastily when they blundered into a strong radar field. These officers stated that the UFOs bounced back out of the radar beam like rubber balls striking a brick wall. There was no turning around or maneuvering. They simply bounced. This, too, is more typical of something living, in the presence of an unpleasant stimulus, than of a machine.

In the same article, Mr. Edwards related a similar incident at a New Jersey cape that occurred in 1950.

"When UFOs were seen there repeatedly, the owner of the only house on the cape, a very famous writer, notified the government. Next morning he and his family were surprised to find three radar towers, mounted on trucks, hidden in the brush along the beach. Why were they there? One of the officers explained that radar made the saucers stagger, and they hoped to be able to bring one down if they could get enough beams on the disc at close range. The radar *did* stagger the discs later that day but did not bring them down. The discs sped away and the radar trucks were withdrawn a few days later."

There are significant conclusions that may be drawn from this encounter, dovetailing with what has been advanced theoretically earlier in this book. Once again there was the cyclic appearance of the objects in a certain area, a basic behavioral pattern of animals. This cyclic pattern was well enough established to justify assignment of several radar units to the area. Radar was directed at the objects, and it was known in advance to those in charge of the radar that the pulsed radiation of the radar transmitters made the UFOs stagger.

At this point, let us review from our earlier discussion just what radar is, and let us never forget its nature as long as we delve into the UFO mystery. Radar is a tremendously powerful beam of pulsed electronic energy. Even a portable military radar could have an output power of 30 kilowatts or more. Three radars were employed in this case. Pulsed or chopped electromagnetic energy has a known effect in stimulating living tissue. Carried far enough, this stimulation leads to a *radiation burn*. There are also extensive but as yet only dimly understood interfaces between living organisms and the microwave spectrum.

Returning to the incident at the cape, let us note that were such beams of energy directed upon our earth-built, airborne machines, there would be no known effect upon the functioning of such machines. Radar beams have no effect upon the propulsion or mechanical integrity of aircraft, dirigibles, helicopters, or any other known aerial conveyance, Radars of any power presently known to man cannot cause any known aircraft to stagger. Yet the UFOs in this incident did stagger, visibly and incontrovertibly. Furthermore, the officer in charge of the radar forecast and anticipated this reaction.

Consider the word "stagger" itself. Living experience is our faithful guide here. We stagger if struck by lightning, we stagger if we touch bare house wiring, and if we put our hand into the modulator circuit of a radar set—the source of the pulsed voltage that produces the radar's electromagnetic radiation—we could be sent staggering several yards and possibly would die from shock. Since such reactions do not occur in a corpse, *they must be deemed due to a clash between biological energy and electromagnetic energy.*

All training, common sense and experience in the operation and maintenance of radar leads to great care in avoiding these voltages that stagger. This may be seen as the reaction of a human biosystem to something lethal. Therefore we should anticipate that anything biological—whether known to official science or not—would be similarly staggered by blasts of pulsed radiation. Staggering now appears less as an effect upon some hypothetical power system or engine than as a direct biological reaction to something lethal.

The staggered UFOs did not return to be staggered again, indicating the presence of something akin to memory in their makeup. A radar serviceman who touches a "hot" portion of a radar set has his memory similarly impressed. He avoids repeating a painful experience. Interpreted in this direct way, the evidence implies that the UFOs off the New Jersey cape were *biological*.

The mechanistic tendency is to evade any living element that appears in UFO investigation. Staggering UFOs are explained by asserting that the radars upset the power source of the UFOs. This explanation keeps the reported effects anchored in the mechanical concept of a craft, for it is neurotically important that the mechanist not relinquish the machine concept in favor of the living. The known effects of electromagnetism upon power sources are essentially confined at present to demonstrable ability to terminate the flow of biological power to biosystems. Jets and rockets cannot be stopped with electromagnetic energy.

The biological effects and facts surrounding UFOs continue to mire the evasive mechanist ever deeper, just as a struggling man accelerates his own doom in quicksand. The only kind of energy known to man that has a demonstrably antagonistic reaction to electromagnetic radiation, and especially to atomic radiation, is orgone energy.[2] This revolutionary discovery will be fully dealt with in due course, and in its proper place in my story, but for now it is sufficient to note that orgone energy and electromagnetic energy are antagonistic.

The orgone energy ocean is in all probability identical with the classical "ether" of ordinary physics. The orgone also equates, in its properties and determinism, with the chemical ether of etherian physics as best exemplified in the expositions given by Dr. Rudolf Steiner, Dr. Guenther Wachsmuth and Dr. Ernst Lehrs under the panoply of Anthroposophical spiritual science. The orgone is essentially the medium in which the electromagnetic waves of radar are created, and in which those waves travel in doing their detecting and ranging work.

Assuming for a moment that the UFOs under discussion were spacecraft, we might expect the power system to be upset *if the power source were orgone energy.* The most limited intellect can grasp this. Radar radiation is a specific mode of disturbing the orgone. Such a disturbance could conceivably upset the functioning of an orgone motor. The functional leads to some core truths of spacecraft propulsion are to be found here. The mechanical explanation, by contrast, leaves us high and dry. Our jets and helicopters cannot be made to stagger with radar because their functioning is not dependent upon the ether or orgone—whichever you wish to call it—but upon the interaction of known laws and forces inherent in gross matter and essentially independent of the ether.

The orgone is the specific biological energy of everything living, in addition to being the formative etheric underpinning to physical matter. The orgone is the power source for the critters we have been describing—our denizens of this queer borderland of physical nature into which we are only now beginning to penetrate. Like ourselves, these critters are *orgonotic systems,* and their staggering from blasts of Life-negative electromagnetic energy is as predictable as our own.

In 1971, there was a knotty problem bothering scientists in both Canada and the United States that bears a close functional relationship to radar beams staggering UFOs. Chicken farmers complained that chickens living under or near microwave transmission towers were staggering and suffering from disorientation. Without any knowledge of orgone energy, it is discernible that the electromagnetic radiation was upsetting the biosystem that we call a chicken. They were staggering just like certain UFOs do when hit with a beam of this energy.

Subtler organisms, such as many of our critters undoubtedly are, might be expected to stagger much more readily and to pull away from such painful stimulus. Earlier in this book it was emphasized that man only really became aware of UFOs on a large scale in the modern epoch after he got his hands on radar. In the New Jersey incident under discussion, we have evidence that radar directly effects UFOs, making them stagger.

There may well be other important consequences. Microwave radiation is used today in the flash cooking of foods. That is what it can do to dead flesh. If you, as a functioning biosystem, get sufficient microwave radiation, the exposed part of your body will become inflamed. That is, it will *luminate,* as your own life energy, or orgone, fights the lethal electromagnetic energy effects on your organism. Medical science would call it a radiation burn. If viewed by a sensitive infrared detector, the inflamed area would be seen to be *glowing.*

Surely we can now see the functional connection between radar, and living organisms being made to luminate by this pulsed electromagnetic radiation. The radiation was not in their environment in the previous known history of this planet. Ask yourself how many thousands of UFOs have been *light manifestations* rather than structured objects. Biological UFOs obviously do not like radar, and it is eminently probable that these atmospheric creatures of tenuous composition luminate if they get too much of it—i.e., act as if they had received a radiation burn. Step-by-step the New Knowledge unravels the mystery, where Mechanism in all its overfinanced glory and prestige produces only befuddlement and endless evasion of the essential.

Several distinguished persons have followed Meade Layne's pioneering thrusts into UFO theory, and agree in general that there are heretofore unsuspected organisms in the atmosphere and in space around the earth. Charles Fort broached the idea decades before Meade Layne, and in 1955, a distinguished Fortean scholar in Europe—the Countess Zoe Wassilko-Serecki—updated Fort's theories.

Writing in the occult publication *Inconnue,* the Austrian noblewoman held that UFOs were lifeforms that fed on pure energy as they dwelt in space, constructing bladder-like bodies for themselves out of colloidal silicones. The Vienna publication *Neue Illustrierte Woehenschau* also presented the Wassilko-Serecki theory in its issues of 17 May and 24 May 1959, under the title "Creatures from the Stratosphere." By that time, I had captured these same bladder-like creatures on numerous infrared photographs and films, and given genuine substance to her theory. U. S. magazines and "saucerzines," however, wanted nothing to do with any layman's discovery of a new lifeform, especially if the discoverer were an American.

Some years after the Countess bravely published her theory, the distinguished American scientist and inventor John M. Cage also suggested that UFOs were not machines. He saw them as "sentient life forms of a highly tenuous composition charged with and feeding upon energy in the form of negative electricity." Cage's view was that UFOs were best described as *life fields* rather than ships or objects, and thus more or less got into bed with me.

Despite Mr. Cage's massive talents and demonstrated achievements in science and invention, his views on UFOs got little attention. He had erred in the direction of the truth. The spaceship-obsessed ufology of those days found his opinions unwelcome. The etherian physics to which Mr. Cage devoted himself has never been popular with this world's provers of foregone conclusions.

Significant among those who came to accept that UFOs are critters was the man who touched off the whole powder keg—Kenneth Arnold. As coiner of the term "flying saucers" he has won immortality, but by the early 1960s he was nowhere near as puzzled as he was on 24 June 1947, when he saw a "formation" of these saucers near Mt. Rainier. In November, 1962, his views on UFOs sounded as though they had been lifted bodily out of my 1958 book *They Live In The Sky.*

Writing in Ray Palmer's *Flying Saucers* magazine, Mr. Arnold said:

"After some 14 years of extensive research, it is my conclusion that the so-called unidentified flying objects that have been seen in our atmosphere are not spaceships from another planet at all, but are groups and masses of living organisms that are as much a part of our atmosphere and space as the life we find in the oceans. The only major difference in the space and atmospheric organisms is that they have the natural ability to change their densities at will."

Any pioneering researcher like myself, slogging through countless chilly dawns on the high desert and in the mountains to capture these critters on film, cannot fail to feel gratified at finding the originator of flying saucers in the same bag as himself.

Biologists should be assuming a leading role in UFO investigation. They are the group of scientists most concerned. They should also be able to recognize in my photographic work the germinal phase of a New Biology. Broader-based in bioenergetic reality on the physical plane than mechanistic biology, and *cognizant of a spectrum of density* that is immanent in what I have discovered, such a New Biology will occupy skilled scientists for generations. An authentic exobiology cannot take root in the sterile soil of mechanistic biology, and the existing discipline bearing that name is essentially scientism, not science.

The first kind words concerning my photographs to be published by any reputable scientist fell appropriately from the pen of the late distinguished biologist Ivan T. Sanderson. Owner of degrees in botany, zoology and geology from Cambridge University, Sanderson has clearly recognized the UFO problem as a biological mystery. His views are presented with great sagacity in his book *Uninvited Visitors.*

In the course of his exposition, he deals with my work and photographs. Despite his three degrees from one of the world's leading universities, he imputes to me a claim nowhere to be found in any of my published writings, namely, that I took my photographs with a special camera that I built myself.

In *They Live in the Sky,* as well as in this book, I have been at pains to emphasize that the cameras, filters, lenses and films used by me in my field work are standard, catalog items available through any photographic dealer. Never have I claimed to have built a special camera. I don't know why Sanderson decided to invent one for me. He wanted to publish some of my pictures in his book, but dropped the idea when I requested the normal ethical right to see his captions to those pictures. While such remarkable errors made me wonder about the rest of his book and research, when one saw Ivan Sanderson's great tenderness and good feeling for animals, all was forgiven. He could not have been other than a fine fellow.

A few paragraphs after inventing my special camera for me, and after pointing out that experts have not been able to prove my photographs to be fakes, he says this in *Uninvited Visitors* concerning what I have captured on film:

"If they are real, we face a particularly unpleasant situation, for it can mean only that our atmosphere is literally crowded with (to us) invisible objects, ranging in size from a cookie to a county; and if so, we would appear to have no way of getting at them, as of now. But, there is a further much more unpleasant aspect to these photographs.

"Be they Unidentified Aerial Objects or Unidentified Aerial Phenomena, they don't look like machines at all. *They look to a biologist horribly like unicellular lifeforms, complete in some cases with nuclei, nucleoli, vacuoles and all the rest* (emphasis added). Some are even amoebic in form. What is more they appear as completely opaque, mildly diaphanous, completely tenuous, or what can only be called evanescent, merging into mere light hazes."

Mr. Sanderson is thus able to see with his eyes what is actually in front of them—a task that the immortal Goethe conceived as being among the most difficult for humans. As the photographer at the time the photographs were made, I knew nothing whatever of "nuclei, nucleoli, vacuoles and all the rest," but as Mr. Sanderson has pointed out, that is what is there on the photographs. Similar forms will be found depicted in major reference books on microbiology.

To this present edition of *The Cosmic Pulse of Life*, I am pleased to add a summary of the original and strongly correlative UFO research conducted by the GRCU group in Genoa, Italy. Under their late dynamic leader, Luciano Boccone, they were able to assemble a compelling and convincing array of UFO photographs of invisible living forms, 20 years after my original discoveries. Boccone's work is summarized in Appendix II, along with examples of GRCU photographs.

This work was carried out without any awareness of my discoveries in the 1950s. A Rumanian research group under engineer Florin Gheorghitza produced similar, completely independent results. I am happy that European readers have this assurance that critters have been photographed in their part of the world by European researchers. Boccone's excellent and lavishly illustrated book, *UFO—La Realta Nascosta,* published by Ivaldi Editore in Genoa, is essential reading for students of this subject.

A significant observation here is that man would have no books on microbiology, or any knowledge of its workings, had he not invented the microscope to overcome the lower border of his optical limitations. The microscope *intensifies* the power of sight. Infrared film *extends* the sense of sight beyond the vibratory range considered normal. In both cases, a range of form inaccessible to the unaided senses appears for study, complete with living organisms. Surely it is the acme of irrationality to accept the microscope, and reject the infrared camera or other sense-extending apparatus that opens a macrotelluric level of life heretofore veiled from human gaze.

Man is busy mounting ventures into cosmic exploration, whose purpose is allegedly to roll back the frontiers of knowledge. The search for extraterrestrial life has high priority in these programs, but the irruption of these heretofore unsuspected and, *in situ,* supersensible lifeforms into human experience is being evaded.

The critters, as I call them, will perhaps one day be better classified as belonging to the general field of macrobiology or even macrobacteria. The more man irritates their plane of life with his ignorant and clumsy electrical activities, the more he will see of them in our own level of tangibility. Ufology can have no coherence or substance without their acceptance as an integral part of its phenomenology.

Before we leave the critters for the time being, perhaps observed and reported activities of UFOs should be related now to our ingenuous probings of that imperceptible yet physical world that is their true habitat. Without burying ourselves in the hard data of specific instances, let us simply recall the following things from the body of ufological knowledge:

UFOs have been seen playing around thunderheads, and playing around aircraft wings as foo fighters and as all kinds of balls and forms. They have been photographed and seen traveling in the atmosphere in luminous shoals, just like fish. UFOs have playfully buzzed primitive peoples at different times, including Alexander the Great's army, which they threw into confusion. UFOs have paced airliners and jet fighters like dolphins pacing a ship. They have chased each other around the sky in a manner consistent with happy animal play as we know it on earth. They have arisen from hiding in remote areas at the approach of man, as though startled, and much as do waterfowl. They have demonstrated suspicions of man, and have made themselves elusive. Man, in turn, justifies this suspicion by pinging bullets off them to see what they sound like, or blasting them with machine guns.

None of these activities are readily reconcilable with the concept of highly intelligent and advanced entities from other planets—the only concept that mechanistically-minded humans are thus far willing to entertain. Biological aeroforms of a subhuman or elemental character are nevertheless typical of a large and growing number of UFO encounters. These critters, in all their diversity, are not only here to stay, but they have always been here—unseen. We of the human kingdom will undoubtedly see more of them as our electrified civilization intrudes ever more strongly on their milieu, disrupting the tranquility of millennia.

An authentic and functional exobiology, which will eventually grow out of a revolutionized educational system, will in due course classify these critters and give them Latin names. In the early 21st century, hosts of young investigators in exobiology will be in full pursuit of the critters of our atmosphere. They will undoubtedly marvel at our stupidity in not stumbling onto such presences far earlier. Objectification of the critters by a couple of borderland adventurers may, by mid-century or earlier, not seem half as strange and incredible as it does now.

Despite their importance to a rational and complete understanding of the terrestrial environment, *the critters neither preclude nor exclude other presences of a far more complex and terrifying portent.* Our photographs already provide evidence of how the vehicles of these advanced intelligences are propelled, as well as evidence of their presence in an adjacent and perhaps interpenetrating density. The 20th century human primitive—with or without a Ph. D.—who cannot face these living aeroforms of an order of creation lower than himself, nevertheless deems himself equipped to cope with a far more advanced visitation.

Scientifically and psychologically, however, man is not equipped to confront entities of a far higher order of intelligence than himself. These entities are clearly masters of the etherian physics that 20th century scientific man has evaded and suppressed. The bold beginning made by Dr. Wilhelm Reich provoked rage and terror, which in turn backed a full-scale legal effort to push this knowledge off the earth. Man is a child, and not even potty-trained.

In all the panorama of theory and surmise about UFOs, there is perhaps no stronger evidence of the essential beneficence of the higher intelligences involved, than that they have refrained from unhinging our best minds by appearing with the wonders of etherian physics at their command. What happens in such circumstances to an eminent Ph.D.? Or to a Nobel Prize winner in physics? Suddenly all the Old Knowledge of which he is a master, is seen as a concatenation of misobservations. Are our finest minds equipped to make any meaningful boarding of spaceships that the Etherean visitor can *think into existence* as we watch?

The critters are the beginning, the opening wedge, to a New Biology, one that will lead inevitably to spaceship propulsion. Through the years that I have toiled in this strange borderland, one lesson struck home to me again and again. I had begun with the desire to photograph spaceships—just as our distinguished men of science wait now for alien confrontation on their own ground. A conference in Washington might well be their ideal.

I wanted spaceships, but I got something else. I actually *resented* the first critter pictures. From the inception of this work it was as though some friendly, superintending intelligence was conducting an educational campaign— watching and helping me handle what I had to find out *first*. The critters fly faster than and easily outmaneuver contemporary aircraft, and they do this on biological power. The key to spaceship propulsion seemed to lie in the same realm.

Pursuit of the critters opened many doors to the New Knowledge. Illusion after illusion was stripped away, and the process was often painful. Some illusions die hard. The grandest illusion of all, inculcated in me since boyhood, was that scientists are objective, unemotional, calm and methodical fellows. My experience proved to me that like all other human beings, their emotional structure tinges everything they do and everything they perceive. Being imperfect emotionally like the rest of us, mechanistic scientists sometimes impress these imperfections on the New Knowledge that breaks in increasingly on the human race.

NOTES TO CHAPTER EIGHT

1. Compare this reported shape with Bravo #1 in this book, which shows a similar object, photographed directly from the invisible state by me, using infrared film.
2. See "The Atom Versus the Orgone" in Jerome Eden's *Orgone Energy,* Exposition Press, Jericho, New York, 1972.

Chapter Nine

THE MASK OF OFFICIALDOM

One must live things to judge them.
— WILHELM REICH

Many people were prompted to try and help me after publication of *They Live in the Sky* in 1958. Interested and kind people who already knew me, and many who sought me out in a spirit of intelligent goodwill, did their best to "promote" me and my work—despite its underfinanced primitiveness. Introducing me to scientists and engineers of their acquaintance was their major effort on my behalf. They were rational people, but unacquainted with the formidable psychological problems attendant on the discovery of biological UFOs, and therefore these efforts all proved unrealistic and fruitless.

The rational aspect of the efforts made to help me was based on what people could observe happening in the U.S. The Pentagon was mindlessly dumping scores of millions of dollars into worthless "make-work" research in many fields. An influential bagman in Washington seemed to be all that was needed, once you had a covey of researchers with official credentials on the payroll or otherwise on tap. Introducing me to steady, established, thoroughly safe individuals was the procedure my friends thought would be most fruitful.

All the well-intentioned efforts made to assist me were barren of results in any immediately applicable fashion. There was nevertheless an invaluable and unforeseen dividend. The efforts involved me in numerous meetings with a variety of scientists and engineers. As a participant once again, I learned for myself at first hand about the sham of so-called scientific objectivity.

Wherever scientific objectivity may turn up in the universe, it is most assuredly *not of this world.* My experiences with human irrationality in connection with radical findings are of minor dimension. Suffice it to say that the persecution of pioneers in cosmic electronics like Dr. Albert Abrams and Dr. Ruth B. Drown, and revolutionary cosmologists like Dr. Wilhelm Reich and Dr. Immanuel Velikovsky, verifies that this irrationalism is amplified to the dimension of the personality it becomes necessary to execrate. The ultimate manifestation of irrationality becomes the murder of the innovator.

By mid-1958, Jim Woods and I had accumulated well over one hundred mutually corroborative photographs of the two basic types of UFOs—vehicles and critters. Motion pictures of the critters whirling around me in the desert dawn decisively crushed the objections raised to the stills by self-styled experts in whom these strange critters seemed to evoke an unbearable anxiety. Seen in its totality, with the underlying reasoning and basic explanation of the experimental methods presented in this book, the collection of photographs made an overpowering impression on all who came with open minds.

There was no question that it all lay outside the boundaries of official science, and violated in the mode of its acquisition all the principles set up by the reception systems for the dissemination of scientific knowledge.

Scientific periodicals returned contributions as though from a sling shot. We had violated most of the rules of mechanistic method, and one rule in particular, for which we would always be rejected by official circles. We had involved *ourselves* in the total process in a deliberate, outrageous and unconventional fashion.

Physics sometimes terms the influence that the individual experimenter has on the outcome of his experiment "the Heisenberg effect." This term commemorates the late Professor W. W. Heisenberg, who drew attention to this phenomenon. In most experiments in mechanistic physics, the Heisenberg effect is small enough to be ignored.

In our work, our sense apparatus was used in an extended, new way, impermissible to a mode of scientific cognition based on evidence received through a single, color-blind eye. The human body force field, or orgone energy field, was an integral part of our work, and success depended upon its methodical manipulation. The total human being was an essential part of the whole happening, and not an incidental effect influencing the results only slightly. To the orthodox mind, we were simply out of sight.

Without an understanding of the marvelous reality of the human visual ray, and direct contact with the living energy ocean in which all these UFO events were occurring, we would never have been able to produce any objective evidence. To this day, no one may say with certainty to what extent my particular personality and extraphysical energies were responsible for the results, since no one else has tackled the problem in the same way. A few people tried, got results, then became frightened, for one reason or another. No one persevered.

Each human is the bearer, at the subconscious level, of a variety of powers that transcend current scientific knowledge. The New Knowledge leaves no doubt that each of us is not a new soul—a brand new entity coming by accident into a world of indeterminate origin—but rather an entity thousands of years old. Given the right external conditions and opportunities, the things we knew and could do in other lifetimes and other lands manifest again as talents, drives and insights.

Our penetration into a supersensible stratum of physical nature invariably proved to be too much for scientists and engineers to whom I was directed by my well-meaning friends. Brainwashing by the universities, the systematic biasing of young minds to a mechanistic world order and an irrational financial system, is the biggest single barrier existing today to the introduction of the New Knowledge. Every time I went to people who were products of this system—at the behest of others—the experience was futile and frustrating.

Where qualified people came to me on their own, impelled by their own forces and interests, the results were quite different. Friendships were established that have endured ever since. When these people of novel bent broached these matters to fellow scientists, their experience proved to be the same as mine. They were beaten down by a peculiar, highly emotional and irrational reaction that bore no relevance to the facts being discussed.

Intolerance is the correct description of this reaction. After detailed study of Dr. Wilhelm Reich's work many years later, I came to understand what

upset these people and why. The intolerance is the reaction of a neurotic personality to his own bioenergetic movement. Neurosis is born of chronic
bioenergetic inhibition. When the individual's bioenergy moves in response
to stimuli that are lawfully rooted in basic life processes—in this case the living creatures of our atmosphere—the neurotic individual clamps down on
this movement. He literally cannot stand it.

Such reactions were observed by me for years, and some of the most highly qualified men I talked to reacted most violently. These reactions were
utterly incomprehensible to me at the time because I had been brainwashed
with the legend of scientific objectivity in my own education. There is no
more fatuous myth permeating our modern life than the one that tells of
detached, unemotional men of science. I have seen Life scare them stiff!

When we deal with the scientific work of Wilhelm Reich in later chapters,
all this will be further elucidated. Having seen this queer reaction scores of
times, among all kinds of people whose views on extraterrestrial life are
already concreted within the mechanistic framework, I believe it to be an
integral part of contemporary human attitudes and reactions to life in space.
Ashcans to the moon at $30 billion per voyage, do not move the bioenergy.
Critters from our atmosphere do. People can tolerate the former, but not the
latter.

Persons who can tolerate new work of this kind, usually have only one
basic question if they are without technical expertise: 'What do the *physicists*
say about this?" Reference is automatically and always to some form of
authority, even though such authority may have no relevance to this work,
which is new and revolutionary.

There are no authorities on this work, not even me.

Almost two decades after I started, I know only that my ignorance of what
lies behind and beyond the realm into which I have broken is appalling.
Pandora's Box was by comparison a picnic basket. Experience has taught me
that the most brilliant men are, from the cosmic point of view, still only on
the fringe of understanding. Scientific cognition began for man only yesterday.

In the 1950s I was an idealist about the world of science, not then having
learned to distinguish between science as an investigative instrument and
method—which is marvelous—and the character structures of the individual
human beings who turn science into scientism. Because of my naive idealism, I was persuaded to seek the approval of physicists, chemists and others
who were qualified in their own formal fields. These people had either
assumed authority in the UFO field, or had authority thrust upon them by laymen who deferred to them in technical matters. This assumption or assignment of authority had persisted despite three decades of zero progress by
formal science in the UFO field, and the unavoidable conclusion that something is lacking in mechanistic scientific cognition.

Irrational reactions toward New Knowledge are not confined to my work
on the part of the conventionally-minded. Highly-qualified scientists who
have broken into the New Knowledge invariably have the same problem.
One might cite here the work of Dr. Harold Saxton Burr with his colleague
Dr. L. J. Ravitz at the Yale University Medical School.[1] They discovered the

Life-fields of which everything discussed in this book is in some way a manifestation—including the critters. Prejudice and resistance attended this epochal work, although it was done in an august institution of higher learning.

The UFO project at Colorado University demonstrated little concerning UFOs, but was rich in lessons concerning human beings. The characterological barriers to straight mechanistic investigation of UFOs were fully exhibited. The project also showed that the Establishment can buy convenient verdicts from teams of scientists—which is corrupt and wrong by any normative standards. Genuine scientists resist corruption, and it was such men within the Colorado project that triggered the controversy that erupted.

The Colorado project was staffed with scientists qualified in their disciplines at the best universities. Men who had studied UFOs avocationally were ruled out of the teams, so that experience, familiarity with the subjects and any insights accordingly won might not contaminate the pristine product the Establishment had ordained should come forth. The pasteurized, thoroughly safe men who gathered at Colorado were to scrutinize and evaluate UFO evidence.

Soon they were accusing each other of duplicity and engaging in squalid quarrels through books, magazine articles and other media. There were firings of top personnel and naked threats of professional vengeance made against those scientists who declined to see the Colorado project run quietly along the rails laid down for it by the Establishment. The largest civilian UFO group in the U.S.A., Donald Keyhoe's NICAP organization, withdrew its support when evidence NICAP made available to the project was not properly utilized.

For years Major Keyhoe had called for a government UFO project, staffed with top scientists, as a necessary and desirable official approach. Numerous other UFO investigators, with similar high expectations from scientific objectivity, similarly urged just such an official approach. When at last these importunings were answered, the character structures of the participating scientists made a public shambles of their objectivity. Whether or not the individuals involved could *bioenergetically tolerate* UFOs seemed to determine which of the two camps they joined. The anti-Life faction won control.

UFOs are pervaded with *life*. At Colorado, some people could tolerate life and movement, including their own corresponding organ sensations. Other people were anti-Life, or Life-negative. Any well-authenticated body of UFO evidence has life in some way immanent in its substance. This is true even if one can stretch no further than the seedy extraterrestrial intelligence hypothesis. Even this is life of a new kind. Some neurotics find this highly disturbing, and the disturbance makes them act irrationally.

Some of the Colorado scientists stood up for Life. The rage and threats against them, well-documented in the whole lamentable episode, sprang from frightened, irrational men, their scientific detachment blown to the winds by their terror of their own biological movement. What happened to the honest and upright scientists at Colorado had happened to me many times and many years prior to Dr. Condon's project.

Having experienced this irrationalism among scientists at first hand over a long period, I was able to predict the outcome of the Colorado project with precision. Barely a month after the USAF announced the contract, I wrote an article for the *Journal of Borderland Research,* predicting that Emotional Plague would break out despite the august university atmosphere. I said that nothing would come of it but chatter and bewilderment. The *Journal's* editor at that time, my good friend Riley Crabb, was jampacked with articles and did not use my piece, but later said he regretted not having scooped other periodicals. In the UFO field, most publications opined that the scientific approach would win out. So it will, when we understand and reform human character. My negative advance judgment on the Colorado project was based on bitter experience with a wide variety of troglodytes.

In a typical instance, a close friend of mine who was an independent UFO researcher, persuaded me to meet with a group of scientists at a suburban home in Los Angeles. I will call the hosting engineer George Broughton, although that is not his real name. He has always maintained his goodwill toward me, and was not responsible for what happened. His interest in UFOs hurt him economically in many ways, and I refrain from identifying him because of his Federal business connections and obligations.

Broughton had an electrical engineering degree and a sizeable plant doing defense work. UFOs intrigued and enthralled him. He was fascinated by my photographs and the possibilities they opened. He could also see the commercial future in mastering the energy form that propelled the discs. The biological UFOs he took in his stride. "The photos speak for themselves. They eliminate a hell of a lot of confusion about these things," he said. Broughton was a straight, bright, unblocked individual who could tolerate Life and the New Knowledge, even if the latter had not yet been brought down into hardware.

The men he invited to his home that evening were of a different stripe. So was his wife. She greeted me with ill-concealed hostility and abruptness, embarrassing him and establishing an antagonistic atmosphere. She hated UFOs passionately, and hated her husband's interest in them. To the kindly Broughton's dismay, the proceedings were taken over by a bumptious biologist, who said to me before we were even introduced, "Has any of this crap you're putting about been proven?" Without apprising Broughton, the six men present had evidently agreed to give me an uncomfortable evening—to put me down.

When I tried to give a basic account of the methods employed to attract UFOs and photograph them, the biologist kept interrupting with "But you *can't* do that!", dragging me into arguments on side issues and quibbling over methodology. My host protested in vain. Two other men present, both aerospace industry engineers, hectored me relentlessly.

When the time came, I opened the case in which I carried eighteen enlargements of my best photographs. I handed the "amoeba" photograph shown in Alpha #1 to the biologist, and held four successively captured shots of the same critter in my hands. He looked as the "amoeba" and *immediately looked away as though needles had shot out of the picture into his eyes*. He handed the photograph back to me quickly. "I don't see anything in this picture," he said. I invited him to look at the sequence of the pictures, so he could follow the obvious expansion of the object, but he flatly refused. "No *true scientist* would ever look at such stuff," he said.

Years later, I understood what Dr. Velikovsky encountered in his struggle for a new cosmology, when eminent, internationally famous men of science put him down in scientific journals. The tirades against Dr. Velikovsky were characterized by the writers stating that they had *not* read his *Worlds in Collision.* Not to read Velikovsky was an article of faith among his critics. The scientific world is riddled with, and addled by, such irrationalism.

In two gruelling hours with the six men at Broughton's house, not more than a dozen intelligent questions were asked. Most such questions came from Broughton. The other men all spoke quickly, with an evident need for reassurance from the others. In those days, I did not understand what was happening in a behavioral sense to those men. I did understand that objectivity, calmness, fairness and rationalism were not there. Something bordering on panic was present in that room. Normally calm and objective men were experiencing *anxiety.*

Striking indeed was the compartmentalization of their knowledge. All of them were sort of walled off from each other. The electrical engineer grasped the photo evidence of quasi-electrical propulsion, but dropped out in dealing with anything obviously biological. The physicist understood the plasma effects around some of the objects, but recoiled from even the tentative idea that organisms and vehicles were somehow functionally related. All of them were basically kind, normal men, husbands and fathers, well-educated and respected, but in terms of both scientific training and character structure *they were not qualified to deal with what I had uncovered.* They were tragically emblematic of the flight to scientism when the New Knowledge breaks through.

They were a microcosm of the whole scientific failure and default on UFOs. Getting these intelligent men to accept that I had photographed these things when they were *invisible* proved impossible. In that respect, they were all totally blocked. When I went home, I was deeply troubled by the multiple examples of irrationality that had given me one of the most uncomfortable evenings of my life.

How could men of science be so emotional, so intolerant, and so strangely fearful? That question arose again and again. Why the antagonistic reception to something innovative—something that was throwing light on the greatest mystery of all time? Until these experiences with workaday scientists crowded in on me, I had always felt that science was a pure instrumentality, untainted by emotion. At first hand, I was learning a central fact of 20th century life: science can never be any better than the character structure of those who labor in its service. If the scientist's character structure is mechanistic, so will be his perceptions, reactions and conceptions. Their direction can only be toward the crushing out of Life.

Next day, Broughton called me and apologized for his guests. We had lunch, and he told me that after my departure, his friends ridiculed him for his support of me. He made light of it at the time, but had been hurt by their behavior. He told me that in his opinion I was too far ahead of my time, and too mercurial for orthodox scientists to follow. Ten years later, I met him by chance in the Mandarin Hotel in Hong Kong, and his first words were, "I'll never forget that awful night at my home..." Nor will I.

Another similar incident illustrative of orthodox reaction to work like mine, took place soon afterward. A close friend of mine, graduate of one of the finest U.S. universities and well-connected in the Defense Department, persuaded me to try and show my work somewhere in official quarters. Earlier I had placed everything at the disposal of the USAF, but there was simply no interest in such work, even though the USAF was dumping money into so-called "think tanks," wherein bright boys were paid to *speculate* about spacecraft propulsion.

A negative reaction from the Federal government does not necessarily mean that what one is offering is no good. The defense services in all countries are markedly resistant to new concepts of all kinds, even when they originate with responsible and qualified people. Innovators who are too daring and too forceful have to be torpedoed, as was the case in the U.S. with General "Billy" Mitchell. Irrational opposition from high-ranking professional officers to new conceptions is a central element in military history.

Each world war has proved that only a handful among the hordes of professionals is able to handle high command successfully. People still trust the military professionals despite their repeated failures even to provide the proper patterns of weapons before major conflicts erupt. My service-connected friend suffered from overconfidence in the defense establishment. He made the arrangements for me to visit a Defense Department research office in Los Angeles to show my films and photographs. Still naive and idealistic, I felt that with his introduction there would be no repetition of that awful affair at Broughton's. After all, I told myself, that was essentially a social affair, and this was a business visit to a Federal office.

The physicist who received me opened our conversation by stating that the official position was that these things do not exist. When I sought to explain the simple methods used to obtain the films and photographs, the physicist kept looking at his watch. He had no interest in a specially prepared folder of fifty photographs I had prepared for this day. In my galloping naivete I offered to leave him my negatives for analysis, but he said it was not necessary.

The motion picture films were disposed of by an intelligence officer, who then projected them upon a dirty and pockmarked wall, winding the lens in and out of focus as the film was shown. Despite this chicanery, sufficient footage came through clearly—dirty wall notwithstanding—for it to be a convincing introduction. The physicist left the room, red-faced and angry. The intelligence officer handed me back the film. "Now you can say, if you are asked, that this office has seen your film and pictures." He then walked away.

Persons who doubt that such reactions occur, should undertake some serious time to the advocacy of any pro-Life viewpoint. Thanks to the findings and discoveries of Dr. Wilhelm Reich, the etiology of the physicist's irrational fury is today well recognized in orgonomic psychiatry. Today's ecology advocates are encountering kindred irrationalities, in their efforts to prevent man from suffocating himself in his own effluvia. To me, in those days, as a young man unlearned in the ways of neurotic reaction, I was thunderstruck that a man who had earned a Ph. D. degree would blow his mind and temper over a piece of scientific film.

These reactions typify untold dozens that I experienced through the years. Whenever I made the mistake of approaching some established, orthodox-minded person, usually at the behest of a well-meaning third party, I could depend on such reactions—in greater or lesser measure. There was also the studied vacuity of official, government reaction, as a counterpart to the individual emotional volcano.

The late Senator Clair Engle of California was a licensed private pilot, and also had a strong personal interest in UFOs. We had a brief correspondence, as a result of which he sent some of my material to the USAF. The reply which came from the Air Force credited me with having photographed "space ectoplasm", a term never used and a claim never made by me. The USAF view was that the things I had photographed had nothing to do with the UFO phenomena. And having made the pronouncement, the matter was dropped without further communication or inquiry into what would nevertheless be an amazing discovery in space, even if it didn't have anything to do with UFOs. Let all others who photograph pulsating, glowing discoidal forms be hereby warned.

The dynamic activity of the dawn period, when the bulk of my early photographs and films were made, is not something invented by me. Rather it is a phenomenon independently observed by scientists, as a result of radar propagation anomalies that have been observed and recorded at this time of day. In layman's terms, a "propagation anomaly" is an irregularity in the normal way in which radar signals are transmitted and received.

One of the anomalies that bothered radar scientists in the late 1950s and early 60s was the reception of strong radar returns from objects that are *not ocularly perceptible.* "Angels" was the term applied to these objects that seemingly are not there, except that radar says they are. My photographs are of objects that are also not seen with the naked eye—except sporadically and occasionally—at which time they become UFOs.

An Air Force project conducted by the Cambridge Research Laboratories attempted to explain why service radars detected strange clumps of targets in the pre-dawn period and also before sunup. The findings, theories, conjectures and speculations of the baffled scientists were presented by Vernon G. Plank in two papers published by the Geophysics Research Directorate of the Air Force Cambridge Research Center. Paper No. 52 was "A Meteorological Study of Radar Angels," and Paper No. 62 was "Spurious Echoes on Radar, a Survey."

Anyone who thinks that my finding invisible flying objects in the earth's atmosphere at dawn, and photographing them, is farfetched should read the alternative explanations in these two enlightening treatises. Imaginative explanations for the dawn "angels" are offered that are far wilder than mine, and for these explanations no proof is adduced. Flocks of birds, heated pockets of gas, clouds of insects and other way-out notions were offered as examples of what the angels might be.

My friend Bob Beck, a former engineering test pilot and an instrumentation specialist, read the monographs and expressed my feelings precisely when he returned them to me.

"Those guys," he said, "are really *reaching*."

Some young scientists who had been concerned tangentially with the angels project, and who knew that they were "reaching," were shown some of my photographs by an electromagnetic interference expert who had an interest in my work. These young men could see a promising and mighty solid explanation for the angels that they had been dogging.

True to the spirit of science, they wanted to corroborate my findings with more orthodox methods and apparatus. I was ready to assist in any way open to me, in complete anonymity and without any desire to receive official credit, funds or recognition. For me, it was enough that these young men should be able to pull off the corroboration of what I had done.

When this proposal got up to the level where heavyweight scientists hand out funds and approval, the whole idea was killed. That life might be present in a heretofore unsuspected form was evidently once again going beyond what the older men in science could tolerate. A couple of ardent young spirits wanted to pursue the venture on their own time with their own funds, but were denied use of service apparatus and told to drop the approach completely. The angels continue their anomalous manifestation, but it is enough to give them a name—angels—and leave explanations for later generations. If it is alive, then *don't touch it.*

There were many other instances involving scientists and engineers who tried to get attention to my work. Hope was always held out, despite the radical nature of my findings, that a more conventional empirical approach might verify my findings in an acceptable fashion. Nothing ever came of all this pulling and hauling. My personal policy was to greet all aid with goodwill and high hope, always assist to the maximum, and to stand aside completely if someone should insist on personal credit.

Every individual who tried to work within any kind of official framework, either in government or in private industry, got set down hard by his superiors when he broached biological UFOs. They experienced frequently the same Emotional Plague reaction that I have already described. In every case, they were men with families to support, and were honest, open-minded and fair men. The threat of economic reprisal—open or tacit—always caused them to recoil, and for that they cannot be blamed. In a hardware culture, living needs are necessarily secondary to the primary mission of selling junk to the government. All these well-motivated efforts were thus either stillborn or strangled at birth.

These examples paint a general picture of scientific reaction to this approach to UFOs, when it was advanced in the 1958 period and in the years immediately following. Considering that the publications, bodies and individuals approached were essentially products of and enslaved by the mechanistic world conception, the reaction is really not surprising in retrospect. Among genuinely friendly, openminded people of science with a feeling for the novel, however, the reaction was not greatly different. Their cutoff point simply came later.

A close friend of mine enjoys the friendship and confidence of many of the top men in science in the eastern U.S. Many of them are world famous for their discoveries, inventions and achievements. They meet periodically to hash over the psychic, the occult, the new and the way-out, seeking stimulus and diversion from the workaday concepts and methods by which they make their livelihoods. All of them were fascinated by my photographs, which continued to accumulate into the Bravo Series, some of which are shown in this book. The Bravo Series includes a remarkable set showing USAF jet fighters, armed with *Sidewinder* infrared homing rockets, chasing my critters above me over the Mojave desert.

These men could tolerate the photographs and the findings. They were delighted at the unequivocal proof that the USAF has lied about its pursuits of UFOs. Helping the work, or introducing it outside their secret circle, was something else.

They quickly joined the ranks of the "if only" men of my experience. "*If only* you had taken this in stereo, it would be incontrovertible," said one famous man. When stereo photographs were made, there were new evasions for new reasons, not the least of them *that the objects appeared to be in two different places at the same time.* The new problems were thus thornier than the old, something to which biologist Ivan Sanderson has referred in his account of my work in his book *Uninvited Visitors.*

The package could not be made neat and unobjectionable. Disturbance or destruction of the illusion of neurotic security under which scientists labor — no matter what their eminence — was simply unavoidable. The old criteria of reality were under fire. As greater ramifications appeared out of my efforts to shape the phenomena to what the mechanistic mentality would tolerate, my need for assistance rose sharply.

Revolution leaped out of the photographs. There was no way I could soften or lessen its impact. All the men of novel bent and goodwill eventually quailed before the challenge. I ended up with friends, sympathizers and even admirers, but without the support I needed. Being told I was fifty years ahead of my time was no consolation.

Fairness and chivalry constrain an understanding of these reactions among active scientists. Their professional tribalisms have not only form, but also force. Families have to be supported and responsibilities discharged that require maintenance of professional integrity and standards, regardless of the fascination a scientist may have for off-the-track findings and borderland phenomena. Retribution attends such dabblings if they become too vigorous. This kind of intellectual and economic blackmail is largely responsible for the general failure of organized science to tackle UFOs as serious business. Only hardy mavericks buck the system. Borderland roustabouts like myself cannot hope to alter the course of such a high-powered juggernaut.

The same kind of hedging and resistance extends into ufology, a subject where involved parties like to think of themselves as being open-minded. Ufology has nevertheless developed its own tribalisms. Official ufology appears to have no intention of departing from its prime dependence upon the ETH, or extraterrestrial hypothesis. While reasonable and probably in some ways true, the ETH is nevertheless inadequate and is unproved to this day.

Opposition to the New Knowledge approaches to UFOs takes the form of evasion of anything not readily classifiable as a spaceship. The more "solid" such a ship may appear, the more solid the attention directed to the incident. That unseen intelligences had given me basic information that I had used to obtain my photographs—to bear into a dynamic borderland of energy and force—was used as sufficient reason to discount my findings.

The irrationality of this posture was never apparent to diehard ETH advocates. Extraterrestrial spaceships were here, according to them. All evidence, in their view, pointed to this. Yet when a biological communications system provided information and methods for making these photographs, originating with entities who stated that they rode in ships from other dimensions, official ufology could not tolerate the new realities.

Ufology also hid behind the mask of officialdom, croaking down through the years for a government investigation of UFOs. When the investigation came, the qualified scientists who ran it turned more than half a million dollars of Federal money to crushing ufology. Twenty-first century students of 20th-century mass psychology will have mordant comments on the way *fear of life* powered this irrationalism. The prized, cherished, adored and worshipped *objectivity* does not exist among mortals.

Only a few periodicals remained opened to my contributions on UFOs, and beyond these it was as though an invisible wall had been erected against dissemination of my views. Editors of general interest magazines and men's magazines, to whom I was steadily selling military articles and stories, quickly fired back UFO pieces as "too wild," or routed them immediately to a tame consulting scientist for the death blow.

One compiler of a book of UFO photographs was told by a young lady who knew of my work that he should include something of mine. At her behest, I went to see him with a satchelful of UFO photographs. These pictures showed a wide range of forms and shapes of unseen things in the air. *He already had many of these same forms in sketches provided by persons who had made sightings.*

These sketches were on the desk in front of us, in paste-ups for the book he was compiling. In any other field than ufology, my photographs would have been like a gift from heaven for anyone compiling such a book. They objectified photographically what independent observers had already sketched, from sightings made in many parts of the world.

Like the biologist at George Broughton's home and dozens of other people in the interim, the book editor looked at the photographs quickly and then *averted* his gaze, like a Milwaukee matron finding a French postcard. He handled the piles of pictures as though they were physically hot. He couldn't stand the sight of them. He was uncomfortable and anxious in the presence of this pile of proof that fitted in so well with what he planned to publish. Finally he got it out.

"These are very interesting," he said, "but of course, *they don't have anything to do with flying saucers.*"

As he spoke those words, the discoidal, glowing shape of the amoeba stood on the top of the stack. This incident more than any other, coming as it did after years of similar reactions where the associations of everything were less direct, convinced me that there is something fundamentally wrong with the way a great many human beings perceive. This man was obviously *blocked.* So were all the others who had acted irrationally.

By contrast, I noticed the way in which free spirits tackled the examination of the photographs. There was no looking away. They went right into the pictures with their gaze, pored over them, and inevitably compared them with microscopic forms. People like these seemed to be more alive and direct than those who couldn't stand to look at the photographs, and I always seemed to have a deep and immediate contact with them.

Official science and official ufology still wear the same mask to this day, although their irrationalism is no longer the enigma to me that it was in earlier times. The etiology of this irrationalism is understandable—and therefore ultimately conquerable—because my individual pathway led me to three giants of the New Knowledge. All three of these titans encountered similar irrationalism and rage in making their contributions to human advancement.

From these three titans I learned in fortuitous sequence the things I needed to know in order to understand something of what I had stumbled upon. I also learned how to carry it forward. The three avatars were Dr. Rudolf Steiner, Dr. Ruth B. Drown and Dr. Wilhelm Reich. Their work, discoveries and inventions permit us to begin a new, Life-positive science. Through Steiner in particular, we can understand—and therefore counteract—the hidden powers from beneath man, whose earthly works include the truth-killing mask of officialdom. Let us meet these three titans in turn, in the same sequence as I did, and we will begin to understand the Battle for the Earth that has already begun.

NOTES TO CHAPTER NINE

1. See *Fields of Life*, by Dr. Harold Saxton Burr, Ballantine Books, N.Y., 1973.

Chapter Ten

STEINER—PRODIGAL TITAN

Wisdom is crystallized pain.
—DR. RUDOLF STEINER

Dr. Franklin Thomas handed me a medium-sized green book one day in the summer of 1957. He had already effected my rescue from involvement with the psychic phenomena attending UFOs—my false start as an investigator—and was directing my curiosity into new channels. He was phasing me into a healthy approach to the laws and forces that stand behind UFOs and kindred phenomena, as well as behind our own manifested world. My old teacher wanted me to begin training myself to see nature in a new way.

"This book is what you need now," he said. "Learn its lessons well. Follow where it leads you." The book was Dr. Ernst Lehrs' *Man or Matter,* and proved to be exactly what I needed. *Man or Matter* is an introduction to a spiritual understanding of nature, based on Goethe's method of training observation and thought. This book changed my life. From its profound depths came the inspiration and guidance for all my practical work, which began with the securing of objective evidence of the presence and nature of UFOs. Today I still lean heavily on this book in my weather engineering research, which involves manipulation of the primary energies behind physical nature.

Dr. Lehrs' book also provided me with a comprehensive introduction to the first of three titans who were to influence me decisively—Dr. Rudolf Steiner. Genius is a word that has been debased by copywriters, Hollywood flaks and public relations people who have made an industry out of exaggerating the importance of business executives. Therefore it seems appropriate to define the word genius before applying the term to Dr. Steiner, who ought to be a basic standard by which genius is measured.

George Bernard Shaw said that a genius is a person who, probing deeper and seeing farther than his contemporaries, has a different set of ethical valuations to them. The genius is able to give effect to this insight in whatever manner suits his or her talents. Steiner proved by the scope and sweep of his work, as well as by the magnitude of his legacy to mankind, that he was a universal genius. One of the most remarkable men ever to walk the earth, he vivified and renewed everything he touched—and he seemed to touch everything.

Steiner's impact on human culture is only just beginning. The passing decades verify his prescience in all fields of human knowledge. A century hence, he will probably be accounted the major avatar of the New Knowledge in the Western world, and the youth of the future will find in his work the answer to all their idealism, as well as the keys to complete renewal of human life and culture.

America stands largely unaware of Steiner, many years after his death in 1925. This is an unfortunate circumstance, since Steiner's insights into coming crises in the life of humanity have been forcefully borne out in the United

States, riven as it is by tumultuous social forces clearly beyond the comprehension of those in authority. Disquiet over dehumanization, for example, is endemic throughout America, and it was Steiner who foresaw such consequences of rampant technology. He also provided workable indications by which man might maintain both his humanity and his technical progress.

Examples have already been adduced in this book of the phenomenon of *sequestration*, by which Life-positive personalities and developments are walled off from human attention, rather than openly contradicted or empirically disputed. Sequestration will be better understood when we get to fuller grips with Wilhelm Reich's bioenergetic and biosocial discoveries. Sequestration undoubtedly has occurred with Steiner's work in America, where there are grave needs that it can meet.

Once I asked a Californian college professor who rode with me in an airliner what he thought about Rudolf Steiner's work. He described him as a "second rate philosopher." On closer questioning, it turned out that this man—charged with the education of young people—*had actually read not one article or book or lecture of Steiner's*. The philosophers of American academia just accord among themselves that Steiner is second rate, and it is appalling how much has to be lost to our youth because of this "rote" thinking. Let us review briefly Rudolf Steiner's epic career, and wonder that the New World has hardly been touched by this titan of the New Age.

Austrian-born in 1861, Rudolf Steiner was educated in Wiener-Neustadt, and later at the Institute of Technology in Vienna. He turned from technical studies in early adulthood to the study of history and philosophy, and received his Ph. D. from the University of Rostock. His Ph. D. thesis dealt with Fichte's theory of science. Later this thesis was published in book form under the title *Truth and Science*. He who would understand the impasse science has reached over UFOs can do no better than study this book.

Steiner's knowledge of basic natural science was to serve him well in giving substantial new impulses to many different scientific disciplines, and he paid frequent tribute to the achievements and to the spiritual significance of mechanistic science. He did not accept the mechanistic world conception as being in any way adequate for modern consciousness, and we have seen this mechanistic world conception break down completely in the UFO problem decades later. In the development of his Anthroposophical world conception, Steiner provided something far more serviceable, practical and pervaded with Life.

Steiner was to become one of humanity's major teachers, and his gifts in pedagogy first manifested when he successfully tutored the retarded child of a wealthy Vienna businessman. The child went on to become a medical doctor. After the First World War, and a giant's work in other fields, Steiner's teaching experience and educational insights led to his establishing the first Waldorf School in Stuttgart. The methods of education developed in this school from Steiner's indications and guidance soon spread internationally. Today there are approximately eighty Waldorf schools all over the world. Many more are in prospect.

At the age of twenty-nine, Steiner began six years of work at the Goethe Archives in Weimar, a labor to which he was summoned as an outstanding scholar. There is little grasp in the modern world of Goethe as other than a poet-philosopher and dramatist. He is best known in the U.S., as well as in his native Germany, as the author of the classic *Faust*. This aspect of Johannes Goethe is rather less than half of him.

Goethe was a natural scientist of consummate skill, insight and versatility, who made far-reaching observations and discoveries in many fields. He probed into and contributed substantially to such seemingly unrelated fields as botany, meteorology and optics. When men of science rid themselves of their traditional and doggy thralldom of Newton—a process that has begun with the Space Age—Goethe will be accorded high station in the history of science.

Goethe provided the break-in to the new modes of cognition and observation that support the New Knowledge in all its diversity. Steiner's task at the Goethe Archives was to edit the massive scientific writings of his predecessor. This work involved comprehensive and sustained study of Goethe's writings. He had to "get inside" Goethe. Steiner's scientific training, combined with his background in philosophy and history, enabled him to grasp fully the significance of Goethe to the modern world. Goethe's significance was lost to his own time, other than in the arts.

Goethe provided the pathway to a new mode of *cognition*. The need for a new way of knowing things was already there, at the dawn of modern material science, but this need was buried under the mechanistic avalanche of the 19th century. The success and progress of mechanistic science trapped man into thinking that all could be known by these methods, an illusion that persisted until the UFOs brought about a staggering moment of truth.

Steiner wrote his classic *Goethe's World Conception* out of his experience at the Goethe Archives. This won him recognition as an authority on Goethe. Such recognition was less important than Steiner's actual achievement, which was to open to modern Man with modern consciousness a modern and dynamic pathway to observing and understanding natural processes. The need that existed for such a Life-positive approach in Goethe's own time has been amplified through the intervening decades to become crucial in our time. The deadness, sterility and persistent drive toward explosive substance that characterize our present world, are the inevitable consequences of expunging life from our way of looking at natural processes.

Steiner produced a series of books that formed the foundation of what is now known as Anthroposophy, described by Steiner as "a path of knowledge that would lead the spiritual in man to the spiritual in the universe." *The Philosophy of Spiritual Activity* is his major work, and the serious study of this remarkable book opens new doors to the student. In particular, the student finds his way to acquisition of the *mobility of thought* necessary to correctly observe and comprehend living processes.

That same mobility of thought is now essential, about a century after Steiner's book appeared, to deal adequately with the cosmic phenomena known collectively as UFOs. This is but one indication of the powers and insight of Rudolf Steiner. He was in his own time a man of the future, and that is one reason why he is being increasingly "discovered" by the questing and restless youth of today. That his major book has such a direct bearing on the UFO problem illustrates that Steiner's work is *practical* in the fullest sense. Impractical people naturally cannot grasp Steiner, and clutching their eminent qualifications and degrees, remain bewildered and defeated by the UFO problem.

By the end of the 19th century, Steiner's universality was coming into full bloom. He was a scholar of distinction, and an author of strikingly penetrant quality. He became a discerning editor, notably as the guiding spirit for the *Magazine for Literature* in Berlin. During his Berlin experience he also lectured at a workers' educational institute, and was thus involved at first hand in the social problems of his times. Steiner did not belong to the academic brotherhood of the Ivory Tower, but rather plunged down with increasing force and effect into earth life.

Those who conceive of metaphysics and occultism in Oriental terms— highly popular in the U.S. today—will possibly be surprised by Steiner's life. There was no withdrawal to mountain tops, no retreat from the world as maya or illusion. The range and scope of his activities continually increased. The versatility of his endeavors never carried him from the basic thrust of his life—the pathway to new means of cognition and observation opened by Goethe. Steiner cleared, widened and lengthened that pathway through his own dynamic achievements, and his personal unfoldment is lucidly described in his autobiography, *The Course of My Life.*

He was a man who moved among, and met personally, such contemporaries as Haeckel, Nietzsche, von Hartmann and Grimm, all of whom have enjoyed fame as major figures of Western culture and wide readership in the U.S. Steiner nevertheless remains a sequestered figure, destined it seems to be carried forward by the force from beneath—the force of modern youth— rather than by academic acceptance or comprehension. The accounts Steiner gives in his autobiography of his encounters with these major personalities and their schools of thought partially explains why he is almost unknown in America.

He founded the Anthroposophical Society. This became the center of the cultural renewal he set on foot, and a point of focus for those who began breaking new paths in various disciplines of science by following his indications. A characteristic of virtually all the great Anthroposophists is that they share Steiner's ability to meet the world as it is. They qualify themselves in their disciplines at leading universities. With their abilities fortified and fertilized by Anthroposophical conceptions, they then dedicate themselves to the challenge of human renewal. We will cite some of their contributions in due course.

Steiner's literary activities extended to drama. He wrote four plays dramatizing the cosmic mysteries, and with the discovery of the orgone energy by Wilhelm Reich in 1939-40, these plays have achieved a new dimension of authority and significance. Steiner's mystery dramas deal with the machines invented by Strader, one of the leading characters. These inventions are intended to transform social life by the free use of primary power by everyone.

Human reaction to the discovery of such devices is poignantly dramatized by Steiner, and much of what he dramatized has already been acted out in real life, by living humans of the 20th century who tapped a force corresponding in essentials to that discovered by Strader. The orgone energy motor of Wilhelm Reich equates well with the "Strader machines" of these plays, and the evasive, smart-alecky dismissal by official science of the orgone discovery is well delineated in Strader's conflicts with a Life-negative know-it-all who could have come directly from our finest universities.

In a startling sequel to his play writing, Steiner designed the Goetheanum in Dornach, Switzerland, as a new kind of theatre for the performance of these mystery dramas and kindred productions. Introducing an innovative style of architecture, inspired by Goethean concepts of living metamorphosis, Steiner produced in the Goetheanum a structure of vibrant, organic beauty.

Destroyed by fire in 1923, ten years after its creation, the partially-wooden building was replaced with a new Steiner-designed Goetheanum on the same site. The new structure used reinforced concrete. This building stands today. World headquarters of the Anthroposophical Society, the Goetheanum is a structure that fairly breathes Life from its organic form, and few people who see it ever forget its impact on their own inner forces.

Steiner not only developed a new style of architecture, but poured his vivifying genius into such details as the colored windows of the building. He invented a process for grinding designs into glass, and plant dyes for staining. From these original and inventive labors he could turn in an instant to lecture to the workmen who were building the Goetheanum, meeting them on their own level but elevating them through his great instructive powers.

He was a *Western* genius in that he provided a concrete physical basis for his innovations. His work was realistic in the truest sense. People were always given something that they could genuinely grasp, use and carry forward out of their own strengths, forces and impulses. When he opened new pathways in painting and sculpture, for example, the basic technique for their pursuit and extension was also provided by this remarkable personality.

Steiner's contributions to the arts have hardly yet been felt in the world, but they did not stop with drama, literature, architecture, sculpture and painting, or with a new approach to the design of jewelry, or with his indications for the renewal of the acting art. He brought into being *eurythmy*. A new art of movement of the human form, eurythmy is sometimes referred to as visible speech. This mode of enlarging the capacity for movement of the human form — making it a vehicle for the expression of heretofore untapped qualities of tone — is not only a new art form, but a significant therapeutic innovation. Curative eurythmy is, in my view, a safe route for the amelioration of muscular armoring, and its potential is vast.

One may watch a stage full of small children at a Waldorf School performing in an eurythmy presentation, and literally feel the force coming off them thirty feet away. Professional eurythmists leave no doubt in the spectator that their art is a spiritual breakthrough of the first magnitude. Body, soul and spirit speak as a functional unity through eurythmy. Those who are about to give up on mankind should first expose themselves to a performance of this new art.

The Rudolf Steiner who elucidated and carried forward into modern form the scientific insights and observational innovations of Goethe would have a claim to world attention on that basis alone. As a philosopher, writer and dramatist in his own right—contemporaneous with such important figures as Haeckel and Nietzsche—Steiner rates study on that basis also. As the founder of the Anthroposophical Society, he fashioned an international movement that spread worldwide. With Anthroposophy came a renewal of education, so that humanity's problems could be tackled at the roots—with the children. Waldorf education takes account of the spiritual realities of the developing child. Virtually all education now moves slowly in the direction to which Steiner gave impetus.

Scientist, philosopher, writer, founder of a new education, editor, architectural innovator—that is a formidable range of achievements. Certainly there is enough innovation there to warrant attention to Steiner. Yet he was all these things at once, and *more*.

As a lecturer, he was probably without a peer in Europe in his own time. He gave hundreds of lectures, to all manner of people, on a positively astounding range of subjects. In his versatility and vivifying power he must be accounted unmatched, from that day to this. He could move from lecturing and inspiring ordinary workmen, to lecturing audiences of physicians on a full range of medical subjects. Study of his lecture series published as *Spiritual Science and Medicine* makes it clear that modern medical research will be a full century exploring his indications in pharmaceuticals alone.

From audiences of doctors, he could move to audiences of teachers, mathematicians or artists. That they were qualified professionals in their fields dismayed Steiner not at all. The results were always the same. Everyone present would be put to their intellectual limits to follow him, and kindled in their depths by what he had to say. He was the expression and embodiment of the new kind of consciousness and the *new mode of mentation* that man must acquire in the Cosmic Age that opened with Sputnik.

There was nothing that could be broached to him of which he did not seem to stand in total command. He was fully familiar with the "state of the art" in diverse fields of knowledge, and had indications for decisive advances in all of them. Specialists with lifetimes of experience in such seemingly unconnected fields as economics, agriculture and electricity found Steiner able to guide their most advanced research.

Typical of this incredible technical and philosophical versatility is the account Dr. Lehrs gives in *Man or Matter* of his first meeting with Dr. Steiner in 1921. Electrical engineer and mathematician Lehrs was toiling at that time with his Ph.D. thesis. He was involved with experimental research in high frequency phenomena, then a new field. In his brief first conversation with Dr. Steiner, in a crowded hall, Lehrs referred to his technical problems, not for a moment thinking that Steiner would be familiar with—or interested in—such a specialized and new branch of physics. Dr. Lehrs writes of this incident on page 21 of the English edition:

"Judge of my astonishment when he at once took out of his pocket a notebook and a huge carpenter's pencil, made a sketch and proceeded to speak of the problem as one fully conversant with it, and in such a way that he gave me the starting point for an entirely new concept of electricity. It was instantly borne in on me that if electricity came to be understood in this sense, results would follow which in the end would lead to a quite new technique in the use of it. From that moment, it became one of my life's aims to contribute whatever my circumstances and powers would allow to the development of an understanding of nature of this kind."

Lehrs' experience was typical of Steiner's impact and influence on the soul that was ready for what he had to give. Those who were not ready for what he had to give often compulsively denounced this God-like man, and Anthroposophy was frequently pilloried in many tongues from European pulpits. Clerics often detested his contributions to a modern understanding of the Christ mystery. His restoration to the Christian ethic of the indispensable knowledge of reincarnation and karma, expunged by the early Church fathers, excited considerable clerical hostility.

Steiner in no sense attacked prevailing religious systems. His teachings were exclusively intended for those for whom their time had come, and he was neither proselytizer nor polemicist. He regarded both mechanistic science and formal religion as *transient* in the evolutionary sense, and each as one-sided. He held that this mechanico-mystical splitting of man would be overcome in due course by the forces of change inherent in the evolutionary process. This transient and essentially contrived splitting process has been sharply delineated, as well as brought to a head, by the UFO problem.

In his writings and lectures, Steiner addressed world economic problems. On a completely new basis, he outlined his "threefold commonwealth" approach to man's socio-economic and politico-economic problems and their role in the human future. He elaborated with precision and originality the need for a threefold social organization that would reflect the threefold nature of man as a being of body, soul and spirit. Revolution in the old political sense has demonstrably been no more than an interchange of inadequacies and evils. As with cognition itself, in economics there must be a completely *new* way. Steiner broke the path.

This incredible man poured his vitalizing, modernizing power into all these diverse and seemingly unrelated fields. He broke through all the artificial, divisive boundaries between disciplines and demonstrated the unity to be found amid their diversity. Philosophy, natural science, economics, sociology, speech, music and drama, cosmology and cosmogony, occult science and Christian theology each received from Steiner their strongest forward impulse in centuries. Agriculture, astronomy, astrophysics, mathematics, medicine, therapy, nutrition, optics, biochemistry and education were similarly broken free from old traps and intellectual habits.

Steiner appears on the broad canvas of his times as a one-man university—a single human being who yet carried all within himself. He is undoubtedly the most astonishing and penetrating generalist of modern times, beside whom even generalists of formal repute appear as mere children in their conceptions. Within each speciality upon which he touched, however, and usually at its most advanced point, Steiner could decisively influence top specialists with new indications for their individual endeavors.

Steiner's free and fluid movement among all these assorted disciplines—covering the whole field of modern knowledge—was living demonstration that the new modes of cognition and observation of which he wrote and spoke were both realistic and practical. The new ways, for him, were far advanced beyond mere hypothesis. Steiner was living proof that man can indeed find within himself the keys to his modern dilemmas.

Steiner's mode of mentation, and his observational and cognitional powers, showed that modern man need not be overwhelmed by technology. He provided the answer to the exponentially proliferating details of mechanistic science, as well as to the ever-multiplying welter of *ad hoc* theories that threaten to overwhelm the discursive human intellect, data retrieval systems notwithstanding. Particulars could result in humanity's end. Access to the universal laws, the womb from which the particulars are born, assures that man will not lose his humanity. Steiner has shown modern man how to win such access, and with it, more true freedom than humanity has yet known.

Some of the leading thinkers, writers and workers in scientific fields who were influenced by Steiner have shown the same ability to advance and support new conceptions. Worthy of note among them are Dr. Hermann Poppelbaum, with his monumental contributions to a new zoology; Dr. Eugen Kolisko, a medical doctor who moved with Steinerian fluidity into geology, cosmology and such startling non-medical fields as the influence of the moon on agriculture; Dr. Georg Unger, eminent mathematician and sometime visiting fellow at the Institute for Advanced Studies at Princeton; Dr. Ehrenfried Pfeiffer, the driving genius behind the now internationally flourishing biodynamic gardening and farming movement; Theodor Schwenk, whose new approach to the creation of flowing forms in water and air will surely come of age in the next few decades; and Dr. Guenther Wachsmuth, secretary and biographer of Rudolf Steiner, whose comprehensive presentation of modern etherian physics dovetails beautifully with both the UFO problem and Dr. Wilhelm Reich's discovery of the orgone energy. There are also many other eminent contributors, whose work is to be found in Anthroposophical literature. Those cited above are the best known and most widely acknowledged in their individual contributions.

Steiner died in 1925, in his sixty-fourth year, laboring until the final days of his life in the service of mankind. He had broken through all the established and purported boundaries of knowledge, and given humanity a new beginning. Subsequent events have verified that he was something more than just an exemplar of the new humanity, destined to pass half-forgotten into history.

The School of Spiritual Science that is maintained at the Goetheanum in Dornach has carried forward and extended Steiner's work. His students have continued to progress along the new pathway he broke. He thus appears virtually unique among modern avatars in that his work has gone forward decisively under the energies of those he trained and guided in the New Knowledge. The more critical the human condition becomes, the more widely will Steiner be recognized, and his wisdom sought.

In America, there is a marked tendency on the part of those scientifically trained, or even scientifically-minded, to recoil from the use of such terms as spiritual science. This is the legacy of two centuries of mechanico-mystical splitting of man. The division between knowledge and faith, which its victims fail to recognize arose at a quite definite historical moment and produces this irrational, spastic aversion to the use of the term "spiritual" as an adjective with the noun "science." Visions are instantly conjured up of various religious systems, such as Christian Science, Divine Science and Religious Science. There is therefore a tendency to equate a School of Spiritual Science in far-off Dornach with some kind of religious cult. Such notions spring not from true knowledge. Investigation reveals such notions as baseless.

Through lectures, seminars and schools, through laboratories, a clinic and other facilities, the School of Spiritual Science applies Anthroposophical knowledge to the practical problems of contemporary life. This work is direct and realistic. As a prime example, the problem of cancer therapy has been tackled at the Clinical Therapeutical Institution at Arlesheim. Steiner even provided indications for the chemotherapeutic conquest of this scourge.

Steiner's indications for renewal of the performing arts are pursued in the departments of speech, music and eurythmy. His mystery dramas and other classic plays are regularly performed, and the school of recitation and declamation produces speakers and performers who carry the New Knowledge into these oral performing arts. Eurythmy performances are given throughout the year on one of the largest stages in Europe, and eurythmists travel abroad to teach this life-giving art. Students come from all over the world to Dornach, and range from idealistic young people to highly qualified, senior professionals in the sciences.

A research institute is maintained with laboratories for biology and biochemistry, food, medicine and physiology. Biodynamic farming and gardening knowledge is provided and taught. In the U.S., a biodynamic farm is maintained in Spring Valley, New York, where practical results of the new ways may be seen by all. An observatory is at the center of the sections on mathematics and astronomy. Steiner's extensive medical indications are pursued through physicians' courses, and pharmaceutical research and clinical tests are conducted at Arlesheim.

The special care of handicapped children has been pursued with devoted diligence by Anthroposophical physicians and workers. They have thus faced squarely, and with remarkable success, a pervasive practical problem from which modern society essentially prefers to remain aloof. Anthroposophists do not withdraw to the mountain tops to wail about the world. They work.

The pedagogy section of the School of Spiritual Science provides training for teachers at Rudolf Steiner schools. Conferences and seminars are regularly held on Waldorf education. The education of handicapped children has received a new impulse through Steiner's indications, and the special needs of such children was made the subject of dedicated, original research. This work has gone out into the world to help the afflicted.

Schools of painting and sculpture pursue new lines. A school of jewelry based on Steiner's cosmic insights into the art of gems has produced remarkable innovations in a field thus far wholly decorative and not cosmically attuned. Sections of the school are devoted to sociology, literature and philosophy.

Regular publications of high grade by scholars, researchers, investigators and thinkers appear under the school's imprimatur. A bookshop offers comprehensive Anthroposophical literature to accommodate any interest. The overall impression is that of a thriving practical center. The Goetheanum is a place frequented by people who have their feet on the ground. They face life in all its fullness, and there is about it all an atmosphere of dynamic spiritual health and positivity.

The Anthroposophical Society is thus not a European tong, or secret in any of its aspects, but a public organization. When he founded the Society, Dr. Steiner excluded any kind of dogmatic position on any issue. One could thus hardly conceive of any society more ideally exemplifying true spiritual freedom. The qualifications for membership? One simply believes that the existence of the Society is justified, and that joining is a good thing!

The Anthroposophists are a union of human beings with a desire to further the life of the soul, individually and in society, through a true knowledge of the spiritual world. The science of the spiritual world opened by Rudolf Steiner is essentially the methodology needed by modern consciousness for exploration of the supersensible realms. This exploration is pursued by methods that are as exact in their way as those of true natural science. No one is fitted to judge the validity of these methods without serious study and personal application. The acquisition of *exact spiritual perceptions* is a task calling for diligence, persistence and patience.

Since Anthroposophy has produced results that serve every human being, racial, religious, national or social differences are not accorded any significance in the Anthroposophical Society. A social life rooted in true brotherly love would be the inevitable social result of fully applied Anthroposophical knowledge. The Society does not consider politics among its tasks.

In this world, there is plenty of bluster and lip service about freedom. One good way to find out what a gathering of free people is really like is to forget the endless, neurotic babble about freedom, and simply meet a group of Anthroposophists. I have traveled all over the world, and known men and women of numerous nations and tongues in a full spectrum of social situations. The Anthroposophists as a group are the most human people I have known, and are easily the most solid and sound of all those seeking a pathway to the spirit.

Anthroposophy has been in existence for almost a century, as of this writing. A typical American inquiry is therefore: "Why hasn't it caught on?" In a culture erected upon the product-design concept that "if it doesn't sell it's a failure"—with mass appeal justifying anything—it is true that Anthroposophy may seem unimpressive. Steiner's work has not "caught on" like Billy Graham, the Beatles or the astrology revival. The reason is that Steiner's work is no instant panacea, and can offer no opportunity for taking an emotional bath in one's own illusory ecstasies. Such a serious pathway of knowledge cannot be merchandised. When the individual is ready, he will find Anthroposophy, and its value to the serious student of UFOs is broached here as one of the main purposes of this book.

Many persons reading this book will be hearing about Dr. Steiner for the first time in their lives. They cannot help but reflect that it is a long way from Steiner, the Goetheanum and Dornach, to the lone wolf investigator photographing UFOs on the high desert of California. I had never heard of Steiner either, until the man who saved my life gave me a gentle bump in the direction of his work. Those who had trodden the Steiner pathway were those in my experience who were instantly capable of understanding—in the full light of consciousness—what my strange research with infrared photography has opened.

Fatiguing explanations were unnecessary with these people. They knew. The elemental, borderland realms into which I had broken were not new to them. I was astounded at the mobile way their minds functioned. I wanted to be like them, and I wanted myself to be able to see nature from the same pro-Life, spirit-permeated viewpoint that they had made their own. Their tolerance and kindness towards me were heartwarming. They understood, and they accorded me a natural respect both for myself and for what I had been able to do in objectifying the unseen.

As the years passed, and Anthroposophy proved to be a safe, sound and healthy pathway, one thought returned to me repeatedly. This thought arose out of the stuff and substance of experiences with the strange, unseen space denizens that my cameras had recorded. I thought often of Rudolf Steiner. Especially did I think of him when speculation appeared in the media about what space entities from other worlds would be like.

These speculations have often concerned themselves with the standard of knowledge that such beings would possess. Mechanistic scientists and mathematicians made the usual linear projections of what purported "advanced beings" would be like, and what they could be expected to know. I always thought of Steiner as a sort of example of what the possessor and bearer of advanced knowledge would be like, particularly in his incredible universality.

Here was a truly advance being, who had walked the earth among us in our own time. He had shown us how to reconnect microcosm and macrocosm. He had given clear indications for the renewal of human life and culture, and for the New Sciences to be wrought out of the old and moribund conceptions. All this was offered without one backward step from facing the world as it is. My question was: "What spaceman could conceivably bring or offer more, and if he did, how much could humanity, how much *would* humanity accept?"

There was more in Steiner's work than our brightest minds could handle in the 20th century. We have been given the pathway to 21st century science, and to a new cosmically-connected culture. Few have thus far listened, even fewer have studied the work of this titan—the blueprint he gave us for the future is as real as this book in your hand. One cannot but wonder at the vacuity of people who imagine that they are going to acquire more by telepathy from alleged spacemen than Rudolf Steiner has already provided.

In the UFO field, the pursuit of messages from spacemen has yielded, in some cases, surprising technical information. My work began that way. Canadian scientist Wilbert Smith, former head of the Canadian government's *Project Magnet*, also got workable technical data and new scientific information. George Van Tassel got a heavy yield of new data pertaining to the science used by the Etherean entities with whom he had contact. Daniel Fry was able to use data so derived—from his own sources—in the design of electronic apparatus. There have been others. The important element that must not be overlooked is that this information is *limited.*

The technical data usually stands valid but isolated. There is no readily discernible connection with extant technology. The conclusion is inevitable that the blockages to further such information lie in man—in his present modes of cognition and mentation. This is probably the reason that more data does not pass. Man does not think in a fashion adequate to handle what is *normal for any intelligence advanced beyond our own.* Steiner's example gives practical verification of the difficulties attending the introduction of New Knowledge, even when it is dovetailed to existing forms and norms.

If you want to gain some insight into the kind of mind, knowledge, compassion, understanding and insight that we might anticipate finding among truly advanced intelligences, then by all means begin your struggle with Steiner. You will be put to your limit, no matter what your eminence. You will find yourself immeasurably strengthened in dealing with earth life as it is. Friends, scientists, educators, physicians and researchers whom I have turned on to Steiner have almost unanimously found this to be true. My own pursuit of three simultaneous careers draws its energies from the same vast well.

We have now met the first of my titans. Let us turn to some of the indications, theories, findings and conceptions of Steiner and his followers—as they influenced my UFO research. Pride of place goes to the Steinerian approach to etherian physics—the physics of the ethers. Ethers, in this meaning, is rightly plural. Without this knowledge, earthmen can have no comprehensive access to the scientific principles that underlie the strange aeroforms in our skies.

Chapter Eleven

COSMIC BLUEPRINT

A man's mind stretched by a new idea can never go back to its original dimensions.
—OLIVER WENDELL HOLMES

In attempting to provide a thumbnail sketch of the spiritual-scientific conceptions of Dr. Steiner, and of those who followed the pathway he broke, one is sharply conscious of a certain impertinence. Steiner's work is destined to change the character of modern knowledge, as well as expand the means of its acquisition. These are weighty matters. There are at present in print, in the U.S. and in the English language, at least 250 separate books and collections of lectures by Dr. Steiner.[1] Let the following pages therefore be regarded as an outline sketch of those aspects of Steiner's work that enabled me to approach UFO phenomena in a new way.

Anthroposophy provides insight into the *changes of consciousness* that have taken place during human evolution, in various regions and periods of the past. Popular conceptions of ancient man present him as just like us, but lacking automobiles, electronics and similar modern appurtenances. Our belief today is that people did not really *know* anything before the rise of modern science and its measurement techniques.

Prior to the development, in the immediately recent past, of modern natural science, human beings held completely different views of cosmic and planetary evolution from those that are considered scientifically valid today. In earlier cultures, extending as far back as written records take us, men expressed their conceptions of cosmic workings in the "mythologies" that are an integral part of the cultural history of early humanity. Scientists lend no credence to these views of early men. Evolution and creation as seen by the ancients arose out of religious ideas and inspirations influencing humans in those early epochs.

Steiner has pointed out in many of his lectures, that all these varying cosmologies and cosmogonies—regardless of the geographical region, ethnic rooting or time period in which they were developed—attest to a mode of consciousness in their formulators markedly different from that of modern man. Spiritual beings were held to stand behind natural processes. These spiritual entities were deemed to be interwoven with earth life in a way that modern scientific consciousness can only regard as absurd.

Dr. Steiner recognizes and accepts the power and clarity of modern scientific thinking. At the same time, he repeatedly states that the particular type of brain-bound consciousness that has given rise to present material science, is a *transitional* form of consciousness. In short, changes of consciousness in the past evolution of mankind are a fact. Changes of consciousness in the future are a certainty.

Breaking through all prior limits of knowledge drawn by earlier philosophers, Steiner himself demonstrated in the breadth and depth of his own life's work what the scientific mind of the future will be like. The scientist of tomorrow will have *exact perception* of the spiritual worlds, and of their

111

processes and denizens. Present-day material science may be considered a sort of prototype of a future science that will understand the laws of the spiritual worlds with the same precision and clarity of thought now attending its physical endeavors.

As we move toward the increasing encounters with otherworld intelligences and forms that are arising out of UFO manifestations, we do well to remember two things concerning the orthodox knowledge with which we are presently attempting official investigations of UFOs:

1. Modern natural science has been on the world scene only about a century and a half.

2. Until the advent of modern natural science, the mythico-religious conceptions of creation were the only conceptions entertained by humans. Their history extends back for millennia.

When technical data has been offered by non-terrestrials in contact with earthmen, as in the Van Tassel communications mentioned earlier, this technical advice has two main characteristics:

1. Advanced scientific concepts are employed that indicate technological mastery of etherian physics. These concepts have been largely execrated by official science when advanced theoretically by highly intelligent earth men. To the "spacemen" they are facts, far into technical application.

2. The entities involved speak of spiritual laws and forces with the same precision that we can apply only to "hard" technical fields like chemistry and optics.

The importance of Steiner's pioneering work—showing humanity the pathway to such superior cognition—should already be evident. He showed us how to get to where the UFOs and their controlling intelligences already are. *We have to do the job ourselves.* We do it in freedom.

Steiner gave mankind the basic means and methods for acquiring the powers of cognition necessary to extend our scientific understanding of creation. The extension must inevitably be into the supersensible realms that stand behind the manifested world. Such glimpses of the New Knowledge in technical application as the UFOs have given us, leave little doubt that 21st century science is going to be radically different to that of today. *So will be the consciousness of its scientists.*

The time to set such changes on foot is now.

The contemporary deracination of any kind of spiritual or soul force as active in earth life is a recent thing in the evolution of humanity. Through all prior time, man maintained cosmic connections. The impact of UFOs on scientific consciousness—their arousal of technical curiosity and their inaccessibility to existing methods—has been immense. The UFOs have confronted us with the need to change ourselves. The direction of change will be as Dr. Steiner indicated—the systematic acquisition of superior cognitive and perceptive powers—and by the means he defined.

Thousands of young science students in the past took "trips" with LSD and other psychedelics, blasting open centers of higher perception with this powerful drug. Out of this pathological, but nevertheless devastatingly real contact with mind-stuff and emotion-stuff, has come a general impulse toward spiritual renewal. The science student who has seen the working of the ethers while under drugs becomes a candidate for their theoretical study, and for the intelligent application of his will to his own inner development. The strong souls will develop in themselves the same perceptual and cognitive powers as Dr. Steiner.

Modern materialistic science and its specialized consciousness have reached back into earlier earth life in the course of scientific investigation. Writings, drawings and records arising from human *inner* wisdom have been classified as mere superstition. This attitude stems from the impossibility of our viewing the cosmos with the consciousness of a Chaldean star-worshipper or an Egyptian priest. There is justification for the attitude of modern science, but it is essentially one-sided, and that is its weakness.

Cosmologies predating modern science had Man as the center and purpose of earth evolution. He was seen as worked upon by spiritual beings, powers and forces in an upward path to perfection. Modern materialistic science, by contrast, regards man as the end product of a series of biochemical accidents and arcane distillations, and there is as yet no coherent theory of evolution capable of withstanding the mounting torrent of scientific discoveries. Science is adrift in this field largely because it has not yet learned to read correctly the facts it has found. As with UFOs, the problem is not a lack of hard data, but a question of perception and consciousness.

The ancients conceived of the world and man as being built up over the eons by divine workings. In the sciences of the future, we will regain a similar knowledge of human origins, according to Dr. Steiner. This future science will develop its understanding of evolution by fortifying its external work with exact supersensible perceptions, scientific discipline, and a firm knowledge of the four basic ethers whose formative powers rule physical creation.[2] Ancient men felt themselves intimately membered into these etheric weavings, and out of this relationship arose what we regard today as ancient mythologies.

Modern materialistic science extinguishes all such intuitive feelings in its votaries. The systematic biasing of otherwise free minds is the bedrock of higher education, which George Bernard Shaw appropriately described as a "budget of lies". The religious conceptions that survived the rise of modern science were relegated to the philosophical domain of faith—the bailiwick of ministers, mystics and the multitudes who followed them. As science multiplied its power of investigation, the religious field came to be regarded as essentially for the misguided and ill-informed.

For the old conceptions of divine workings, science substituted the Kant-La Place theory of evolution. Synthesized out of the conceptions of philosopher Immanuel Kant (*Natural History of the Heavens*) and the theories of Napoleon's astronomer Pierre La Place, this was an approach to evolution wholly in tune with the juggernaut of Mechanism. Stripped of all soul, spirit and kindred influences, the Kant-La Place theory presented a lifeless, sterile picture of the origins of the earth.

A glowing, revolving gas ball is deemed to have slowly condensed, under this theory, spinning off the planets into their orbits. The whole thing eventually settled into the configuration we see today. Verifiable mechanical laws rule the whole happening. Soul and spirit become superfluous to what are regarded as the hard, physical facts. Life on earth is a later, seemingly accidental addition to the basic sterility of the earth's origin. To this day, scientists have been content with this theory. As this is written, it is being crammed anew into the heads of yet another hapless generation of youngsters.

Evolution thus came to be regarded scientifically as something starting from inert substance, proceeding from below upwards. The basic astronomical mathematics of Johannes Kepler were an indispensable step in the consolidation of these modern scientific ideas of evolution, but were divested of Kepler's conception of the universe as the body of a gigantic spiritual being. Any suggestion of spirit in the new concepts was left with the mythological notions of ancient men, amid the dust and junk of long-dead civilizations.

The searingly hot Kant-La Place gas ball is a mechanist's delight. Physical man or other life—either as an extant entity at the time or as a prior contributor to evolution—was totally eliminated. My discovery of the plasmatic bioforms—heat critters—undoes the idea that great heat and life are incompatible and paves the way for a new empirical beginning. Since Kant-La Place, however, notions of when and how life on earth started have, in formal circles, occupied some of our brightest minds. These varying notions have also triggered some of the fiercest arguments between scientists.

The sterility that is seemingly inseparable from that glowing gas ball, and the mechanical notions of evolution constrained by its acceptance, have blocked generations of thinkers from reaching any valid conceptions about the origins of the earth. Dr. Steiner's indications, based on exact supersensible perception, open all the necessary doors. What he has provided is in harmony with basic natural scientific findings. All this, in turn, reaches with illuminating power into some of the central riddles of ufology.

Dr. Steiner stated in his numerous lectures that lime formations are deposits of living organisms. Shellfish, snails, polyps and similar organisms have *deposited* the lime formations of the earth. Dr. Steiner made this observation about a century ago. Geochemical research in modern times progressively confirms Steiner's statements. In a relatively recent incident bearing upon ufology, the origin of other earthly matter in life processes was elevated to an even more startling eminence. We reach forward from Dr. Rudolf Steiner in the first years of the 20th century to Ivan T. Sanderson's *Uninvited Visitors* for one instance worth a thousand.

Biologist-geologist Ivan T. Sanderson was among the scientists interested in ufology who believe that before we seek to understand extraterrestrial life, we should try to better understand terrestrial life. That modern science is woefully derelict in this sphere, is evident from the incident he cites on page 99 of *Uninvited Visitors*.[3]

"A scientist in Germany was assigned after World War II to make certain surveys of the deep salt mines in Bavaria, where the Nazis had stashed art and other treasures that they had plundered from museums and collections all over Europe. This man, a biologist in its widest sense, happened to become interested in the salt itself, which was laid down under some overwarm sea about 200 million years ago. He took samples and analyzed them, and one of the things that startled him was that living things turned up in the resultant solutions. Naturally, he assumed that these were the product of contamination during the taking of the samples, or later in his laboratory, since all air is choked with such living things as spores and bacteria.

"However, he refined his techniques with every conceivable precaution, and still the living things appeared, until finally he was forced to the conclusion that they were resuscitated generations of creatures that had lived in the ancient sea and become fossilized along with the salt, to rest for millions of years before being 'brought back to life.' This bizarre experiment was subsequently tested by others, starting from scratch and using both their own and the original methods. Always the same answer."

Limestone and subterranean salt, oil, coal and slate deposits are all produced out of living processes. Then why not the totality of the matter of which the earth itself consists? To the objection that azoic rocks have never been found to contain fossils, there is the work of British biochemist Morley-Martin as an answer. Under conditions of total sterility he proved in the 1930s that *animal forms exist in and can be awakened from azoic rocks*. Such devastating findings wreak havoc upon the neurotic security of those whose whole world conception rests upon the opposite. Like Steiner, Reich, Drown, Abrams, Velikovsky and other genuine scientists, Morley-Martin has been the victim of obscurantism.

Recognizing the epoch-making nature of Morley Martin's discoveries, the late Meade Layne compiled an introductory brochure on the English biochemist's work, and published it in July 1950 under the imprimatur of Borderland Sciences Research Association. Entitled *The Morley-Martin Experiments and the Experiments of Dr. Charles W. Littlefield*, this valuable document is currently available from BSRF.

Morley-Martin, by a series of manipulations in total sterility, caused to emerge in microscopic miniature, vertebrate forms such as we know, together with others which no longer exist. Morley-Martin's vigorous condensations and chemical transmutations reproduced these creatures, in his own words, "in the way they probably came out from the firemist or the gas of our nebula." Under the microscope, trusted witnesses reported seeing vertebrates take shape within the protoplasm, with the outlines of limbs and claws following, then the heads and eyes. One crustacean developed its legs and then walked off the field of the microscope.

When Morley-Martin died in 1938, he had published only one small brochure, *The Reincarnation of Animal and Plant Life from Protoplasm Isolated from the Mineral Kingdom*, which appeared in 1934. Renown awaits the enterprising young biochemist who treads this same pathway today, for the lesson of all such work—Morley-Martin's experiments, those of Dr. Littlefield and the earlier Andrew Crosse—is that the opposite of life is not death, but *latency*. Biogenesis is not the mystery that obscurantist high priests of official science would have their fellow humans believe.

Human experience and experimental results have thus advanced slowly to those conceptions of eternal life on earth first put forward in modern terms by the amazing Steiner. The inert matter of the earth has been precipitated out of the life processes of the earth itself, just as inert minerals appear when we kill any living organism. That is, the corpse becomes of the substance of the earth itself. The unanswerable question is really no longer "When did life appear on earth?", but rather "When was there not life on earth?" The continued and dogmatic teaching in the universities of a world conception rooted in sterility is a fraud against the young. Ultimately, such lies will lead to revolt.

The concept of life being ever-present on earth is in no way upset or disturbed by geological evidence that certain mineral substances were once liquid. As long as we are strapped to mechanistic conceptions of the earth's origin, we consider this onetime liquefaction of presently dense substances to have been due to colossal temperatures. Mechanistically, there is no way around this.

In our pro-Life viewpoint, we read our lessons directly from nature. We know that in biological processes within living organisms, substances are liquefied which, if they were external to the organism, would be solidly mineralized. Such processes go on continually in our own bodies, and especially in our digestion. All manner of "solid" substances are dissolved in protoplasm and in blood. The peculiar power and character of biological processes is exemplified in the vulgar television commercial that shows fabric in which a hole has been burned by "concentrated stomach acid." The acid does not burn through the stomach, but is held in check by biological processes that are, at best, poorly understood.

Hydrochloric acid is formed in external nature by massive processes of volcanism—a happening that approaches in temperatures those applicable to the Kant-La Place ball-of-hot-gas theory. In the individual human being, this acid is formed without our even being aware of it, at a modest 37° centigrade! Why do learned men *avoid and evade* the clear testimony of all manner of liquefaction in living processes? Wilhelm Reich has provided the answer to this chronic evasion of the essential and we will deal with this in due course.

Given the undoubted ability of organisms to liquefy otherwise solid substances, it is but a short step to grasping Steiner's account of earlier epochs of the earth. One goes back not into sterility, but into an earth *far more actively permeated with life than it is today*. The farther one goes back, the more the earth seethes with life. Today's earth has grown cosmically old and mineralized, like a human octogenarian. Steiner takes us back in time to progressively more active life forces, and hence we can entertain the idea of a liquid state and farther back a gaseous state as a teeming center of life, whose nature is quite different to present earth life.

This "backward progression" allows soul and spirit forces to enter the evolution of an earth constituted differently from the present earth. We can easily conceive of an epoch where all would consist of *energy only*. In modern times, we have a crude counterpart of this type of life in our plasmatic creatures that dwell in the atmosphere. My pictures are *heat pictures*, and they show living organisms.

With Anthroposophical knowledge, one works back through ever more active life processes. Soul and spirit are admitted. Ancient men discerned their workings in earlier epochs, and recorded their perceptions in what we now discard as myth. Kant-La Place starts with the moribund, cosmically old earth of today, and attempts to go back into deadness to find life. In Steiner's conceptions, life and change are endemic. The mechanists are the misguided men who deny life and try to kill change—especially change in their own conceptions.

Radical as Steiner's conceptions may seem, they actually do no violence to the facts natural science has determined concerning prior evolution. This must be seen as distinct from the judgments and theories attached to those facts. When Morley-Martin brought critters to life in miniature out of azoic rocks, it is only the *prior assumption that those rocks predate life on earth that is demolished.* Knowledge and understanding are thus advanced by such original empirical findings.

That particular discovery, given new force and merit by the correlative Bavarian salt-cave discoveries mentioned earlier, permits science to go forward and embrace new truths. The men who are troubled and disoriented by such findings are those whose neurotic security is anchored in the old theories and the old knowledge. Their eventual passage from the scene is a certainty.

Meanwhile, such people control natural science and its reception systems. They ensure that the living, vibrant youth embarking upon higher education is biased and deadened beyond redemption. The price of making a living in science is compliance with established practices and knowledge. Although sound when properly applied, this principle has been misused and is now a major instrumentality of intellectual corruption.

At every hand, therefore, the real innovators, the real discoverers and trailblazers work outside the scientific establishment. They have no option. Much of the fury of today's youth arises from an intuitive sense of right and wrong about what is taught. The discoveries made by the scientific mavericks of the 20th century to the present are positively staggering. Every one of them was a hammer blow at the foundations of the old order.

Consider the dating techniques possible with radioactive carbon, for example. Immense authority attaches to these dating techniques. When Dr. Wilhelm Reich discovered the orgone energy, he also discovered in subsequent work that concentrations of this energy were capable of so decisively altering the half-life of radioactive substances as to corrupt the dating process completely. An earth more alive than at present would be an earth more intensely permeated with orgone energy—life energy—and thus all those lovely linear projections of the mechanists back into the past go straight down the drain! Reich's work in this area will be enlarged upon later, but it is cited here as an example of yet another empirical assault on scientific views that are mere *beliefs.*

Basic natural science will inevitably follow the lead given by the maverick innovators, equipping itself gradually with the new modes of thinking and cognition that at this time are congenital talents of the mavericks. A new empiricism will, in time, spring up to verify and objectify the fruits of such expanded cognition. Dr. Steiner was a sort of archetype of the scientist of the future, with the old primeval wisdom living within him in harmony with modern natural science. Man needs both scientific method and new access to cosmic processes in the age that came upon him with the UFOs.

The primeval wisdom should not be taken lightly or dismissed as mere myth. Archaeological research continually makes startling discoveries of the human past. Aided by such modern techniques as paleomagnetism and archeomagnetism, the human past is continually being moved farther back. Powerful cultures are uncovered by modern techniques. In 1967, for example, the Hohokam culture in the American southwest was unearthed. Suddenly, it was learned that Americans before Christ had learned such amazing arts as the acid etching of ornaments and artifacts, and had created masterful irrigation systems. Every decade verifies the presence of intelligent humans on earth farther and farther back into the remote past, before written records.

The artifacts of such remote cultures bespeak a more spiritualized man than walks the earth today. These men of old were not sundered from cosmic forces and powers, and they precipitated their perceptions of these cosmic forces and beings into the media then available, such as clay and wood. Today, we will proceed with similar precipitations, except that our media will be films of all kinds, videotapes, dvds, and other wonders soon to be born. In all these spheres of thought and work, Steiner's indications earn new authority via scientific investigation. On every hand, what comes forth has intimate relevance to ufology.

Functioning with a different mode of consciousness, ancient human beings perceived and thought differently from us. We cannot think their thoughts today. A significant point must now be advanced. In this book we have shown how modern photochemistry, in combination with some knowledge of etherian physics and occult science, has objectified a range of form adjacent to and probably interpenetrating our own. This range of form contains typical UFO shapes, as illustrated in Appendix I.

Surely it is not a wild surmise or a preposterous notion that ancient, closer-to-the-spiritual humans might have directly perceived such forms—lost to untrained modern consciousness and perceptions. There is powerful evidence that ancient people did perceive such forms. The evidence consists of an indelible and comprehensive record of their perceptions—side by side with other records that in themselves verify what accurate observers and highly attuned artists these ancients were.

This evidence is in the form of cave drawings found in southwestern France and Cantabrian Spain, dating from 30,000 to 10,000 BC. There are some seventy-two of these painted caverns, and the hundreds of drawings of mammoths, bison, horses, ibex and other species are unequivocal proof that the Stone Age men who carved, painted and sculpted these forms were vibrantly functional in those aspects of soul and spiritual life that sustain and inspire art. Truly, it is an astonishing record. Amid the animals appears a full spectrum of UFO shapes that might well have been lifted directly from an official NICAP publication.

The distinguished French mathematician, Aimé Michel, internationally renowned also as a serious UFO researcher, has dealt with this incredible paleolithic art exhibit at length, in an article entitled *Paleolithic UFO Shapes*. Published in the November-December 1969 issue of the *Flying Saucer Review*, the article is required reading for all serious students of ufology. Particularly is its content required by those who naively believe that UFOs have been involved in earth life only in relatively modern times.

Pointing out that we have no direct knowledge of the 700 or 800 human generations responsible for this art, the brilliant Michel emphasizes that these masterpieces are still today numbered among the high peaks of world art. Independently of what I have set down here regarding the perceptions of ancient men, Aimé Michel describes his own reactions on visiting the caves:

"Entering the cave at Lascaux is like *visiting the Parthenon or Sistine Chapel* (emphasis added). Every time I have been there I have seen the same impressive spectacle: the crowd of visitors, chattering, passes into semi-darkness. And then the walls are lit up, and suddenly all is silence: 15,000 years after the deaths of those who painted them, the pictures still produce the effect of a sublime presence that imposes respect for its genius, and prompts us to meditation and contemplation. The men who fashioned these masterpieces may lie beneath the dust of the centuries; but at the very first glance we understand that, if there is something in us that escapes death, that something is there, on that white stone, and it will remain there until the end of the world."

Aimé Michel is that rarest of scientific thinkers, the formally trained and accomplished man whose feeling world remains vitally alive. Thus we may see from the above quotation that the distinguished Frenchman was able to come into full contact with the spirit behind this Paleolithic art. The scientist in him, however, does not dream or lose vigilance for a moment. Michel immediately raises—as a scientist surely must—the question of the degree of confidence that can be placed in the accuracy of these cave drawings. Since the animal forms are as well known to modern man as to the paleolithic artists—and they appear intermingled with the UFO forms—the accuracy of both is self-evident. Says Michel:

"A mere glance gives us the answer. It is art of an admirably representational quality. The painters at Lascaux, Rouffignac and Altamira possessed a sureness of vision and of execution that is comparable to the great painters of the Renaissance, and that, by its realism and its movement and at times even its humor, is greatly superior to all that has been bequeathed to us by the Classical and Near Eastern worlds of antiquity."

The *unknown* objects portrayed by the paleolithic artists can therefore be regarded as accurate recordings of what those artists perceived. Discs, doughnuts, large fusiform shapes accompanied by lines of small discs *a la* Adamski, ventlas with ladders, ventlas trailing vapors, and objects closely resembling America's own Apollo Lunar Landing Module (circa 1969) are all there. In short, it is a collection of shapes indistinguishable from many UFOs reported and photographed in the 20th century.

That ancient men saw and drew UFOs can therefore hardly be doubted, although as Michel points out in his article, it is only the UFO explanation of these otherwise unknown forms that is fantastic to the orthodox mind. These retarded spirits prefer eternal enigmas to explanations that will bring better understanding of life processes. Better sterility and deadness, with its illusory safety, than life from space perceived and recorded by ancient men.

This collection of paleolithic UFO art, and its clear implications, leads us back to one of the original postulates in this book. When heretofore invisible organisms and other objects started to become visible sporadically with the electricalization of civilization—and especially with the advent of high-powered pulsed radiation—we were forced to consider a cardinal point: if invisible but present now, then *why not invisible and present always?* Expanding human consciousness and changing human cognition will lead inevitably to the acceptance of this central truth. By contrast, the ships-from-other-planets theory, with its two-plus-two reasoning and no demand to stretch either consciousness or our means of knowing things, hardly seems worthy of the enlightened human mind. Paleolithic man knew UFOs were here.

Dr. Rudolf Steiner has also provided a comprehensive outline of etherian physics—an area of thought without whose content ufology cannot possibly move forward. Major mysteries of terrestrial life, as well as fundamental relationships between man and microcosm and the macrocosm of which he is both part and product, may be gradually understood via etherian physics. Without such knowledge, UFOs will continue not only as a mystery to mankind, but also as a prolific source of illusion.

The most succinct and comprehensive presentation of etherian physics from the Anthroposophical point of view has been compiled by Dr. Guenther Wachsmuth, Rudolf Steiner's biographer for the final twenty-five years of his life, and also his secretary. Dr. Wachsmuth's monumental book, *Etheric Formative Forces in Cosmos, Earth and Man*, was published by the Anthroposophic Press of New York in 1932. One of the masterpieces of esoteric writing, the acquisition of this book is worth whatever effort the student must expend. As might be expected, it is out of print, although thus far no government has ordered it burned. Systematic study of this book will provide the keys to the shapes and forms of *all natural objects—including UFOs*. Fortified by this knowledge, the ufologist will be able to "read" correctly the signs in the skies, and in this way find his way around popular dogmas.

In order to make the subtle but important distinction between the ethers and the formative forces, researchers should also acquire and study *The Four Ethers*, by Ernst Marti, MD.[4] This small but invaluable book supplements, augments, extends and clarifies the presentation of Dr. Wachsmuth.

True to the innate proclivities of mechanistic thought—which embraces immobile, stationary and fixed entities with strangling enthusiasm—most conventional thought has held the ether to be stationary. Ether is thought of only in the singular in official science. In more recent times, concepts of a mobile ether have gained wider currency, with this medium seen to participate actively in processes of motion essential to any valid conception of cosmic workings.

The motion of the ether is beyond question to anyone who has extended his powers of cognition beyond the arbitrary boundaries set by conventional philosophy. Such persons, including myself, perceive basic etheric workings directly in the perception of physical force fields around living organisms and in such obvious external happenings as the flimmerings that are perpetually present in both the day and night sky. All such persons accept out of their own experience that what Dr. Steiner and Dr. Reich say of the ethers is true, namely, that they move a *priori*.

The ether moves, without doubt, and acceptance of this premise has opened to man the control of planetary weather, of which more will be said later. Steiner and Wachsmuth present the ether as *seven-fold*, and four essential ethers reveal themselves in physical space-time processes, according to these two brilliant men. The four ethers are:

1. Warmth ether
2. Light ether
3. Chemical ether (AKA sound ether and number ether)
4. Life ether

These four ethers have evolved phylogenetically out of each other, so that we may consider that warmth ether has evolved into light ether, light ether into chemical ether, and chemical ether into life ether. Empirical findings arising out of Wilhelm Reich's discoveries already partially confirm this. For example, orgone energy—which may be equated with the chemical ether— can be made to luminate *in vacuo* and in nature through weather control apparatus. Orgone accumulators literally create warmth "out of nothing". Thus, light and warmth are contained in orgone energy, i.e., chemical ether. Etherian physics is thus already demonstrating empirically its own theoretical basis.

Thus it is seen that the characteristic of the mutual relationship between the ethers postulated by Wachsmuth and Steiner—that the more highly evolved ether always contains in itself the properties of the ether(s) from which it evolved—has already been externally demonstrated. Each "new" ether, however, is a new formative force with its own innate activities and properties. Understanding this relationship, and living with it, naturally requires a mobility of thought which contemporary higher education usually extinguishes.

Life ether contains in itself the other three, chemical ether contains in itself the preceding two, out of which evolution has taken place. On page 40 in his book, Dr. Wachsmuth states: "Only when, through having penetrated into the substance-world, it has been modified, may a higher ether, for instance, be reduced as it were to the action of a lower. Warmth ether from which the other ether forces have evolved, has in turn come into being out of purely spiritual states outside of time and space."

This at once gives us a theoretical link to earlier states of evolution, and to the gradual condensation of our world through countless millennia. The glowing gas ball of Kant-La Place, as it were, had an illustrious and long prior etheric history. Dr. Wachsmuth's statements regarding evolution of the ethers themselves are open today to direct empirical proof in some areas, such as is provided by Dr. Wilhelm Reich's orgone energy accumulator. By perpetually producing heat "out of nothing" it directly contradicts the Second Law of Thermodynamics.[5] From the point of view of etherian physics, the orgone accumulator squeezes the warmth ether out of concentrated chemical ether. Incomprehensible to the old knowledge, Reich's invention yields to the basic theories of etherian physics.

Ingenious and enterprising young men will carry all this forward. Unlike the Old Knowledge, the comprehensive structure of etherian physics provides them with a chart for their experimental work. There can be little doubt that the science of tomorrow will penetrate these principles and bring forth a new technology—indeed, it has already begun.

The four ethers as carriers of formative forces, have the following characteristics:

1. *The warmth ether* manifests in heat, which etherian physics regards as the fourth state of matter,[6] and not a mere mode of motion affecting the other three states. My photographs of the atmospheric bioforms could not have been made without *heat-sensitive* film. We have recorded animate things extant and conscious in this fourth state of matter. These objects were photographed *in situ*, and are an adventurer's contribution to a valid new conception of heat as the fourth state of matter. The warmth ether tends to produce objects of spherical form, i.e., it shapes matter into globes. I have more than my share of globular critters on both stills and movies.

2. *The light ether* reaches the direct physical perception of man as the normal light of the terrestrial environment. When light ether acts unimpeded in nature, it produces triangular forms. In field experiments with the cloudbuster, triangular forks of lightning have been produced on many occasions, in another break-in to the experimental side of etherian physics made possible by Reich's discoveries.

3. *The chemical ether* or sound ether, also known as the number ether, is active in chemical processes of all kinds, and also transmits sound to the human ear—the tones of the audible world around us. Sound has tremendous formative power, well illustrated externally by the Chladni sound figures. In these experiments, dust laid upon a tightly stretched diaphragm adopts varying forms in accordance with varying tones.

Physically audible tones nevertheless are an infinitesimal part of the total tone activity vibrating everywhere in space, the waves produced by the forces of chemical ether. Sound or chemical ether gives form to the natural things of our world. In nature, the chemical ether, when working unimpeded, produces half-moon forms. Centripetal in action, it is interlocked with all the phenomena of cold and contraction.

4. *The life ether* is the most highly evolved of the four, and also the most complex and diverse in its essential qualities. Life ether is radiated to the earth by the sun. Like the chemical ether, it is a suctional, centripetal force—a force that draws inwards. The formative power of the life ether, when working unimpeded in substance, manifests in the square shape.

Each of the four ethers manifests naturally in its own color. The colors are:

>Warmth ether—Red
>Light ether—Yellow
>Chemical ether—Blue
>Life ether—Purple Violet

Each of the four ethers has a natural spatial tendency. These spatial tendencies are:

>Warmth Ether—Radial, expansive, centrifugal
>Light Ether—Radial, expansive, centrifugal
>Chemical Ether—Suctional, contractive, centripetal
>Life Ether—Suctional, contractive, centripetal

Each of the four ethers manifests in its own natural state of matter. These states are:

Warmth Ether—Heat
Light Ether—Gaseous or aeriform
Chemical Ether—Liquid or fluid
Life Ether—Solid

Anyone who has wondered why or how different colored flowers of the same type come into being should inspect not just the color of the flowers, but the shape of the leaves. The formative signature of the appropriate ether will be found in the differing shapes of the leaves, i.e., the yellow flower's leaf will have a more triangular form than the red flower of the same kind. Try it and see.

Before proceeding further with this sketch of fundamental etherian physics, a practical observation may be made here that is not without its fascination. The majority of the photographs of UFOs made by me through the years, using heat sensitive film, reveal spheroidal objects. In the 1958 motion pictures made by Dr. Woods and myself, the highly mobile forms captured were essentially spheroidal. Thus do we see how etherian physics, even in this elementary state, provides new tools with which to evaluate and study UFO photographs *from any source*. They are, after all, and as Meade Layne said over four decades ago, "ether ships" that we are pursuing—those of them that are ships.

Another point arises from the obvious and lawful connection of UFOs to etherian physics. A purported photographic expert who is without expertise in etherian physics should recognize that if he wants to evaluate photos of ether ships, he must acquire and apply a knowledge of etherian physics. His valid conventional experience is not undermined by such knowledge, but rather, made amenable to both extension and expansion.

Etherian physics sees the earth as a gigantic living organism. Like all living things, the earth "breathes". The planetary breathing cycle is diurnal. Chemical ether is inhaled and exhaled by the earth, with marked and measurable effects upon the atmosphere. Many of the central mysteries of meteorology are bound up in this breathing process.

The earth and its atmosphere consist of four spherical, concentric and interpenetrating spheres. In each of these spheres, one of the four kinds of ether is predominant. Light ether and warmth ether function mainly in the atmosphere and heat mantle of the earth. The chemical ether and life ether predominate in the liquid and solid matter of the earth respectively. These distributions of forces and their influences are the normal, primary terrestrial state.

Let alone, the earth would quickly stratify these ether spheres and tend toward fixity. The sun is the chief force in chaoticizing these ethers. By constantly roiling these ethers and the formative forces of which they are the carriers, the sun produces their dynamic intermingling and interpenetration, out of which comes life with all its rhythms and diverse forms.

During the night, the earth reestablishes something akin to the normal primary distribution of the ethers. Contraction, stillness and chill characterize nocturnal conditions. Human life sinks to its lowest ebb during the night, especially between midnight and dawn. The human being thereby reflects, in his inert and unconscious organism, the quiescent condition of that portion of the planet upon which he is living. Then comes the morning, the sun's influence, and the remanifestation of the life forces.

The old cosmologies saw everywhere in creation the deeds of all manner of divinities, the spirits of the ethers. The immortal Kepler, kinsman and antecedent of the modern etherian physicist, saw the universe as the body of a gigantic spiritual being, so that to Kepler it was a living cosmos. Nineteenth-century science expunged from its rubric all such concepts. The physical-material was considered sufficient, so immense was its impact on the human mind with the rise of measurement. Goethe's conception of the earth-breathing process, physically verified in the morning and evening "double wave" of pressure at the barometer, was among the brilliant insights into natural workings ignored at this time. The stampede for mechanical interpretations of nature trampled under all pioneer efforts at functional enlightenment.

The major etheric force involved in the breathing processes of the earth as a living organism is the chemical ether. We shall find ourselves intimately concerned with this ether in understanding UFOs, not only at this break-in stage, but also for many decades to come. Apparatus has been designed by certain earthmen to concentrate and manipulate this particular ether, and we had best therefore understand its basic action in the daily life of the earth.

Many mysteries of the earth, atmosphere and weather become accessible through this knowledge and the apparatus based on its application. Beginning at dawn, the planet earth exhales the chemical ether into the atmosphere, the latter being, as we have seen in our preliminary tabulation, the normal sphere of activity of the light ether. All the sensory testimony of the dawn, sunrise and immediately post-sunrise period confirm the exhalation process. A cardinal one: water vapor arising from the surface of lakes and rivers.

This exhalation process continues until the noon period, by which time the exhaled chemical ether extends up into the heights of the light ether sphere. Barometric pressure reaches an obvious minimum. The earth organism, like anything living, then inhales the chemical ether back into itself, completing the inhalation process just before sunrise. The moisture-producing, fluid-influencing chemical ether is responsible for such seemingly disparate phenomena as morning and evening fogs and mist, fluctuations in soil humidity, barometric pressure changes, increases and diminutions of potential gradient, and the rising and falling of plant sap.

All these happenings, and many others, are not isolated, discrete phenomena. Man has made them so. In truth, these phenomena are *functionally unified* by virtue of being produced by the macrotelluric breathing process. Those who think today in terms of "antigravity"—a term that exemplifies man's brain-bound, point-based consciousness—may see plant sap rise and fall daily in billions of plants. Relating this enormous lift to a force capable of lifting a spaceship should not be beyond the capacity of even a modest mentality.

Most of the riddles of atmospheric physics can be understood through the earth-breathing process, and the role played in this process by the chemical ether. Dr. Wachsmuth summarizes these electrostatic, fluidic, barometric, thermic and biological factors on pages 45-56 of his book, written in 1932.

In the interim, virtually all of this has been made empirically accessible through the discoveries of Dr. Wilhelm Reich. Dr. Wachsmuth's book should be studied today with this revolutionary fact in mind.

Dr. Reich discovered the chemical ether as a physical force in 1939-40. He called this force orgone energy, but etherian physicists will find in Reich's empirically-established properties of orgone energy the traditional properties of chemical ether. The blue color, the affinity for the fluid state of matter and the ability to produce thermic, barometric, electrostatic and biological effects are all there. What Dr. Wachsmuth has written has therefore long since come down to the empirical level, and the scientist who does not today understand at least some of these functions is simply out of date. A new generation of young physicists, strong enough to break with the Old Knowledge, will explore these things fully in the 21st century.

The mechanistic notion that the earth is a dead ball of matter has paralyzed creative thought in official science for generations. The earth is a living organism. Breathing processes are carried out diurnally, as a terrestrial, organic function, only partly dependent upon cosmic intervention. The endless cycle of fixity and chaoticization is what gives rise to, and also sustains, the life processes of the planet.

The side of the earth remote from the sun naturally contracts, and the four etheric spheres tend to stratify. Striving toward the sun, living organisms from the plant kingdom upward to man contribute to a mighty etheric torque that literally turns the dark side of the earth back toward the sun. Under the sun's influence, the etheric spheres that became stratified during the night are once again brought to chaos. This is a crude sketch of the etheric formative forces and their diverse roles in living organisms, including the planet Earth itself, and the basic source of the earth's rotation.

The earth may thus be seen rotating behind an etheric envelope that moves around the earth slightly faster than the earth itself. Dr. Wilhelm Reich, in his discovery of the atmospheric orgone energy, independently determined the presence of this basic west-to-east movement. He called it the *orgone envelope* of the earth. In his time, he was called mad and paranoid, but not long after his death cameras aboard the satellites were photographing the envelope, while mystified scientists speculated on what it was.

Reich likened the earth, in its spinning progress through the heavens, to a ball being spun by a stream of water moving faster than itself. From this perceptive description—originating wholly in his own needle-sharp observational powers and unconnected with any esoteric sources—Dr. Reich opened the pathway to weather control as a reality of *this* age. This work is now far along toward providing and perpetuating a life-giving environment, but everything that is done to press it forward is at the risk of the life and liberty of the pioneers. America is a hostile environment for such work, not innately, but for reasons that will become all too clear in the next few chapters.

The etheric earth is the driving power, and induces the earth's rotation. Out of the etheric formative forces and their workings on matter come the colors and forms of all living creatures. From the highest to the lowest, all are ruled by these forces. The ethers and the formative forces give rise to the grand design of nature. Artists are the only general class of humans to have at least partial access to this grand panorama. The full perception and understanding of these forces require the use of untapped, unutilized and latent powers of the human mind.

The scientist of the future will perceive, understand and work with all of it, just as Steiner indicated. Living processes will be paramount. On this account, it should be no surprise that the basic technical breakthrough into this realm came from a scientist who was a biologist in the highest and best sense—Dr. Wilhelm Reich. He found the pathway from psychology to biology, and in the process made the study of the etheric formative forces possible empirically—the anchorage for the new kind of thinking that is upon mankind.

The scientist of the future will not direct his energies to determining what a rose is, or of what a rose consists. Rather will his labors be invested in determining how the rose arises. The question of the future is not "What is a rose?" but "How does a rose come about?" Man is about to ask, at long last, how a rose achieves its particular form and color. The scientist of the future will "read" the expression of the formative forces in all living things—as a determined handful of them do today. The challenge of etherian physics is immense and thrilling.

Ufology will undoubtedly be a key aspect of tomorrow's sciences, since it is a bioenergetic wellspring from which much New Knowledge will flow. UFO investigators of our own time cannot pass beyond the bewildering, enigma-ridden periphery of their subject because they have not thus far tackled etherian physics. Most evidence of UFO presence gathered to date can be interpreted accurately only with an understanding of etherian physics. Here lies the key to spaceship technology and to the new communications—the means by which man will enter into commerce with otherworld entities.

Gifted people have already penetrated far deeper into some of these mysteries than most people interested in UFOs even suspect. The reward of these bold ones has been uniformly depressing: they have been branded as charlatans, crackpots and quacks. Dr. Steiner's work—the miniscule portion of it described herein—helps us to understand and accept on new ground the work of these innovators.

The loneliness of the gifted individuals who left the beaten path and went off on their own across new frontiers has been appalling. Later generations will appreciate their sacrifices. In the new realms they penetrated—usually unbeknown to them—lurk the subhuman, nonhuman and superhuman intelligences that now reach into and play upon human evolution with ever-increasing force and purpose. We will show in the ensuing chapters that man will need more to survive this challenge than the fraud and flummery of contemporary scientism.

NOTES TO CHAPTER ELEVEN

1. Interested readers should write for a current list to St. George Book Service, P.O. Box 225, Spring Valley, New York 10977.
2. As a bold commencing effort see *The Loom of Creation* by Dennis Milner, D.Sc., and Edward Smart, published by Neville Spearman Ltd., London.
3. Published by Cowles Communications, Inc., New York, 1967.
4. Published in 1984 by Schaumberg Publications, 1432 South Mohawk Drive, Roselle, Illinois 60193.
5. The Second Law of Thermodynamics is a statistical law, inasmuch as no exception has ever been found to it until Reich devised the orgone accumulator. The rooting of the Second Law is now no longer statistical, but neurotic, since the orgone accumulator proves its basic contentions false in certain arrangements of materials.
6. For a highly intelligent and comprehensive approach to heat as the fourth state of matter, see *Man or Matter*, by Lehrs and *The Warmth Course*, by Rudolf Steiner, Mercury Press, 241 Hungary Hollow Road, Spring Valley, NY 10977.

Chapter Twelve

ETHEREANS

Helpful and promising data from any source should never be rejected a priori through prejudice and ignorance.
— MEADE LAYNE

Etherian physics is dealt with in great detail in the works of Dr. Rudolf Steiner and his followers. The Anthroposophical conception of these things is comprehensive and clear, and is expressed suitably for modern Western consciousness. The empirical pathway for its steady and systematic verification is open. A New Technology is inevitable. All this is realistic in the highest sense, overlaid as it is by the necessity for those who assume its duties and challenges to engage energetically in the work of the world.

Striking, marvelous and revolutionary as all this is, Anthroposophy is not the only approach to these matters. Nor have Anthroposophists ever claimed to be the exclusive ushers to the future. The same principles have been approached in other times and in other ways by other people. The search for knowledge of the microcosmic-macrocosmic relationship is as old as humanity, and objective evidence of this search abounds in archaeological findings and relics.

Each epoch of human evolution has seen this knowledge carried forward synchronously with the changes in human consciousness wrought by evolutionary development. Theosophists, Rosicrucians, Masonic orders, Qabalists and kindred organizations have guarded this knowledge through the centuries. Possession of this knowledge was thus confined to a few souls in each epoch who could meet the combined challenge of its acquisition and further advancement.

The secret imparting of this knowledge, under confidence enforced by oaths, has now essentially passed away. Initiation follows a modern pathway. An example is that the Anthroposophical Society is a *public* organization. Today's aspiring and determined student may secure all the esoteric knowledge he can absorb. He has only inner limits. While finances may block or disqualify young men and women from university education, no such barriers exist between the young and the New Knowledge.

A greatly expanded consciousness is the characteristic of all those who have earned knowledge and insight through esoteric training and study. A few years serious study of Qabalism, for example, suffices to expand the consciousness of the student far beyond that of ordinary men. This expansion of consciousness is *permanent and dependable*, and in no way dependent upon drugs and chemicals. He thus becomes an extraordinary man.

Such an extraordinary man, and a Qabalist, was ufology's most outstanding theorist to date, the late Meade Layne, M.A. We have already referred to his work many times. Let us turn now to his writings on UFOs, aided by what we have learned of etherian physics in preceding chapters. In this way, we may begin to win insight into the nature and purposes of the entities who are behind the mysterious flying discs.

128

Meade Layne's personal background in etherian physics stemmed from his deep studies of modern esoteric Qabalism. Specifics of the Qabala and their implications for the physics of UFOs will be dealt with more thoroughly when we meet the incredible Dr. Ruth B. Drown. Suffice it for now to say that Qabalism, proceeding from a base in the intuitive ancient wisdom of Chaldea, provides its students with a safe guide for penetrating the microcosmic-macrocosmic mysteries. The whole system rests upon a precise mathematical basis.

Meade Layne was fully familiar with these principles, and able to apply them, when a remarkable trance medium named Mark Probert swam into his ken in the middle 1940s. Here we take an even sharper departure from comfortable paths of conventionality. Mediumship must be broached.

Execrated by official science, ridiculed by academia, derided by official religion and a source of anxiety for virtually every neurotic person, mediumship is one of Western culture's most violently rejected paranormal phenomena. In one form or another, mediumship is certain to increase in the coming years in connection with the UFO incursion. Therefore it must be understood—without fear or favor. The case of Mark Probert opens the best possibilities of mediumship.

When Mark began jabbering in strange tongues in his sleep, his wife Irene became perturbed. The nocturnal dissertations continued, without Mark being aware of what was happening. In the true sense of the term, when he went to sleep, he was "out" for the night. When he went "out", others came "in" and made use of his physical facilities to make their ideas known in this world.

Mark Probert was duly introduced to Meade Layne, in anticipation that the knowledgeable Qabalist and teacher would help Mark get control of a situation that was not without its alarming aspects. Meade Layne began working with Mark. Tape recordings were made of the nocturnal ramblings. After much diligent research, it was established that in his sleep, Mark was speaking coherently, and with great precision, in an ancient Himalayan language of which he had no conscious knowledge whatsoever.

Under Meade Layne's guidance, various *discarnate entities* were identified as speaking through Mark. The major personality was the Yada Di'Shiite, whose Himalayan discourses had first caused consternation in the Probert household. Others manifesting were Professor Alfred Luntz, 16th century astronomer Ramon Natalli, and an Oriental sage named Rama-La-Ko.[1]

Competent esotericists like Meade Layne, when they work with mediums, seek to bring the faculty of mediumship under the control of the medium, and to systematize the manifestations. Mediums usually resent such directive assistance and go their own ways. In Mark Probert's case, he cooperated fully. A panel of excarnate philosophers and scientists was gradually set up, and a high yield of useful information subsequently obtained from these personalities.

This panel of excarnates became known as the Inner Circle or the Mark Probert Controls. They are called "controls" because during the time that they utilize the medium's physical body to express themselves, they are "in control" of that vehicle. Experience overwhelmingly confirms that communication with the entities behind the UFOs is by variants of this mediumistic transduction between different planes of consciousness and density.

This is one reason why our many thousands of radio stations do not hear "spacemen" talking to each other, or to earthmen. Communication is essentially *biological* and bioelectric. Human experience with radio communication, which is merely a precursor to biological communication, is less than a century old. While government agencies in the U.S.—egged on by malefic corporate forces—have destroyed and crushed all significant efforts at investigation of biological energy and the instruments designed to direct that energy, the Soviet Union continued to promote such investigations. The validity of telepathy as a means of communication was established by Soviet parapsychologists, and its expansion was highly promoted. The U.S. was left far behind in this field, and became increasingly vulnerable to the psi technology Russia successfully developed.[2]

The panel of philosophers and scientists manifesting through Probert were from many ages and epochs in the earth's history. By intelligent, searching questioning—based upon his own strong foundation in Qabalism—Meade Layne obtained from the Mark Probert Controls the basic "etheric interpretation" of the aeroforms or flying discs. This work was done in 1946-47.

Data on UFOs were then scanty. Much has happened in the interim. Up into the early 21st century, nevertheless, the etheric interpretation remains essentially valid, as well as extensible to cover newer data. Since the entities assisting from "the other side" manifested through Mark Probert in *full room light* and could be directly questioned and cross-examined, it was mediumship of a high order, and the validity of the data received provided its own verification of the quality of the contact.

Meade Layne founded the Borderland Sciences Research (Associates) Foundation to ventilate all manner of borderland phenomena. The idea was to deal with things that official science declines to approach. The etheric interpretation of the discs was the major initial contribution of BSRA to contemporary thought. Establishment ufology, of course, does not acknowledge the etheric interpretation of the aeroforms, preferring the comfort of mechanistically-oriented naivetes, applied on an *ad hoc* basis to individual UFO happenings.

Human thought must enter boldly and creatively upon the subject of disc occupants, whose nature, characteristics, origin, loyalties, technological knowledge and *relationship to mankind* continue to be the most pressing problem in ufology. External, objective and verified evidence concerning these entities, arising out of encounters between disc occupants and human beings in various parts of the world, does not square with the naive expectancy that we confront beings who are either simply or exclusively humanoid.

Hairy dwarfs; sticky, jelly-like amorphous beings with intelligence and motility; stinking, bouncing, towering monsters; little green men; humanoids with the appearance of metallic asparagus. These are but a few of the bewildering range of creatures who have appeared to human gaze after dismounting from UFOs—often with aggressive intent or violent reaction to human presence. Mechanistic projections and speculations concerning these weird entities are uniformly inadequate and colored by the pressing, inner desire to have the "spacemen" appear as both friendly and superior to inhabitants of earth.

The ships-from-other-planets hypothesis has crowded out all other hypotheses because of its simplicity and easy reconciliation with prevailing modes of mechanistic thought. The extraterrestrial hypothesis (ETH) causes the least disturbance to the mechanistic world-conception of the modern human. Meade Layne, amid all this intellectual ferro-concrete, was the first ufologist to emphasize that "there is another possible source, and we are convinced of the true source" for these strange craft. His etheric interpretation drew the proper theoretical picture for a public that has been brainwashed with the simplistic ETH notion.

Meade Layne held that the origin of the discs was *Etheria*. In this contention, he was fully supported by the Mark Probert Controls. Let us have Meade Layne answer for us in his own words, quoted here from *The Ether Ship Mystery*, the obvious question that follows the broaching of Etheria.

"Etheria is *here*... if we know what that means! Alongside, outside our world. Because our world, that is, the so-called dense matter of the objects in our world is a *rarefaction*. It is spaced out like a vast net—a net with enormous meshes. Imagine if you will, a net of wire with meshes a mile wide. Would not wind and water flow through that net as if it did not exist? A little friction, very, very little. On the strands of those meshes we live. That is the so-called dense matter of our world. We look out across the mesh and do not see anything in it, or hear or feel anything in it, and so we call it *empty space*. Meaningless words in the abysses of folly! We see, hear, feel the other strands in the mesh, and call our world solid, firm and material. Call it a real and substantial thing. But the 'spaces' of our mesh are a thousand and ten thousand times greater than the substance of the strands.

"What is in these spaces of the mesh of our 'matter'? Space? Space is not nothing. Space is stuff, is matter. What kind of matter? The matter which makes up the ethers. More dense—not less, but more dense by far than the rarefactions of our world. The matter of the etheric world.

"Inside the molecules, inside the atoms... other atoms. Still other atoms inside of those, or ten thousand Chinese eggs, each inside of another. If you think this is imagination, or that there is no room for these interpenetrating universes, ask the physicist and the mathematician. They will tell you there is nothing in the omniverse *but* room; and room, again, is space, is stuff, is basic world stuff, mind-stuff, life-stuff; and there is no emptiness anywhere. And that is why Etheria is here. But it is also everywhere. All heavenly bodies have an etheric realm."

So much for Etheria, and the mind-bending potential of admitting Etheria to ufology. Let us proceed to Meade Layne's views on the craft originating in Etheria:

"The etheric ships are of many types, not all of them discs by any means. Some are long and cigar-shaped; many are shaped like balloons or dirigibles; some have wings or wing-like attachments. Many of them appear to change shape while in the air. Their color varies also and has also been seen to change. Some of them resemble fireballs, others are accompanied or followed by smoke, vapor or luminous appearances. Some are translucent, or seem to glow from within.

"They travel at all speeds, from zero (stationary) to thousands of miles an hour. They also vary widely in size, from a few feet or yards in diameter to nearly incredible dimensions; we have reports considered trustworthy of at least one craft ten miles in length. They have a curious way of disappearing suddenly when in plain sight, and also of rapidly dwindling in size without withdrawing in distance. It is well to bear all these items in mind, since any single explanation must obviously cover all of them; it is not enough to invent some theory for the disc shapes alone and then imagine that the whole matter is accounted for. Now, under the 'etheric explanation', which we are presenting here, our understanding is briefly this:

"All of the types of craft just mentioned come from the same source. All are constructed by Ethereans. Their variety arises from the fact that many of them are experimental, and that different types of them serve different technical purposes. These craft are built of matter of the etheric plane, and are imperceptible to us, since they exist in what we call 'empty space'. They enter our plane, or our level of perception, by a process of materialization. The extremely dense matter which fills in the pattern of etheric stresses is replaced by matter of a kind that is visible to us... or better, its vibration rate is lowered until it becomes visible.

"It is a little like slowing down a very fast fan until one can see its blades. This is closely similar to what happens in the phenomenon of apportation (where objects of any kind or size are 'miraculously' transported into and out of sealed rooms, sealed containers and the like). Many people, and most scientists, take no stock in this alleged phenomenon, but it is well attested and cannot be surrendered on that account.[3]

"Neither the pattern nor the molecular filling is destroyed. The object becomes invisible because atomic motion is speeded up, and it can pass through other matter, e.g. a wall, because of this same fact. This is equivalent to an increase in density when an object on our plane is being apported from one place to another. To the etherean, an object materializing in our plane would dematerialize on his; to us it materializes in our field of perception, and dematerializes when it returns to its etheric state."

Meade Layne also states that several of these ships have fallen, and their substance has been carefully studied. He wrote: "The outer skin of fallen discs, however, is harder than steel, and of an unidentified substance." To this day it is not publicly known whether or not a complete disc was ever captured. What has subsequently become known beyond any dispute is that pieces of a UFO are in the possession of the U.S. authorities. Their analysis is as Meade Layne stated all those years ago!

The material in question came from a UFO after being shot by a USAF jet fighter. The well-known Canadian scientist Wilbert Smith, who headed the Canadian government UFO project named *Magnet*, made these revelations before his death. Mr. Smith further stated that he had been loaned, and had personally handled and analyzed some of this material before returning it to a super secret authority in Washington, D.C. Smith's description of these happenings appears in a tape-recorded interview published in the Frank Edwards' book, *Flying Saucers: Serious Business*, and more latterly in *The Boys From Topside*,[4] edited by Timothy Green Beckley.

Proceeding with his description of Etheria and Ethereans, Meade Layne writes:

"It is our information that certain types of ether ships are and have long been used for interplanetary travel. Those who persist in the belief that they come from visible planets or our system will eventually seize upon this as a confirmatory fact. But the ether ships do *not* originate on visible planets... interplanetary travel is simply one of their innumerable uses, the fact should not mislead us. The craft are made in Etheria, and Etheria is *in* and *around all visible objects*. It is also the home and motherland of our own bodies. We ourselves are ethereans by ancestry and descent, for Etheria is the home of the human race. That we should be so ignorant of it is strange indeed."

As the story of my experimental work unfolds into the connections I was able to make between my objectification of unseen UFOs and the discoveries of Dr. Ruth B. Drown and Dr. Wilhelm Reich, the reader will find Meade Layne's insights all the more remarkable. His theories and findings long predated everything I did.

Furthermore, when I began my work that led to the discovery of the invisible fauna and forms in our atmosphere, I knew nothing of Meade Layne and his writings. When my work cut across his at various points, and friends told me of BSRA and Meade Layne, my prescient old teacher, Franklin Thomas, artfully steered me away from Mr. Layne's work. This guidance resulted in my being able to confirm with my experimental work much about UFOs that Meade Layne had hypothesized. In particular, the *invisible* was empirically established as the source. Both Meade Layne and myself were richer for our first meeting coming later rather than sooner.

Etheria being invisible, unheard and untouchable to ordinary human beings, it nevertheless precipitates various artifacts within range of our senses at various times. The electromagnetic works of man undoubtedly aid these appearances. In the UFO phenomena in all their diversity, we have a wholesale irruption of the workings of Etheria into our world. We ignore this at our peril. Similarly at our peril is a continuance of the scientific cowardice that has prevented until now—the work of bold amateurs excepted—any academic approach to the extension of our thought processes. Such extensions are mandatory as the reality of imperceptible worlds within worlds around us, in us, and beside us in space becomes steadily more evident.

Meade Layne expresses the problem thus:

"If there are colors we cannot see, sound we cannot hear and dense and solid bodies which we cannot discover by touch, it is certainly not too difficult to conceive of a world which, though possessing sounds, colors and material objects, is yet imperceptible to us. It would be as unknown to us, by direct perception, as a man is to an ant, or a bird to a beetle. And if an ant or beetle, or a fish in its element could be questioned and could make reply, they would probably be deeply skeptical about the existence of men and birds... unseen and unseeable entities alleged to be living right alongside of them in the same world. The ant world is the next hill; the fish world is a limited area of sea; and though it happens that man can peer in on both of them, they have no way of discovering him. At this point, the fruitful principles of analogy suggest that possibly *there are invisible beings who can peer in on man also*. He may be the crux of creation (or cross of affliction of it) but is not necessarily its crown. And though reputed to be a little lower than the angels, this earth manifestation of him at least may be a little higher than a caterpillar, in the eyes of unknown but observing intelligences...."

These views, arising out of the communications of the Mark Probert Controls and their synthesis with Meade Layne's esoteric knowledge, correspond in all essential details with what my plunge into telepathy had yielded on the same matters in 1957. At that time, I had no basis in occult science on which to judge what was imparted. I was largely bewildered as a result. The ethereans who found me, at that time, open to their modes of communication, *described themselves as ethereans*, invisible to us under normal circumstances, and *in no sense excarnate humans* or "spirits". They are not of this world of ours, but they can and do penetrate into this density at will. They are not something brand-new in the life of the earth, but rather, have always been present in Etheria—around, beside, and inside our own vibratory level. Like the unseen plasmatic critters of our atmosphere, they have *always* existed, unseen by man.

As Meade Layne pointed out in *The Ether Ship Mystery*, we can conceive of matter being compressed to invisibility, becoming so dense in the process that the rarefied matter of the earth plane becomes like a net with vast meshes that present no impedance to the passage of the denser substance. Here is his approach to these considerations of density:

"Let us suppose we have before us two small cubes, one of cork and the other of lead, each of the same size, one inch in each direction. The lead cube weighs more and has more mass, but it has the same overall volume as the other cube. Now imagine that we can exert a very powerful pressure on all sides at once, of this lead cube, so that it is compressed. That means that the particles of lead are forced a little closer together. Imagine this pressure continued and increased until the cube becomes half an inch, a quarter inch or a 16th of an inch in dimensions. Imagine it decreasing to pinpoint size, and then to the point where it becomes microscopic—invisible. All of the original cube is there; all of its mass and weight; only the volume has changed. Suppose this ultramicroscopic cube is hurled against the cube of cork. Will the latter stop it? This enormously heavy and extremely small particle will pass through the loose structure of the cork as if it were not there at all. The cubic inch of lead, by virtue of becoming thousands of times more dense than the cork, by the process of reducing its volume, penetrates the lesser density with ease."

Layne points out also that through such a crude picture-thinking, we can begin to conceive of density as a function of the vibration rate of the material particles involved. Compression and increased density naturally infer a speeding up of vibratory activity. Thus the tremendously dense lead cube of ultramicroscopic size has enormous mass and very high velocity of its particles. "These characteristics will make it penetrative," states Meade Layne, "so that it will go through the cork or any other less dense matter like wind through a wire mesh."

Authentic communications from ethereans, such as the early technical information received by George Van Tassel, support with much data what Meade Layne set down in his epochal treatise. The principles coincide with what Qabalism has taught to its initiates for centuries, they coincide with the statements of the Mark Probert Controls, and they coincide with what the ethereans say about themselves. Further coincidences too comprehensive to permit dismissal exist with the cosmo-conception presented by Dr. Rudolf Steiner, and with the basic New Technology discoveries made by Dr. Ruth B. Drown and Dr. Wilhelm Reich—among others.

That all this is beyond the limits that official science can tolerate is obvious. That it is by no means beyond the creative imagination of earth people— or in any way beyond their latent powers—is equally obvious. After all, humans have assembled these theories and begun building the New Technology in our own time. They must begin to think of new forces in earth life such as the UFOs in new ways. That is a demand of our age.

Among the men of science who have not recoiled from the obvious etheric origin of the aeroforms is the late Edward S. Schultz, of Buffalo, New York. He has contributed considerably to the advancement of these concepts through BSRA and otherwise. In a commentary to Meade Layne's original work, Ed Schultz has this to say on pages 23-24 of *The Ether Ship Mystery*:

"The science and engineering of Etheria is doubtless subject to continued progress and improvement, but (we may conjecture) at a faster rate than ours, since they must long ago have evolved out of such nonsense as technological security and secrecy.

"We must not harbor the illusion that the four etheric states are ghostly and unsubstantial. They are a good deal more substantial than our own so-called dense world, and it is we who are the ghosts of the etheric realm…

"There is reason to suppose, also, that the lowest of the etheric states, just beyond our gases, serves as a zone of demarcation between the planes and that there is some overlap of phenomena of the etheric and of our world of matter in this zone. Etheric matter will interpenetrate all things of our world (etheric doubles)."

Schultz obviously is referring here to the warmth ether—the heat of our plane. By plunging persistently into the fourth state of matter—heat—with simple devices for objectifying any unseen activity in the borderland realm to which Schultz draws attention, we did indeed find and prove the existence of just such an overlap. Further work by younger men with greater resources will, in due time, overpower even neurotically-derived resistance to these findings. Etheria is here, right now, and access to it *wholly a function of tuning*.

The ethereans are clearly masters of the biological energy tuning processes to which we are just gaining access on this earth—over the dead body of official science, as it were. This ability to tune in directly on the organism of any desired carthling pre-empts use of electromagnetic radiation for communication by the ethereans. Not only is EM radiation an irritant to living organisms, with innumerable biological interfaces which our wisest scientists have only begun to discover, but it is monstrously inefficient. The ethereans can get in touch with us bioenergetically and individually, *if they so desire*.

Certain etherean entities evince deep interest in the course of human development, but they do not intervene with stock solutions to our problems. Nor do they attempt to unload on resistant terrestrials technical principles for which humanity is not ready. Any person or entity able to look in on our plane, as Meade Layne suggested they may do, would immediately discern that humanity murders its benefactors and rejects their gifts. They will go as far, it seems, as making it clear to us that there is an immense job ahead in having earth evolution fulfilled in freedom. Forces both incarnate and excarnate are determinedly subverting such fulfillment, and in view of humanity's resistant ignorance of their machinations, may well succeed.

The role of the earth's etheric double in underpinning all the phenomena of earth life was sketched in the previous chapter in Anthroposophical terms. One of Mark Probert's Controls, a discarnate 16th-century astronomer named Ramon Natalli, dealt with this same question of the etheric doubles of planetary bodies relevant to UFO phenomena. There is a remarkable coincidence between the two descriptions, revealing the high intelligence of Probert's "Inner Circle". Natalli states:

"There is around every planet and every body in the vast heavens, an etheric world. These bodies are governed by those in the etheric. Every planet, including the earth, is under the control of its etheric counterpart. All these strange sky craft come from the etheric region of some particular planet. The disc type come from the etheric region of Venus, and the design of the discs is copied from Venus with its outer vibrating disc or ring... all such craft must pass through the etheric region of the earth, and must have permission of the ethereans to pass through. Each comes from its own particular rate of vibration, which is an unseen pattern of the seen body.... They are not solid; their solidity is formed according to where they are going and what their mission is. They make their vibration conform to that of the body they approach.

"Their matter is so formed that it can be driven, by operative force of the mind, at speeds which exceed that of light. But there is not 'matter' or 'stuff' as you use the words, which passes over the enormous gulf of space. For such speeds, *here* and *there* do not exist; space-time is a continuing *here-now*... but you must have your laws and theories, of course!"

So much for Etheria and the ethereans, who will become more familiar to us as we learn, with the aid of new inventions now on the earth, to tune in on them and their realms. This will be an integral part of the earth's evolutionary struggle, in which the ethereans have come in recent decades to exhibit increasing interest. Factors and forces involved in contemporary life are moving earthly affairs toward an obvious climax.

During my 1957 experiments with telepathy, the ethereans with whom I had commerce conveyed to me, in a simplistic yet emphatic fashion suited to my ignorance of things cosmic, basic facts concerning the forces at work within the UFO phenomena that are anti-evolutionary and anti-humanity. This was, and is, unpleasant information. Nevertheless, I have found it in the intervening years to be all too true.

Unmasking and confronting these inferior forces, that is, forces *from below*, has been just as important a part of my work as my ventures into practical etherian physics. Ufology will never be anything more than a hollow shell, an illusion and a miserable fraud, unless we recognize the outlines within it of a massive evolutionary struggle. Along with the essentially benign Etheria and its ethereans, there are other forces impinging upon human life in decisive fashion. These include the malefic unseen forces from beneath, who have made uncomprehending humans their playthings. The time has come for us to meet The Boys Downstairs, often known as the Black

Gang, and to unmask some of their workings so that we may read them directly from life around us.

NOTES TO CHAPTER TWELVE

1. Tape recordings and transcripts of these communications are available from the Inner Circle Foundation, PO Box 16476, San Diego, California 92176, sister company to The Book Tree, publisher of this book. 1-800-700-8733.
2. Anyone who doubts the scope and magnitude of the Soviet effort should read *Psychic Discoveries Behind the Iron Curtain*.
3. Numerous verified incidents of teleportation have occurred since Meade Layne wrote these words. See Gorden Creighton's article "Teleportations" in the *Flying Saucer Review* for March/April, 1965. Also see the story of the "Chascomús, Argentina Teleportation" in the same publication for Sept./Oct., 1968. Dr. Gerado Vidal and his wife were teleported 6400 kilometers from Chascomús to Mexico— automobile and all!
4. Published by Saucerian Press, Inc., Clarksburg, West Virginia.

Chapter Thirteen

THE BOYS DOWNSTAIRS

We have stacks of reports about flying saucers. We take them seriously when you consider we have lost many men and planes trying to intercept them.
—Gen. Benjamin Chidlaw, USAF
Commander Continental Air Defense[1]

The assumption that entities "from space" are necessarily benign is one of the prime humbugs in conventional ufology. Mechanistically-minded humans are reluctant to associate superior technology with inferior motives, despite the lesson of our own world that man's ability to survive is threatened by the *misuse* of technology. In this world, we have directed the finest scientific manpower and immense resources to weapons development. These men of science have demonstrated the ethical benefits of higher education—as it is presently conveyed to the world's young people—by designing devices to destroy millions of human beings at a stroke. Were an individual to come forward today, in any major land on earth, with an invention that could exterminate the entire populace of another country at one blow, he would be granted unlimited resources for its perfection. That is the way earth life is. Men plan and perform acts of cruelty never seen among the beasts of the earth. Realists accordingly see clearly a distressing partnership between superior technology and inferior motivation, right here in this world of ours.

In other worlds, around us, beside us, inside us in space, the same relationship is likely to exist. In these other worlds, there are not only the ethereans to whom we have been introduced by Meade Layne, but also beings who are undoubtedly more rotten, decadent and degenerate than modern man has allowed himself to become. These entities are in a large measure responsible for pulling man down. They oppose his evolutionary development, and have involved themselves intimately in UFO phenomena. They are the beings of death and destruction, of splitting and rigidification. They are the "boys from downstairs".

The NICAP organization, exemplifying establishment ufology, still holds that there is no convincing evidence of hostility from space, even though untold hundreds of aircraft have disappeared without trace and frequently under sinister circumstances—since the end of the Second World War. NICAP founder Major Donald Keyhoe detailed one such instance in his *Flying Saucer Conspiracy*, where a covey of U.S. Navy aircraft disappeared in the "Bermuda Triangle" area, leaving no trace. A Martin flying boat sent out to investigate similarly vanished without trace.

Another example of violence that has been permitted to slide conveniently into the limbo, appeared in Leonard Stringfield's CRIFO *Orbit*, published in Cincinnati on 4 November 1955. One of Mr. Stringfield's most grievous shortcomings, in the eyes of those who insist that all spacemen must be friendly, was his ability to look facts in the face—and print them! Here is what Mr. Stringfield printed in that singular issue of *Orbit*:

138

VIOLENCE IN RETROSPECT

Beyond estimation is the number of aircraft, military and otherwise, that have been lost to the UFO. Surely the earliest date for such occurrences goes back farther than the classical Mantell "death chase" of 1948. In this category, we refer once again to the files of Robert Gardner, which reveal for us this unusual incident dating back to 1939—just before the outbreak of WW II.

CASE 107, between San Diego and Honolulu. Late summer, 1939. At 3:30 p.m., a military transport plane with thirteen men aboard left the Marine Naval Air Station in San Diego for a routine flight to Honolulu. When three hours at sea the aircraft was in dire distress.

Mayday calls were radioed back to the base, then suddenly nothing more was heard until the craft came limping back and executed an emergency landing. The first men to reach the craft were shocked by what they saw—all thirteen members of the crew were dead, save for the co-pilot, who managed, miraculously, to steer his charge in safely. Three minutes later he was also dead.

Examination of the bodies showed remarkably large, gaping wounds, not unlike those received by the surface of the craft, which indicated the impact of missiles. A second amazing discovery was the service pieces, .45 Colt automatics, carried by the pilot and co-pilot, had been emptied and their shells found lying on the floor of the cockpit. Lastly, and possibly akin to saucer phenomena, was the characteristic rotten egg odor which pervaded the chamber's atmosphere. It was also learned regarding the incident that personnel who handled parts of the aircraft showed a mysterious skin infection. Security measures, Gardner was told, immediately blanketed the affair and cameras restricted. Corpsmen were barred from removing the bodies and the job of identification and diagnosis was limited to three medical officers only.

Hostility has been woven into UFO sightings all along. Keyhoe's description of the awesome Bermuda Triangle incident, and his presentation of the Kinross AFB plane kidnapping are merely a couple of instances among many. Similarly typical are those cited by me in *They Live in the Sky*. The amazing thing about all these incidents without exception—is the evasion of their study. In due course, we shall attain a little insight into the means by which this strange blindness and evasion is induced in vulnerable humans by the Boys Downstairs—the perpetrators of these criminal assaults on aircraft and human beings.

Where others in the UFO field in the 1950s avoided this issue of hostility, I touched it fully. Only a few investigators like Harold T. Wilkins and Leonard Stringfield were willing to entertain the idea that not all entities from space were benignly motivated. Analogy will be found a valuable tool in reaching an understanding of a key UFO issue: that all who ride in spaceships may not be intent upon humanity's salvation. With his warrior's intuition, the late General Douglas MacArthur warned of the necessity for human beings to unite against possible attack from other worlds.

Now that we are in possession of some understanding of the crucial role of *invisibility* in UFO phenomena, as well as some technical rudiments of the New Knowledge, the conception of inferior (that is, from below or beneath) beings manipulating an uncomprehending mankind can be broached. As we do so, we should bear in mind that the exposure of this particular activity has been set on foot by the ethereans, who have intervened, it seems, to arrest or slow this process by such exposure. While we need to understand the reality and methods of our foes, therefore, we should remember our friends—those ethereans whom Meade Layne characterized as the Guardians.

Let us turn to our first analogy. Malfeasance, embezzlement, theft, bribery, corruption, deception, premeditated murder and all other left-handed activity on this earth shares a common functioning principle: *keep the victim unaware that he is being preyed upon.* Criminals and crooks of every stripe, from the pickpocket all the way up to the super-rich manipulator in his fortress of power, utilize this principle. They proceed with their depredations and enjoy their particular conception of success because their activities are either unobserved or, in the case of the big-time operations, stand beyond the comprehension of most of their fellows.

The pickpocket depends on legerdemain, the thief on stealth, the embezzler on misplaced trust. In the case of the real rulers of this world—those to whom politicians are manipulable entities of existence—every device of deception is employed to divert attention from their machinations. While American idealists have waved flags, submitted to confiscatory taxes, and shipped their sons abroad in every generation of the 20th century and beyond—to be mangled in the maw of a foreign war—the rich and the super-rich are otherwise busy. They plunder the Republic and its ideals, multiply their wealth exponentially, and make the avoidance of all taxation their religion.

Ferdinand Lundberg's monumental book *The Rich and the Super-Rich*[2] has provided an appalling introduction to American legalized criminality—extant today on a staggering scale. Detailed official hearings on television, of the Watergate and Iran scam type, have shown publicly what has long been hidden: corruption on a prodigious scale. Men are in charge of public affairs—and men are in charge of such hearings—who will deliberately evade the truth. This truth-hatred must be understood as the signature of the controlling invisible forces. Does this seem far from UFOs? Not if you are seeking to prove that most of this planet is already under illegal control of various kinds, functioning through legitimate governments. Obscurantism on UFO matters is government policy, and has its roots in this same *illegal control* of human affairs and also of individual human beings.

There are essentially two factions at work in UFOs. There are the ethereans, who enlighten and guide, and the Boys Downstairs who confuse and control mankind. The earthly instrumentalities of the Boys Downstairs include those who are inspired—by whatever means suit the particular individual—to kill every rational approach to the UFO subject. Governments and their satellite scientists adequately demonstrate the strangulation of rational inquiry.

Analogy may further elucidate these processes. When the embezzler is discovered, he is no longer trusted. He has been exposed and trust in him is ever afterward withheld. A pickpocket caught in the act not only goes to jail, but might well be smashed senseless by his victim. Exposure cripples the functioning of the under-handed activity, regardless of its magnitude. The top-to-bottom rot of the American government, illustrated by various televised official hearings, has revealed criminal and amoral activity on a scale that shocks decent Americans. More than ever before in history, decent Americans are realizing that their government is riddled through with evil. As with the pickpocket and the embezzler, so with the highest echelons of rule—under-handed workings depend for their continuance upon remaining veiled.

In the UFO field, broaching the idea of *unseen beings* is the beginning of the exposure of those beings, and their workings. The ethereans frankly state that they, themselves, are invisible. So are the Boys Downstairs, but humans lacking discrimination persist in the belief—despite all evidence to the contrary—that *there are no invisible beings*. When a man like myself takes photographs of various invisible organisms and constructs, using the simplest apparatus, such findings are sequestered in favor of what is termed "harder" evidence.

There is no taste in ufology, down to this day, for the implications of changing physical density and transitions of substance. "Scientific" things are sought, and the manifestation most thirsted after is a sighting of a "structured" object. For these people, happiness in ufology comes with the killing out of soul and spirit from the subject—of any living element whatsoever. Neurotic comfort attends the anchorage of the subject within the mechanical rubric, even if the cost of this pathological comfort is what we have already paid—over forty years of zero progress. The subject has become *rigidified*, and thus bears the signature of the Boys Downstairs.

This rigidification draws the necessary veil between the malefic workings of the unseen predators and those upon whom they prey. The refusal and reluctance to break out of this straitjacket, on the part of scientists, philosophers and investigators of all kinds, permits the unseen manipulators to continue. An awakening is positively necessary if the Boys Downstairs are not going to win out. Just as awakened Americans would certainly take back their country from its own native-born despoilers, *if they knew about the whole horrible panorama*, so will awakened human beings take back human evolution from the clutches of the devil.

Who are these unseen beings? What is their origin, and what purposes are they serving? How can they be combated? How are they involved in UFO phenomena? When we tackle these questions with the resources of spiritual science and etherian physics, we will soon find that we no longer have to answer that embarrassing question that always comes from the back of the hall at UFO lectures: "Why don't they come down and say hello?" Those of them that have come down to say hello—in their own way—are far removed from those dreamers' conceptions of spacemen that were current of yore. The days of golden-haired titans with no dental caries belong to a quaint, bygone epoch.

What is to be set down here is required, by the limitations of this book's physical size, to be a sketch. Once again the sketch must be almost presumptuous in its simplicity in dealing with profound esoteric events. Readers who wish to expand on what is said here, or to penetrate into it more deeply, should refer to Dr. Rudolf Steiner's classic *Outline of Occult Science*, or to his series of lectures on the matters in question, *Lucifer and Ahriman*. Given in 1919, these lectures were published in 1954 by the Rudolf Steiner Publishing Company of London.

Study of this titan's work will prove rewarding. There are three tremendous events, according to occult research, that have influenced or will influence, the course of human evolution more profoundly than any others. These events are the incarnation in the mortal body—walking the earth among men and as men—of three mighty spiritual beings. These beings are Lucifer, Christ and Ahriman. Lucifer and Christ have already incarnated. Their influences in human culture are traced by Dr. Steiner. Everything he writes in this regard may be read directly from life as we know it and from history as it is set down.

Ahriman has yet to incarnate, but he is coming soon. He is the bossman, if you like, of the Boys Downstairs. To the enlightened observer of contemporary human affairs, there is no doubt whatever that human life is being prepared with feverish intensity for this tremendous incarnation. UFOs are a part of the preparation, together with the kind of thinking we have had from official science regarding UFOs and their origin.

Earlier in this book we referred to the powerful cultures, long predating the Christian period, that are constantly being unearthed by modern archaeology. These pagan cultures reveal in their artifacts and mythologies a remarkable rapport with cosmic forces, so that the cultures themselves may be regarded as products of a more profound insight into cosmic workings than is consciously possessed by modern man. Grecian art in all its magnificence is an example of this.

These earlier, ancient people, even those that pre-date the Greeks, were inspired to their great achievements by the incarnation in Asia, in the third millennium before Christ, of the spiritual being known as Lucifer. All too often, primitive theology has identified Lucifer with the Devil, or with Satan. His name actually means "light-bearer" or "bringer of the light."

Lucifer's influence on man, right down to this day in residual form, is to lead man out beyond himself into ecstatic (in the root sense of the term, *ex stasis*—out of a fixed and rigid order) religious ideas and experiences. Lucifer tends to divest man of his earthly anchorage and orientation. Among humans, the mystic is likely to be under Luciferic influence. In more recent times, the irruption of so-called psychedelic experiences and perceptions—the subculture erected around psychotropic drugs—may be deemed Luciferic. The spectacle of scores of thousands of young people led out beyond themselves is verification enough of the source of their inspirations.

The tendency of Oriental cultures to withdraw from the world—the estimation of the physical world as *maya* or illusion—is another residual Luciferic influence. The powerful westernizing forces and influences, reaching their strongest development in Japan and Red China, have not yet swept away the sedimentary Luciferic influence on Oriental peoples and civilizations. This might be anticipated in the general area where Lucifer's incarnation took place.

Lucifer's influence spread westward, and led to the kind of rapport with spiritual powers and forces that characterizes ancient cultures. Men felt themselves membered into cosmic workings, and these were fully as real to them as electric power is to us. The initiates of these great cosmic mysteries were the leaders of ancient men, and it was under their general direction that life maintained its evolutionary course. Hebraic culture literally sprouts into this ancient wisdom, and provided the religious and cultural milieu for the incarnation, more than two millennia after Lucifer, of Christ, the spirit of the sun.

The ministry of Christ and the events preceding and following Golgotha, are to the eyes of the esotericist, misunderstood by modern people. The formal religionist, having jettisoned reincarnation, has to think of Jesus as a man-god. To the occultist, Jesus the man was a human being who was trained to adepthood by the Essenes. This training was for the specific purpose of providing a physical vehicle in which the spirit of the sun could incarnate for a period of three years. During this time, the cosmic mysteries were dramatized for man. This is the essence of the Christ event from the occult viewpoint.

The earthly consequence of the Christ event, which significantly occurred in what we now term the Middle East, was the development of European civilization. Christ's incarnation provided a corrective impulse against the untrammeled spread of Luciferic influence. In short, to prevent man from going out too far beyond himself and thereby failing to do the world's work. The time since Christ has been largely occupied in the creation of European and American civilization.

Preparations began in the middle of the 15th century A.D. for the incarnation on earth, among humans in a mortal body, of the third of this stupendous cosmic triad—Ahriman. What spiritual science tells us concerning these preparations may be verified by life around us, by actual observation of the "real" world that we know, or think we know. Recognizing that there is in the relatively near future to be such a stupendous being coming among us is one of the major aids we have in unraveling UFO phenomena. Without this knowledge, the phenomena remain impenetrable. Danger and disillusionment will certainly attend further arrogant ignorance of Ahriman's workings.

As with Christ and Lucifer, there is no way that mankind can avert Ahriman's incarnation. He is, as it were, designed into human evolution, and in some degree or another is already a part of every human being. Everything that comes from beneath man, everything that drags him down beneath himself and beneath the beasts of the earth in the process, comes from the Ahrimanic powers. Christians of primitive faith would call him Satan—the ruler of darkness. He is polarically opposite to Lucifer, who would lead man out beyond himself. Degenerate impulses, murderous thoughts, greed, hatred, revenge—negative activity of all kinds is powered and inspired by the Ahrimanic forces. The chieftain of all this is about to come among us in the flesh. Dr. Steiner has this to say on page 11 of *Lucifer and Ahriman*:

"If Ahriman were able to slink into a humanity unaware of his coming, that would gladden him most of all. It is for this reason that the occurrences and trends in which Ahriman is working for his future incarnation must be brought to light.

"One of the developments in which Ahriman's impulse is clearly evident is the spread of the belief that the mechanistic, mathematical concepts inaugurated by Galileo, Copernicus and others explain what is happening in the Cosmos. That is why Anthroposophical spiritual science lays such stress upon the fact that *spirit* and *soul* must be discerned in the Cosmos, not merely mathematical, mechanistic laws put forward... as if the Cosmos were some huge machine. It would augur success for Ahriman's temptings if men were to persist in merely calculating the revolutions of the heavenly bodies, in studying astrophysics for the sole purpose of ascertaining the material composition of the planets—an achievement of which the modern world is so proud.[3] But woe betide if this Copernicanism is not confronted by the knowledge that the Cosmos is permeated by soul and spirit. *It is this knowledge that Ahriman in preparing for his earthly incarnation wants to withhold from men* (emphasis added). He would like to keep them so obtuse that they can grasp only the mathematical aspect of astronomy. Therefore he tempts many men to carry into effect their repugnance to knowledge concerning soul and spirit in the Cosmos. That is only *one* of the forces of corruption poured by Ahriman into the souls of men."

We now begin to throw light on the downright *dishonest* attitudes of official science towards UFO phenomena. Official science is the exterminator of soul and spirit in earth life. Such things are not considered "real" knowledge—an assessment that could well emanate from Ahriman's own mouth were he here today in a mortal body. Since he is not, he utilizes the bodies, brains and minds of certain humans who already are here. In due course, we will outline the basic mode of instrumentation by which such control is wielded—and through which Ahrimanic beings not of our density help prepare the earth for Ahriman. One has only to review the angry, irrational reception accorded by official science to the de-rigidifying, innovative cosmogony of Velikovsky to see official science for what it is, a sleazy Ahrimanic fraud on true science. Every thinker, worker or inventor who would de-rigidify is handled the same way.

Drawing further upon Dr. Steiner's insights, we may list some of the ways in which the Ahrimanic powers influence or control human thinking. Critical areas bearing upon Ahriman's incarnation carry the heaviest signature of the Boys Downstairs. The connection of a number of things listed to the UFO mystery will be self-evident.

1. Execration of soul and spirit by official science.

2. Inspiration in the average man of contempt, suspicion and antagonism toward anything that leads to genuinely spiritual considerations.

3. Idolatry of goods and services as all-sufficient for humanity.

4. Perversion of cultural life so that nothing is deemed worthwhile unless it puts food in our mouths.

5. Denial of the invisible worlds.

6. Nationalism as the grand divider of humanity.

7. Ceaseless political party strife in which opposing forces fail to recognize that their ideas are of equal merit and equally justifiable.

8. Dead scientific concepts. Endless thousands of scientific books on library shelves and no cosmic knowledge whatever in the scientists.

9. Dead education, characterized by rote learning and rigidification of the natural biological correspondences that exist in children with cosmic processes.

10. Idolatry of numbers, as exemplified in the computer age with its dehumanization of life.

11. Tedium in outer life, as exemplified by the endless "toys for adults" mode of living in wealthier countries. More cars, more speedboats, more trailers, campers, guns, fishing tackle and aircraft to offset the mounting deadness of work in civilized nations.

12. Obsession with measurement. Official science accepts only what can be counted.

Steiner points out that it will become increasingly difficult for men to relinquish the notion that all knowledge of the external world not acquired through measure and number is delusion. To measure! How much emphasis is laid on that. What does it really mean? In reality, it means to compare something with a given dimension, be it length, weight or volume or what have you. In such measurements, the qualitative element is lacking, a state of affairs the Ahrimanic beings find it propitious to perpetuate.

Dr. Steiner emphasizes that a natural science so based leads only to an Ahrimanic illusion. Man must have natural science, of course, because since the Greco-Latin epoch he has been on the descending curve of evolution. He is a being growing physically weaker, in the sense that he no longer lives on the sense plane with the same intensity as did the Greeks. It is this weakening apprehension of the world that has made necessary the *fata morgana* of official scientific conceptions. There is no greater illusion than that of a cosmos minus soul and spirit—the concept that is relentlessly hammered into generation after generation of helpless young students.

When the listed influences of the Ahrimanic powers are examined, it is obvious that all these effects have accelerated in recent years. Man stands in clear danger of losing his humanity to the juggernaut of technology— although it need not be so. Life is being steadily dehumanized, and had it not been for the revolt in the Sixties of the younger generation, this tremendous Ahrimanic stream might have swept all before it. The Ahrimanic grip on world thought is undeniable, and on official science, that grip is total.

Denial of the invisible worlds is a pillar of Ahrimanic concepts. When evidence is adduced of the presence of invisible UFOs—with its incalculably damaging consequences for extant denial of the invisible—that evidence is sequestered or ignored. Ahriman inspires, through his agents, the denial of invisible UFOs—which would otherwise undo his racket, and expose him. In ufology, it is a pervasive aspect of the phenomena as reported and recorded by thousands of persons. No attention is given to it.

Ahrimanic influences also lead to the ridicule of all experiences with UFOs that in any way touch upon the presence of spiritual beings, or the intervention of the earth's guardians. "Contact" experiences are arranged deliberately by the Ahrimanic powers to have the involved personage report back to his fellow men that *physical* spacemen are involved. Official science then ridicules it, pointing to the lack of evidence. That same lack of evidence then paralyzes official scientific endeavor. Nobody looks in the invisible where the answers are. Despite the illusory round-and-round produced by these engineered "contact" encounters, still man persists in his illusory evaluation that UFOs are ships from other planets and *nothing more*. These notions are all Ahrimanically inspired.

The eye of the investigating scientist remains fastened on the illusion that Ahriman has fashioned for him—the flittering glimpses of spaceships projected momentarily and ubiquitously into human sight. Measuring, numbering, pointer-reading instruments fail at every turn, and indeed, are often completely corrupted by the manifestations. Still the highly-trained scientist cannot relinquish the mode of mentation into which he was brainwashed while still an innocent child. With his illusory objectivity and life-killing "controls", he is only the incarnate extension and forerunner of Ahriman himself, and is all the more pathetic on that account.

The bankruptcy of official science concerning UFOs has its origin in this Ahrimanic inspiration and control. Demonstrate, as I have, that intelligent organisms exist unseen in plasmatic form, and you are ripping away the Ahrimanic illusion. You are exposing Ahriman just as the cop detects the burglar or the auditor exposes the dirty little embezzler. You are breaking his grip on the human race, and his earthly representatives have no taste for facing such truths or mounting the necessary investigations to obtain the verifications for which they clamor.

There are thus these good, solid reasons why science has been so overwhelmingly defeated—and even humiliated in the UFO field. The whole thing seethes and pulses with Life, and cannot be rigidified. What the ethereans bring to our attention in this way, the Ahrimanic powers must smother with engineered illusions. The science that has been bankrupted by this has enjoyed simultaneous booming success in the development of mass murder. That ought to tell every intelligent human being something about the inspiring forces behind it. Ahriman controls official science and the minds and brains of its high priests.

The only mass effort made by human beings to counter the Ahrimanic juggernaut was the Luciferically-inspired "psychedelic revolution", in which thousands of young people, sickened by dehumanization and the murderous nature of the culture from which they had sprung, took to drugs. Without conscious knowledge of what they were really doing, these young people sacrificed themselves to re-awaken awareness of the spirit. Out of the horror of the drugs came, in the long run, a genuine interest in the arcane sciences, and in the New Knowledge as a viable force for human renewal. Nobody else has done much to oppose Ahriman.

One sees preparations for Ahriman's incarnation on every side. Uncomprehending humans bend every effort—or so it seems—to make Ahriman's sojourn among us as easy and beneficial to him as possible. Ministers of religion thunder about the Second Coming of Christ—one of the major inspirations of the Ahrimanic powers. Such claims as the imminence of the Second Coming can only issue from the mouths of people blind to all genuine spiritual insight. The Second Coming is an *etheric* event, and not a materialization in a mortal body.

When Ahriman arrives on earth, he does not proclaim himself as Ahriman, ruler of the nether powers. That men would flee from him if he did is only part of the story. His very essence is *untruth*, the lie his very substance. Therefore he comes as that which he is *not*, namely, Christ. In this guise he steps into a world ready to fall down at his approach, thoroughly prepared for his advent by ignorant, thundering ministers.

Ahriman will wield all the powers of the miracles related in the Bible. He will be able to heal and to control natural forces. He will not only have these attributes, but will also be able to bestow clairvoyance upon people who fall under his sway, giving them visions of the spiritual worlds without any effort on their part. All of them will see differently, of course, but this is part of Ahriman's goal—the dragging down and corrupting of mankind.

The task of people today is to confront Ahriman in the right way, by knowing who he is and what he is about, and not only what he is going to do, but how he does these things. Man himself must balance off the Luciferic and Ahrimanic influences on his own being, pitting one against the other and surrendering to neither. This is how the earth is won for Christ, the Golden Mean between the two extremes. Today these matters have become practical concerns of importance.

Exemplifying the practicality of confronting Ahriman aright, let us briefly consider here, and in detail later in the book, the question of controlling natural forces. Were a being to appear on the earth today capable of stopping a storm, creating rain, or turning the wind around 180° on command, this would reduce terrestrial science to the level of superstition by comparison. Scientists would be overawed by such mastery of natural forces.

No matter how many doctorates graced their walls, or how many Nobel prizes they had earned, they would be *willing to listen to any such being*. They would recognize that their own mode of scientific cognition had been transcended. They would be under the sway of the miracle worker.

The plain fact is that in preparation for that time, humans have already been presented with the basic techniques for controlling the weather through the discoveries of the late Dr. Wilhelm Reich. The feats mentioned above have already been performed with simple apparatus by certain men, including myself, and more is being learned of these manipulations every year. The discovery of the orgone energy by Reich leads functionally to control of the weather, since orgone energy is the prime mover in all terrestrial weather systems. Developing our understanding of orgone engineering will allow us, as earthpeople, to be standing foursquare and in a posture of challenge when Ahriman appears to try and blind us with his science. The more we learn about cosmic forces and their control between now and Ahriman's advent, the stronger we will be to wrest from him for Christ what is ours *if we will but earn it*.

The Ahrimanic influence appears in mass attitudes and in scientific attitudes toward such discoveries as the orgone energy. At the time of writing, not a single university in the United States is investigating the discovery of the orgone, or repeating the published experiments of Wilhelm Reich. Many devote their resources to such incredible trivia as filter tips for cigarettes, the composition of sourdough and the fat layering of female Korean divers. These projects lead only to proliferating detail and piles of books and paper moldering on library shelves. This evasion of responsibility by the universities and their fossilized leaders, is clear evidence of Ahrimanic control of what is probably the most crucial area of modern life—higher learning.

Ahriman's influence always goes towards differentiation—towards multiplication of detail and overwhelming splitting and fragmentation. The near-insuperable barriers between the various disciplines of science are a clear Ahrimanic signature. So is the misuse and perversion of science, scientists and the technology they create.

Noble goals, ideals and institutions are usually subverted by Ahrimanic control of earthly finance. There is hardly anything that begins nobly and well that cannot be dragged down by the subtle stratagems of temptation. This is not metaphysics, but the facts of earth life.

The involvement of the Ahrimanic powers in UFO phenomena is heavy. They inspire and sustain every notion, theory or contention advanced that keeps ufology locked within the mechanistic rubric. The investigator, student or scientist who finds his mind "shutting off" the moment he goes beyond mechanistic concepts is under Ahrimanic influence or perhaps even control. Whether or not they are conscious of these influences on themselves has little to do with the effects.

The Ahrimanic powers were put into a bad bind when man, utilizing the mechanical inventiveness—the cleverness, if you like—that is itself a part of Ahriman's contributions to the human race, developed radar. Man's new electronic feelers then cut across the biological realm that I refer to as "the critters"—the invisible, plasmatic fauna that inhabit the earth's atmosphere and stratosphere. The presence of living things around us, having their existence in the form of heat—i.e., actual biological form in the fourth state of matter—constitutes a body blow to the Ahrimanically inspired, mechanical notions of life and evolution. Life in the form of heat explodes all the beloved illusions earthmen have about life on other planets. Gone forever, too, is the idea that when the earth itself was hot in the remote past, life was not here then—or couldn't be here. Here indeed it was, as the very core of the earth's existence.

The official scientists and the laymen who defer to them were out chasing the illusion of flying saucers—the illusion engineered by the Ahrimanic powers to divert human attention from the *bioenergetic root phenomena* that were pouring in on man. This obscurantism has been successful up to now, but its day is done. One day soon there will be resources and bold young men and women to chase my critters, and enough funding will come forward because someone, somewhere, will see that all this is more valuable to humanity than sourdough. In the meantime, the life-killers and the life-haters, the hideous caricatures of people who crouch in university and corporate laboratories doing Ahriman's work, will be happily sustained by the forces of finance and power who are Ahriman's major earthly minions. Without the resources of spiritual science, no rational explanation is possible for the prodigious funding of destructive activity and the simultaneous pauperization of practically all good works on this earth.

The ethereans by their gentle intervention in mankind's behalf have made it possible for earthpeople, *entirely through their own efforts*, to lock on to the orgone energy as the power source of etherian craft and as the animating energy of the bioforms in our atmosphere. Reich, Steiner, Drown, Abrams and a dozen other gifted heroes of the New Age have won this discovery and many kindred discoveries for all humanity.

The Ahrimanic powers are trying frantically, with book burnings, corruption of the courts and psychic control of officials to kill all this and roll it back. Idealistic, devoted and spiritually hungry young people are not going to be denied, and the youth revolt of the Sixties was only a beginning. In the future, they will secure the New Knowledge to themselves, and when Ahriman comes, he will find a substantial human element confronting him in a condition of preparedness.

The word "control" has been used in this chapter to explain irrational human resistance to the New Knowledge, and to anything likely to aid its introduction. The principle of "extraterrestrials" controlling humans and wielding such arcane arts as telepathy, has been broached in science fiction. Even among official scientists, progress has been made along this line, notably in the Soviet Union. Dr. Jose Delgado at Yale University has demonstrated the principles of controlling humans by injecting minute electrical impulses into specific areas of the brains of monkeys via remote radio transmitters.

Human beings are often controlled from the unseen, as the literature of occult science and psychic investigation reveal. In the case of the Ahrimanic powers, this control is crucial because they are themselves excarnates and nonhumans, with only limited ability to manifest on this plane. Therefore, they must control living human beings already here. Little imagination is needed, having seen Dr. Delgado elicit rage, fear, fright and other emotions through radio-controlled electrical impulses to the brain, to conclude that an advanced technology could exert an advanced control of emotion and reaction.

Psychic control is fully understood in the inner circles of certain secret societies, whose black adepts not only hold conscious commerce with the extra human Ahrimanic powers, but also occupy key posts in government and in financial circles. Some of them are famous in law, politics and finance. They know in advance, by this means of communication, the dimensions and details of forthcoming events, since time runs backwards for their astral informants. These evil men are able to manipulate policy and action to serve Ahrimanic aims. Breaking down the moral, industrial, economic and spiritual strength of the USA was a major goal. Merging the wrecked U.S. republic with the Soviet Union, to create one rigidly-ruled Ahrimanic world, was the original intent. That goal was aborted by the greedy Russians who took over, largely, Soviet policies and intentions since Communism's retreat. At lower levels, the results of psychic control are all around us and easy to comprehend, once awareness is awakened.

Where total control of a human being to achieve a specific end is desired, we could expect such things as a judge in a court of law to insist that he was empowered on his bench to *judge a matter of basic natural scientific research*. We should also expect to see accessible and controllable judges turning criminals loose. We should expect to see the criminal element slowly acquire the ability to overwhelm the rest of the people. All of this is actually happening.

Now we must proceed to those refinements of instrumentation that make such control possible. Their forerunners are on this earth already. The Ahrimanic powers have made every effort to push them off the earth. The instruments that tune directly in on the human vital body were designed and developed by the late Dr. Ruth B. Drown from the foundational discoveries of Dr. Albert Abrams. Both of them were relentlessly persecuted by the enemies of truth.

The time has come to meet the third of my titans. Knowing, loving and serving her in the final years of her life was a rare privilege. You may have heard many of the numerous and elaborate lies circulated about Dr. Ruth Drown. Let me tell you the truth.

NOTES FROM CHAPTER THIRTEEN

1. Comment to Robert C. Gardner at Ent AFB, Colorado, Feb. 1953. Quoted in Leonard Stringfield's CRIFO *Orbit*, 4 November, 1955.
2. Published by Lyle Stuart.
3. Moon rocks hauled back a quarter of a million miles through space at a cost of $25 billion show that in the fifty years since Dr. Steiner's lecture, man maintains his juvenile thralldom of material substance. In more recent years our hugely expensive Mars missions, using robots and computers to determine composition, have continued the trend.

Chapter Fourteen

CRIMINAL OR GENIUS?

A new scientific truth does not triumph by convincing its opponents and making them see the light; but rather because its opponents eventually die and. a new generation grows up that is familiar with it.
— MAX PLANCK

The question of the instrumentation used by aliens from space to contact earthmen—and in many instances to control them—takes us far beyond what official science will tolerate. Establishment ufology is also left far behind, stumbling in the mists of mechanistic conjecture. Doubt and incredulity must inevitably surround such a subject, seemingly imported directly from the pages of science fiction.

The reality of human experience in connection with UFOs nevertheless overrides these vacillations, for comprehensive, participant investigation leaves no doubt that *the aliens are using bioenergetic methods of communication and control that eclipse terrestrial science with its current knowledge of electrical communications.* Such electrical communications as we have— the entire field of radio—are the product of less than 100 years of thought and work along one single line. Only the visionless see us at the end of that line, or hold that there are no other lines.

Bioenergy has thus far received only fringe consideration by modern official research. While interest and effort in this field are now expanding rapidly, our questing into the field of bioenergetic communications should begin with those who have *worked in this field for a long time.* Such persons have worked without official support, sanction or interest. Bioenergetics have one thing in common with UFOs even when viewed superficially: neither field is officially accepted. Our search therefore begins logically enough with work and persons who have been officially *rejected.*

There is no better beginning than with a lady who was cruelly branded "The Queen of the Quacks" by Big Medicine, the late Dr. Ruth B. Drown. My meeting with Dr. Drown, my study of her work and the personal services I was able to render her, precede my introduction to the discovery of the orgone energy by Dr. Wilhelm Reich. These two titans, while contemporaries, never knew each other. Today, it is crucially important that they be brought together so that bioenergetics may receive a powerful forward thrust.

Dr. Reich discovered the orgone energy, and verified its presence visually, thermically and electroscopically in 1939 40. Later he was able to demonstrate its presence with the Geiger-Muller counter, and he also designed an experiment to make the orgone energy luminate *in vacuo.* He worked methodically from observation and experiment, publishing his findings for replication by other scientists. Dr. Reich was a stickler for scientific method, but the phenomenal reach and power of his mind quickly carried his work beyond what official science could tolerate.

Dr. Reich developed badly-needed new patterns of thinking—he called it functional thinking—by which the human mind could train itself to follow living processes. We will in due course detail his life and work against the

UFO backdrop. He made a massive contribution to the knowledge of the future. Ruth Drown, by contrast, *was* the future. The orgone energy that Dr. Reich slaved to objectify and verify was to her a simple reality no more in need of proof than her own heartbeat. A magnificent anachronism, she had harnessed the energy for medical ends long before Reich's 1939-40 discovery along recognized scientific lines.

As far as I was aware up to the time of her death, Dr. Drown did not know the basic determinism of the energy form she was using, in the fashion that Wilhelm Reich established that determinism. Long prior to Dr. Reich's orgone energy discovery in 1939-40, however, Ruth Drown's intuitive genius had enabled her to develop instruments for tuning and manipulating this energy in medical diagnosis and therapy. Two full years prior to Reich's discovery, Dr. Drown had designed and clinically demonstrated a photographic attachment to her diagnostic instrument that made cross-sectional photographs of both soft and hard tissue of the human body. The agency by which these photographs were made was undoubtedly the energy that Reich later called orgone.

Ruth Drown was the natural possessor of a mode of mentation that sent her vaulting over mechanistically derived and fundamentally inapplicable "control" methods so beloved of orthodox workers and thinkers. Such "controls" tend to blur and confuse man's native clarity of mind when taken out of their own field—the field of inert substance—and dragged into areas where they do not belong and do not apply. Dr. Drown's work with the orgone energy is a towering *fait accompli* that may not be gainsaid, any more than may Reich's definitive discovery of the orgone.

Part of my contribution to ufology is to synthesize in a preliminary way the work of these two outstanding pioneers. Free young men and women will in this way find their break-in to the New Knowledge facilitated. While it might be desirable in some ways to deal with the discovery of the orgone energy first, and sketch Dr. Reich's achievements before describing the functional principles of the Drown instruments, that is not the way in which my own unfoldment occurred. Furthermore, the reader has already had more than brushing contact with this energy if he has followed the earlier chapters closely. Dr. Reich was instantly comprehensible to me because Dr. Steiner and Dr. Drown provided the foundation upon which I then stood. The reader who continues to share my adventure as it happened to me will emerge with a clear understanding of where all the pieces fit.

My old teacher, Franklin Thomas, said to me shortly before his passing, "One day, you are going to have to meet Dr. Ruth Drown. She has much of which you stand in need. The time is not yet, but it will be soon." He went on then to introduce me to the general field of radionics and radiesthesia, which might be briefly summarized as the modern rebirth of the ancient art of divination in all its forms.

Locating underground water with a divining rod is the most widely-known application of the art of divining, and it is as old as human record. What is less widely known is that the same principles may be applied to the location of other minerals, missing persons and ships and submarines at sea. Divination can also be applied to the diagnosis of disease and in Europe has received the attention of many eminent medical men.

The distinguished English physician and surgeon, Dr. Henry Tomlinson, has provided a splendid overview of these possibilities and their practical application in his book *The Divination of Disease—A Study in Radiesthesia*, published by Health Science Press of London. In the U.S., however, physicians who wish to seek truth along these lines are subjected to the relentless, mindless persecutions of Big Medicine, so that all of it has been driven underground. What has been set down in this book earlier about the Ahrimanic powers and their control of earthly humans is nowhere better illustrated than in the suffocation of this kind of thought in America—the place in the world where it should have enjoyed the greatest freedom.

Readers prompted to further study of radiesthesia and radionics will find the name of Dr. Ruth B. Drown prominently mentioned in practically all text and reference books on the subject. These references are unfortunately minus any real understanding of how she fits into the history of this field. The version that follows is abstracted from my unpublished biography of this remarkable woman, and should be considered "official" even though merely a sketch of her life and work. The material was personally approved by her prior to her passing. Almost everything else published about her is apocryphal, while Big Medicine's comments on her are simply lies.

As Franklin Thomas foresaw, I did meet Ruth Drown. Mutual friends introduced us, discerning our common interests. Dr. Drown was a licensed Doctor of Chiropractic, and her office and laboratory were on La Brea Avenue near Sunset Boulevard in Hollywood, California. Optimism and positivity pervaded her premises. Heart and soul were lifted and upborne the moment you entered the building.

In her late sixties when I met her, Ruth B. Drown was a slightly-built, slender, gray-haired lady with an indefinable air of authority. She dressed in quiet good taste, but whatever she wore, it could never shift attention from her unforgettable face. Dominated by a hawkish nose, it was a face of stem and even fierce aspect. Penetrating hazel eyes bored into my very core when we met. In all my days, before or since, no one else had ever looked into my depths as she did.

My thoughts turned to the ancient Egyptians and their reverence for the hawk as the only earthly creature that could fly directly into the sun. Ruth Drown looked as though she could do just that. The inner strength of the woman was overwhelming. Her initial survey of my depths was over in an instant, and then there was a smile and softness. We sat down and talked in her office and enjoyed instant deep contact with each other. By the time I departed nearly two hours later, we had joked about each other's hawk noses, and I knew I had found my way to another teacher of surpassing wisdom and greatness.

During her lifetime, Dr. Ruth Drown was one of the most widely misrepresented and vilified women in America. The poisonous rubbish circulated about her in magazines and newspapers was never written by anyone who knew her. Alleged technical descriptions of the Drown work, invariably condemnatory and always inaccurate, were printed in national magazines and published in books by writers who had never even met Dr. Drown, let alone studied her work. This pillorying went on for decades.

Hoodlums in the service of the United States government, hirelings of magazines parroting the anti-quackery line of Big Medicine, and officers of the law and the courts sought to make her life a hell on earth. When finally they brought about her death in old age through a cowardly legal attack that resulted in a stroke, they must have felt something of relief. Thanks to the Women's Liberation movement, today it is unlikely that such a vendetta could be carried out against a solitary woman.

Even within the international radionic field, which is growing apace and is achieving greater respectability as the knowledge frontiers expand, her work has been misunderstood and misrepresented. There is also a notable reluctance among writers on radionics—often born of ill-concealed male jealousy—to acknowledge the magnitude of her achievement in making cross-sectional photographs of both the soft and hard tissue of the human body. No medical researcher of either sex, anywhere in the world, has been able to approach even remotely this staggering technical feat. She did it alone.

Franklin Thomas described Ruth Drown as a "unique cosmic figure" standing far above her contemporaries in achievement, knowledge, skill and ethics.[1] He added the following significant observation: "In Atlantean times, she was a master of these forces that we now dabble with as radionics, and she is in the body at this time to contribute to their rebirth on a higher arc." Franklin Thomas had the clairvoyant ability—as mentioned in connection with Rudolf Steiner—to travel back in time and read the events of other epochs.

Ruth B. Drown was born in Greeley, Colorado in 1892, the third daughter of Annette Beymer Chase and Morton Ellsworth Chase. Both sides of her family pre-date the American Revolution as residents of this country. Nowhere on either side of her family are there any ancestors noted in science or invention. Those who accept predestination will delight in learning that her father was a professional photographer. At a time when photography was still a rare art, he taught Ruth the whole photographic process in a practical way. Decades later, when she invented her radiovision instrument, she knew exactly how to handle the photographic aspect of the device. Predestination was to leave its signature at many points in her life.

Married at nineteen to Clarence V. Drown, a farmer who later worked for Standard Oil and also as a railroad fireman, Ruth seemed well-launched on a career as wife and mother. The arrival of two children, Cynthia in 1912 and Homer in 1917, completed a tranquil domestic scene—or so it seemed.

Seven years after the marriage, inexplicable friction arose between the couple. "Incompatibility" had stalked on the scene. Intolerable before long, the situation made a separation necessary. Decades later in England, one of the circle of unseen mentors who guided her work in the world, manifested to her through a noted English trance medium.

This entity told her directly to her face that her marriage to Clarence Drown had to be disrupted, so that she would go on to discharge the earthly tasks she had undertaken. Her destiny involved something much more comprehensive than just being a wife and mother.

Armed with $800 and bringing her two children with her, Ruth Drown arrived in Los Angeles in 1918. She bought a gasoline station in the town of

Alhambra at the behest of a brother-in-law, who promptly took another job and left her to run the operation alone. In a preview of the enormous willpower that was to drive her through many far larger barriers, she filled gas tanks, repaired tires, checked oil and tended an eating counter.

This was 1918 and long before Women's Liberation. Difficulties were legion for a woman in any type of business. Gasoline retailing was among the roughest, toughest occupations of all, even for men. Prejudice against her as a woman made things even tougher, but it would attend her throughout her professional life as a physician and inventor. The gasoline station was a superb training ground for dealing with prejudiced men.

She ran the service station successfully, selling out in due course to a mechanic formerly her employee. She then went to work in a Hollywood photographic lab, utilizing the skills taught her by her father. This was a happy period in which she dealt with familiar things, but a peaceful life was not to be hers. Step by step she was moving toward her real work in the world.

A lady friend from Colorado, secretary to a Southern California Edison Company executive, invited her to work for Edison. The job offer made little sense. Ruth was to take charge of the addressing machines in the Edison accounting department. The job required mechanical skill and expertise, and the young Ruth Drown had no such experience.

Against her better judgment, she agreed to tackle the job. Physically frail and weighing only 92 pounds, her pre-employment medical examination might well have seen her rejected on health grounds. A kind physician who spotted her Eastern Star sorority pin bent the rules a little, and a few days later her Edison period began. She took over a room full of clanking mechanical beasts that gobbled in paper and vomited addressed bills.

A belligerent young German assigned to show her around made it clear that he didn't think she could do the job. "We got along like two strange bulldogs," she recalled of this time, "and his attitude made me determined to succeed." Buckling down, she put her fierce will to work. Latent mechanical ability burst into expression. She mastered the operation, maintenance and repair of the machines, and was soon in charge of the addressing department. Fifteen girls worked under her direction. The Edison period brought her phenomenal mechanical aptitude to light, and her later work would have been impossible without awakening these dormant talents.

Alongside the mechanical abilities developed during four years with Edison, there appeared the germinal element of her major work. She became burningly interested in the new fad of radio. Building her own crystal sets, she found yet another exceptional gift manifesting—this one rooted in intuition in the highest sense of the term.

Radio stores sold parts from huge bins. Thousands of assorted components were piled together in these containers. Normal experimenters would select carefully the parts for the unit under construction, making sure that specific values or ratings were chosen. Ruth Drown short-circuited all this in a fashion that became typical of her later technical work with life energy. She would simply pluck parts, seemingly at random, from the bins. They would turn out to be exactly what was required, and when the crystal set was completed, there were no surplus parts.

Electrical components and concepts that others could only handle with conventional, step-by-step reasoning and planning, Ruth Drown handled with sublime ease and stunning rapidity. She just *knew*.

When Edison was taken over by the City of Los Angeles in 1923, the young Colorado lady had no intention of remaining in her job. Although there was no alternative employment to her high-paying job, she was seized by a sense of *finis*. A few weeks before the Edison job ended, she knew why.

Walking one evening on Hill Street in downtown Los Angeles, she saw a signboard outside the People's Auditorium advertising a lecture by Frederick Finch Strong, M.D. He was lecturing on the application of radio to treatment of disease. Interested intensely in radio, Ruth Drown felt irresistibly drawn to this lecture.

Dr. Strong was a Cornell graduate who had studied at the University of Berlin and other European centers. He was also a former teacher of bacteriology at Boston University Medical School, and one of the founders of the American Roentgen Society at the turn of the century. This learned and forceful humanitarian was able to stir mighty but heretofore dormant forces in Ruth Drown. As she dwelt upon the potentialities of radio therapy, she felt the electrifying effect of inspiration. She knew intuitively and immediately that her future lay with this new idea.

Dr. Strong was not only a medical doctor, but also a serious student of Theosophy—the approach to microcosmic-macrocosmic relationships founded by Madame Blavatsky in the 19th century. Numerous American scientists today, including many in the aerospace industry, are finding out for themselves the merits of Theosophy as an aid to understanding the universe in this cosmic age of the earth. As a practical researcher in these fields, Dr. Strong was far ahead of his time in the medical field. Worth noting here also, in view of this book's course, is that Dr. Rudolf Steiner was a member of the Theosophical Society prior to founding the Anthroposophical Society—the latter providing a wholly occidental approach to the same principles.

Getting into this kind of work became a matter of urgency for Ruth Drown from the time she heard Dr. Strong's lecture. Destiny once more moved the pieces on the chessboard of life. By "coincidence", she was acquainted with Miss Maude Breeze, secretary to Dr. Strong. Eagerly seeking out Miss Breeze, Ruth Drown asked about the possibilities of working for Dr. Strong as a nurse.

Part-time work at an extremely modest salary was all that was available. She grasped this opportunity and thereby began one of the most controversial careers of invention and innovation in the history of the healing arts. Her strangely broken marriage had led, by a pathway that seems predestined, to a completely new beginning.

Dr. Strong was at this time using the diagnostic method and instruments developed by Dr. Albert Abrams of San Francisco. Big Medicine has rarely hated anyone with the pathological fury it turned on Dr. Abrams. To modern minds familiar with the filth of medical politics, that Abrams was hated and vilified is certain evidence that he had something of value to the human race. Abrams's invention was the much-maligned "black box" or "magic box", concerning which an immense volume of lies has been circulated by Big

Medicine and the many writers it has brainwashed, bribed and bought out-right.

Dr. Abrams was a graduate of Heidelberg University, where he earned top honors and a gold medal. He had become convinced of the vibratory nature of disease as a result of systematic research into bodily reflexes and percussive resonances. Abrams held that histological tissue, as well as each kind of diseased tissue, has a definite vibratory rate. He theorized, and later verified, that subtle radiation coming from the human tissues may be numerically detected and numerically classified. This detection process was known as E.R.A., or Electronic Reaction of Abrams. A distinguished Stanford University medical professor, Abrams devised a new system of diagnosis and therapy based on his original findings with the E.R.A.

Until the 1939-40 discovery of the orgone energy by Dr. Wilhelm Reich, the only radiation of which scientists had been aware was electromagnetic radiation. In Abrams's time, even electromagnetic theory was primitive, but it was the only conceptual peg in physics upon which to hang diagnosis and therapy based upon tissue radiation. Dr. Abrams was in fact dealing with the orgone energy or vital energy, and not electrical or electromagnetic energy.

Terminology inevitably became a stumbling block in this new approach to diagnosis. Development of Abrams's work paralleled the first major and general technological development of radio. All this took place as well at a time when human consciousness of radiation was just dawning. Terms were used in Abrams's work—such as radio-therapy—that were drawn from another technology altogether. The terminological confusion has persisted right down to this day, and through the decades an expertise has been automatically assigned to radio physicists in evaluating radionics. In fact, such people know nothing whatever about the energy involved in Abrams or Drown instruments, and have contributed greatly to the throttling and ridicule of American research along the lines started by Abrams.

In recent years, the internationally recognized Soviet "discovery" of this life energy, termed by Russian scientists *bioplasmic* energy, has eliminated all save pathological opposition to scientific legitimacy. American efforts have been continually harassed by this kind of irrational opposition. Today, America's scientific sultans would rather credit the discovery to Soviet Russia than have to choke on the embarrassing fact that Abrams, Drown and Reich—and a host of lesser figures—were right.

Big Medicine naturally—and spastically—condemned the Abrams work. The British National Laboratories found curious and inexplicable anomalies in its evaluation of the E.R.A., but through incomprehension dropped the work short of a definitive investigation. At the practical level meanwhile, many able and successful practicing physicians became convinced of the value of Abrams's inventions and concept of the vibratory nature of disease.

As a Theosophist, Dr. Strong naturally understood completely the role of the human etheric double, vital body, orgone body, functional body, formative-force body or bioplasmic body—whatever we wish to call this animating sheath that lies inside the physical organism and animates the physical body. He knew therefore that a fundamentally new approach to diagnosis was being born. His opinion was that despite its primitiveness, the Abrams work in the hands of a competent operator gave diagnoses superior in accuracy to any method then in use.

The Abrams system of diagnosis involved three people. First, the patient— or a sample of his blood. Second, a "subject" or suitable healthy human. Thirdly, the operator or diagnostician. By comparison tests and tuning of the vibratory rates of disease as Abrams had established them—using the "subject" as a detector—much diagnostic guesswork was eliminated. The Abrams system utilized one leg of the commercial power supply (110V A.C.), and was affected both in diagnosis and treatment by spurious energies from colors, lights or certain foreign substances in the room.

Treatment required an electrode over the areas of the patient's body that were affected. The Abrams method also treated many patients in one hookup. There was a great deal of groping in what was a completely new field, but there could be no doubt that people were being more accurately diagnosed than before Abrams devised his system. There were also dramatic therapeutic successes occasionally, and these provided the practicing physicians with great encouragement.

When Ruth Drown went to work for Dr. Strong in his Abrams setup, her natural aptitudes were immediately evident. Gifted in diagnosis, she was also outstanding in the general care of patients. They seemed drawn to her by a strange magnetism or instinct. In her care, they seemed to make more progress than with anyone else. This marked ability to reach and to heal others was with her to the end of her days. No matter how beaten down I was by business or other worries in the time that I knew her, when I came into her presence it was like being pumped full of adrenalin.

Her abilities were discussed by Los Angeles doctors using and following the Abrams methods. Dr. Thomas McAllister, an osteopath, asked her to work for him on a full-time basis. Noting the intensity of her interest and her surpassing abilities, he loaned her books and personally taught her from his own medical knowledge. The association with Dr. McAllister was one more of those strange "coincidences" that shaped Ruth Drown's career.

A patient of Dr. McAllister's named Louise Thrall benefited considerably from Ruth Drown's ministrations. Oil had been found on the Thrall property in Kansas. As a financial blessing flooded in on her, Mrs. Thrall felt impelled to share it with somebody, without simply giving it away. To the young nurse whose care she valued so much, Mrs. Thrall gave $5,000 so that she could attend the Osteopathic College at Kirksville, Missouri. This $5,000 was later repaid in fall.

An enthusiastic Ruth Drown attended Kirksville for a year, during which she received the intensive training in histology that is a feature of osteopathic education. This was to have vital value later on, when her invention of the radiovision instrument provided photographs of histological and pathological cross-sections.

After the first year, however, misfortune intervened. Her aging mother's health had begun to fail. This made it impossible for the grandmother to care any longer for Cynthia and Homer during Ruth Drown's absence. She was forced to quit osteopathic college and return to Los Angeles and her children. She had been delayed, but she would not be denied.

In Los Angeles she entered chiropractic college. By nursing for the Los Angeles doctors who were using Abrams's work, she not only maintained contact with this field, but also was able to support and raise her children. She graduated as a Doctor of Chiropractic in 1926, and was licensed to practice in California in 1927.

During her chiropractic schooling, all her spare time had gone into experimenting with new ways to handle vital energy in diagnosis. Like every pioneer doctor who used the Abrams system, she was aware that acceptance of such instruments would require that they be simplified and further developed. Eliminating the clumsy patient-subject-diagnostician triad was a prime problem. Her intuitive gifts and her practical expertise were directed to tuning in directly on the patient, source of the vibratory activity under analysis by the instruments. In *individualizing* diagnosis in the strictest sense of the word, she developed a strong intuitive aversion to utilizing commercial electric power sources in the instruments in any way whatsoever.

Commercial electricity is created by manipulations of inert matter. The young woman experimenter could clearly see that this power source was neither individual nor biological. Her experience had already taught her that the vital energy patterns of every patient were individual and distinctive, affected as they were in each individual by the particular areas of lowered function with which that individual was afflicted.

Experience and intuition thus united to convince her that commercial electric power was in some way inimical to the energy she was seeking to tune and manipulate. A quarter of a century later, Dr. Wilhelm Reich was to find out in the Oranur Experiment—of which more in due course—that a fierce and potentially lethal antagonism exists between life energy and electromagnetic energy. Worth noting also is that Kirlian photography—now becoming widely investigated in our universities—depends upon exciting the life energy with high-frequency energy in order to make the life energy luminate. In this application, the antagonism between the two energy forms is utilized to objectify the life energy, although this simple fact appears to elude most persons doing this work.

These observations will give the reader a better appreciation of Ruth Drown's profound intuition to keep electricity out of her instruments. At that time, and right down to this day in the general field of radionics, there is a widespread conviction that nothing can be achieved in this kind of diagnosis and therapy without injecting the commercial electric supply in some way into the human body. This is an archaic and dangerous holdover from the Abrams days, when even doctors experienced with this work were skeptical that the patient's own biological energy could ever be tuned or detected without electric assistance of some kind. Ruth Drown hung on tenaciously to her own opposing view and kept experimenting.

RUTH B. DROWN

Every spare dollar she had was put in this research. Her mechanical skills were used to fashion panels and in the assembly, machining, fitting and connecting of components and cabinets. She was a dogged empiricist on the pathway of trial and error. A student of metaphysics since 1916, she meditated daily upon the instrument she was trying to make concrete on the physical plane.

By 1929, the first Drown instrument was a reality. Someday this simple, seven-dialed instrument will be ranked as one of the major breakthroughs of the New Technology. All the instruments she designed in the remaining years of her life were only slightly modified from this original 1929 model.

Problems existing in the Abrams system had been decisively eliminated. The means of localizing the source of disease vibration, or the desired tissue on organic vibration was transferred from the healthy human "subject" to the instrument itself. Detection was localized in a simply-constructed rubber diaphragm approximately 2 inches by 4 inches included in the instrument circuitry. Not only was the "subject" now unnecessary, but there were no connections to commercial electricity. The instrument was simply grounded. There were no electron tubes, batteries or other familiar electronic impedimenta. Examination of the instrument's simple construction belied the years of grinding labor, experimentation and prayer that had made it possible.

Ruth Drown introduced the Los Angeles doctors she knew to the new invention. Their enthusiasm was enormous. The development of the rubber detecting diaphragm was a triumph for the young woman. The need to go over the "subject" body with a glass rod and other clumsy aspects of the Abrams work had been overcome. The doctor now sat in front of the instrument and localized the patient's troubles on the detecting pad, conveniently placed for his right hand. A new era in diagnosis had opened. Unerring diagnosis was now a possibility.

Development of the instrument had taken all Dr. Drown's money. Her doctor friends now subsidized production of additional instruments so that six months after the original was shown, the doctors all had copies and the inventor had a second instrument of her own. From this beginning came the hundreds of Drown instruments manufactured down through the decades.

The origin of the *therapeutic* use of these Drown instruments should be recorded, because of its obvious membering into the pattern of happenings that brought Dr. Drown's work into the world. She thought it might be possible to use the diagnostic instrument for treating, but treatment with the Abrams setup utilized commercial power. The structure of this therapeutic system had been swept away by the new invention—divorced now from electric sources. How then, was therapy to be accomplished?

While Dr. Drown was directing much thought to this, a man she had never seen before stepped into her Hollywood office and asked about her instruments. She explained in a general way the principle of diagnosis. The man shook his head.

"No. No. That is not what I am looking for. I am looking for an instrument that will take the patient's own energy and *focus it back into specific areas of his body for treatment*. I am sorry. Good day."

With that, he departed, but he had set off a bomb in Ruth Drown's mind. *Of course*, that was the way it must be done. The theory and the technique came to her now with blinding swiftness.

The energy that is leaving the patient and is used in diagnosis—she reasoned—must return to its source, the patient. Why not focus it directly into the same tissues and organs that the device can tune? Only a slightly different hookup would be required from the diagnostic circuit. Tests proved this could be done. Experience proved the procedure therapeutically effective.

The man who was seeking that strange new therapeutic method never came back. Was he just another "accident" in the chain that made Ruth Drown's unfoldment possible? Who was he? Perhaps we will never know. Surely it is enough that he came that once.

The basic principle of tuning subtle emanations from all substances, be they living or inert, has grown into the worldwide and international radionics movement. Prestigious figures in European medicine turn increasingly to this field today. Even in the U.S., long the scene of mindless harassment of radionics pioneers, a major university finally entered this research. Stanford University in California, where Dr. Albert Abrams once taught medicine, started a program in the 1970's under the redoubtable Dr. William Tiller. Early in 1975, the first U.S. Radionics Research Congress was held in Indianapolis, Indiana. Later on, the U.S. Psychotronic Association was founded.

There have been many mutations of Dr. Abrams's basic work developed in many countries. In this connection, tribute should be paid to numerous European investigators, including the late George de la Warr of England. Mr. de la Warr got his start by building instruments for English friends of Dr. Drown, after the Second World War prevented her from shipping instruments from California. That Mr. de la Warr subsequently designed and used instruments of this type in the exactly opposite way to that intended by Dr. Drown does not detract from the substantial interest that the English worker was able to attract to the entire field.

In the U.S., an outstanding researcher along these lines was the late T. Galen Hieronymus,[2] who worked devotedly—and successfully—for many decades with what he elected to call "eloptic" energy. His achievements are considerable, but cannot be detailed here, any more than can the work of de la Warr or perhaps a dozen other inventors and innovators in America, England, France and Germany.

Dr. Drown remains unique in developing a method of photographing soft and hard tissue in cross-section, as an extension of her basic diagnostic invention. She is thus the only worker in the world to date to be able to objectively verify that her instruments tune in as and how she says they do. She called this device *radiovision*, and it added to the basic tuning of the diagnostic instrument a simple method of flashing light on a photographic plate. This flashing was done in such a way that the light was modulated by the particular tissue emanations from the patient to whom the diagnostic instrument was tuned.

The film was processed in normal chemicals with one incidental step added to produce a polarity reversal similar to the well-known "solarizing" effect. The vital energy patterns precipitated their form into the emulsion in this way, and the finished plate could be examined on a viewer in a manner identical to that used for x-rays.

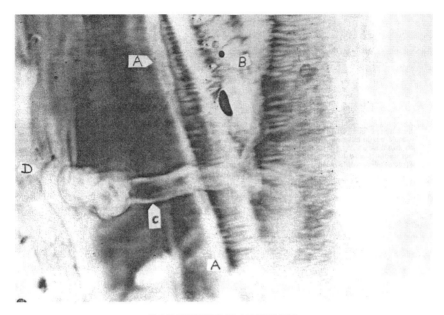

RADIOVISION AT WORK

This Radiovision photo was made of a dental abcess by tuning in with a Drown Radiovision instrument *on the abcess*. The line of tooth structure is shown by the arrows A, with the tooth root pressing into the gums. The photo is *cross-sectional* and arrow B shows veins in the central pulp of the tooth, black areas where blood flow has been sectioned in the veins by the Radiovision tuning. Arrow C shows a sinus of infection from the lower part of the pulp through the gum of the patient. Arrow D shows the gum boil as a bulge in the patient's mouth.

Radiovision terminated argument about whether or not it was possible to tune in this vital radiation from organs, glands and tissues. The photographs spoke for themselves. The diagnostician made the diagnosis, perhaps revealing, say, a tumor near the Eustachian tube of the ear. Such a manifestation would produce the gross symptom of deafness.

By tuning in the vibratory rate of the tumor and the vibratory rate of the Eustachian tube tissue itself, the instrument would then photograph the tumor and the surrounding soft tissue and hard tissue in *cross-section*. The effect was akin to vertical surgical sectioning of the area, so that the doctor was enabled—through this energy photograph—to view one wall of such an incision with all its detail. In one such actual radiovision photograph, the tumor may be clearly seen pinching shut the Eustachian tube.[3]

From 1937 until her death, Dr. Drown made untold thousands of radiovision photographs of her own patients and of the patients of other doctors—M.D., D.O. and D.C.—who came to her for diagnostic assistance. There were numerous cases where medical doctors and osteopaths confirmed the accuracy of both the Drown diagnosis and the radiovision photographs by post-mortem surgery. The dauntless Dr. Drown, in repeated efforts to get attention to her work, went to various cities in the U.S. and abroad and made these photographs *in front of audiences of medical doctors and sometimes of their own personal pathologies*. She even went to Chicago, epicenter of Big Medicine, to demonstrate her work there.

For every critic of Dr. Ruth Drown and her methods that Big Medicine might bribe or brainwash, at least a hundred former patients could be produced who would gladly testify to the benefits of the therapy. She not only met the challenge of introducing her work in this way, but also trained doctors to use her instruments. She would not sell an instrument for diagnosis to anyone who was not a physician. The physician who purchased an instrument was required to take her personal training course as a condition of that purchase. She also edited and published the *Journal of the Drown Radio Therapy* and a magazine, *The Philosopher's Stone*.

This lifegiving endeavor and the success of her practice provoked a decades-long battle with the Ahrimanically-inspired forces of *status quo*. Sending hired liars to her as fake patients was a common stratagem, invariably frustrated by her remarkable intuition. In 1951, nevertheless, her enemies finally got her into Federal court through an action nominally filed by the Food and Drug Administration of the U.S. Government.

The government charged that Dr. Drown had shipped a device in interstate commerce that was "misbranded". The charge was based on a man from Chicago having purchased a treatment instrument from Dr. Drown in Los Angeles, and taken it back to Illinois. This was legally interpreted as "shipping". Just to show you how far they were willing to go, *the instrument* was actually sued under Admiralty Law in other litigation, and ordered destroyed by the court in Chicago on 15 November 1949.

Egged on by Big Medicine, the Food and Drug Administration got hold of this instrument, or one like it, in Chicago. The instruction leaflets were studied by FDA investigators. In 1951, eighteen months after the instrument had been ordered destroyed by court order, criminal charges were brought against Dr. Drown. The leaflet enclosed with the treatment instrument was deemed to be "labeling" under the Food and Drug Laws. Since the government held that nothing could be treated with the instrument, it was mislabeled, and shipping the device in interstate commerce was a criminal act under Federal law.

This action forced Dr. Drown into defending more than a quarter of a century of original research in basic natural science in a court of law. The same Federal statutes and the same stratagem were used a few years later against Dr. Wilhelm Reich and his invention, the orgone accumulator. At the time the government assailed Dr. Drown, she had treated some 17,000 patients with her instruments and her methods of diagnosis and therapy.

During her career, because of the ethical barrier erected against the work by Big Medicine, she had published her findings in the *Journal of the Drown Radio Therapy*, in her books, and in *The Philosopher's Stone* magazine. She had demonstrated her instrument before audiences of licensed, qualified medical doctors all over the U.S.A. and in England. She was also accessible to any person—in or out of science and medicine—who had an intelligent purpose in seeing her. She would discuss her work with any scientist. Few came. They already knew, or were told by Big Medicine and its lackeys, that it was all fake. All of these people must have felt proud, if they survived until the 1960s, to see Russia rub America's nose in its own stupidities by proving that bioplasmic energy *did* exist as a physical natural force.

The transcript of the 1951 Drown trial is an incredible document. Opinion witnesses from the medical profession on the Drown instrument made goats of themselves and could have been destroyed and humiliated by a defense attorney with even rudimentary technical knowledge. None of these opinion witnesses had either met Dr. Drown or studied her work. The University of Chicago, which had forced Dr. Drown to sign a paper in 1949 stating that she would never mention the university in connection with alleged "tests" of her instruments made in Chicago in that year, happily sent a professor to testify against her.

Persons who were present at this obscene legal proceeding report that the judge—a fair and decent man clearly upset by the travesty of justice before him—groaned audibly when a guilty verdict was rendered by the jury of laymen. The judge had done his best to assure justice. The Federal attorney urged the judge to enjoin Dr. Drown against ever using her instruments again. His aim was to destroy her work. The judge declined. He imposed the minimum $1,000 penalty required by law and demurred at imprisonment. The trial had presented abundant evidence of, and unassailable witnesses to, the accused's high ethical and moral character.

Two internationally famous and universally respected character witnesses for Dr. Drown were Dr. Ernest Holmes and Dr. Arthur Young. Dr. Holmes was the founder of the Church of Religious Science; Dr. Young was the inventor of the Bell helicopter. The people who bore varying forms of false witness against Dr. Drown were, by contrast, little men. Their recorded technical testimony condemns *them*, not the woman they sought to destroy. Dr. Ruth B. Drown walked out of court branded a criminal. Years later, every thinking American with a grain of idealism in their soul can see how the once noble American legal system has been almost completely corrupted. Policemen who arrest criminals find themselves *sued by these thugs*. Vicious criminals are repeatedly unleashed from prison on society before their sentences are up. Humanitarians who devote their lives to the service of mankind receive criminal convictions under this twisted system. Still there remain those naive victims of negative superstition who cannot accept that this massive perversion of justice is inspired and engineered from beneath man by the Ahrimanic powers.

In the case of Dr. Drown, the 1951 trial and her conviction frightened off dozens of skilled physicians who were just starting with the Drown work. Building it up in the difficult era from pre-war had been an exhausting task. Just as these labors bore fruit, the founder got into trouble with the law. Most of the promising recruits to the new ways bolted, for fear of their professional lives. They felt the ship was sinking.

The ship did not sink. Dr. Drown reduced her practice and continued her research work. She developed a special modification of her instruments for the location of minerals. Abuse, vilification and indignities continued through the years! What Dr. Wilhelm Reich was to term the "Emotional Plague" of mankind—irrational destructive action on the social scene—

turned its full fury on Dr. Drown. Every avatar—every great teacher of mankind—has had to face the same kind of abuse.

In 1963, the State of California brought fraud charges against Dr. Drown, arising from her diagnosis of an allegedly healthy woman sent to her as a decoy. Then aged 72, she was arrested at her home during lunch and taken to her laboratory. State agents there were already looting her premises of patients' records and apparatus. Television camera crews and crowds of uncomprehending newsmen took part in the trampling of a fellow American's constitutional rights.

This aged and dignified lady was hurled into the Los Angeles County jail. Her arrest and imprisonment were integrated into a triumphant anti-quackery article in *Life* magazine, which probably engineered the whole proceeding in servitude to its massive commitments to official medicine and the drug industry.

Endless litigation now began. The judge involved, who years before had publicly sworn that he would "get" Dr Drown, turned up on the bench to administer justice in the Drown case. A contemporary affidavit by a member of the Los Angeles Board of Education who had heard his threat, resulted in his being pushed off the case for prejudice. The judge would not have gone on his own.

As a result of this final savage assault on her work and the ensuing turmoil, which involved the complete convulsion of her affairs, Dr. Drown suffered a stroke and did not recover sufficiently to stand trial. She died with a giant's work behind her. Shame and dishonor are not on her face before history, but they stain the character and the karma of the little district attorney—himself twice bilged out of medical schools for poor scholarship—who engineered her end.

Twenty-first century science will honor Dr. Ruth B. Drown as we now honor Tesla. She had broken through to the new cosmic electronics. The practical workings, significance and importance of her discoveries and methods will now be described. Let those who crucified this godlike woman say that it is not so. Let them utter their croaks while the wind roars around the mountain tops.

NOTES TO CHAPTER FOURTEEN

1. Biometric evaluation of Ruth Drown, using procedures now widely accepted by large corporations in personnel work, confirms that she bordered on the superhuman.
2. For information on the work of T. Galen Hieronymus contact Advanced Sciences R & D, P.O. Box 109, Lakemont, GA 30552.
3. Published in *Radiovision—Scientific Milestone*, 1961, under the imprimatur of the Drown Laboratories.

Chapter Fifteen

COSMIC ELECTRONICS

It is not in an arbitrary decree of God, but in the nature of man, that a veil shuts down on the face of tomorrow; for the soul will not have us read any other cipher than that of cause and effect.

—Ralph Waldo Emerson

The basic discovery of radionic medicine and the radiant nature of living organisms made by Dr. Albert Abrams will probably be judged by 21st century science as epochal. The subject of continuous efforts at suppression in the U.S., these discoveries have nevertheless survived. Split into many different streams and modes of development—all of them thus far outside official scientific acceptance—they are nevertheless the beginning of the new cosmic electronics. The confusion that often exists within the radionic movement exemplifies the way in which the Ahrimanic powers, when all else fails, split and compartmentalize the New Knowledge even as they have the Old. Mechanistic science is a monument to this kind of splitting action, with its interdisciplinary barriers and multitudes of highly educated people barely able to comprehend one another's work.

The reality of this splitting and fragmentation of new discoveries that are cosmically significant is demonstrated in the case of Sigmund Freud's work. Working a little before Abrams' time, the master of psychoanalysis was in no doubt that his discoveries would eventually open a pathway from psychology to biology. This was yet another possibility for microcosmic-macrocosmic relationships to come into modern human ken. Half a dozen of Freud's followers who thought they were both smart and right, splintered Freud's work. All became famous and respected. None found the psychology-to-biology pathway. That achievement belongs to Dr. Wilhelm Reich, who maintained his direction on the pathway broken by Freud, his mentor.

Similar circumstances and a similar fate attended Goethe's major impulses to scientific cognition at the inception of the age of modern material science. This has been dealt with already in our review of Steiner's work. Goethean conceptions lead to a spiritual, dynamic understanding of natural processes. Following Goethe's methods of training observation and thought would have obviated for mankind the Ahrimanic fragmentation of knowledge that ensued from the mechanistic, lifeless world conception that was impressed on mankind instead.

These examples are cited to illustrate how the right pathway, even when found, can be quickly obscured and lost again through Ahrimanic influences on vulnerable humans. Vulnerability is rooted in spiritual ignorance. As long as humanity remains ignorant of the existence and workings of the Ahrimanic powers—as detailed in Chapter 13, "The Boys Downstairs"— these influences will be felt and wielded in evolution.

Negative superstition sustains these processes and its major instrumentality is academia. What we have here to relate concerning these new discoveries is that instruments and concepts have been poorly received by academia. Ask yourself why this is, when the same segment of society can devote a disproportionate percentage of its energies to fashioning devices of destruction. Your answer lies in Ahrimanic inspiration and control—the control of good men by unseen degenerates. If you have ever wallowed in the philosophic bog separating human knowing from human doing, knowledge of the beings from beneath man brings you to firm ground. Life can then be read with accuracy.

In the case of the Abrams work, there was one individual who followed the initial breakthrough and in due course found the cosmic mainstream. That individual was Dr. Ruth B. Drown. Her achievements do not diminish those of others who devotedly pursued the Abrams discoveries in their own way. Not one of them, however, was able to make the monumental cosmic connections that Ruth Drown established. Her cross-sectional, full-plate photographs of human histological and pathological structures are to this day unequalled. These photographs are staggering verification of the technology from which they are derived.

Almost everyone involved in the radionic field became a party to the general process of sequestration where these photographs were concerned. Sequestration is the process of walling off a discovery by not looking at it or into it, and by verbalizing it away when it arises to consideration through its own native buoyancy. In extreme cases, as with Wilhelm Reich, the scientific literature that would guide men to these principles is burned—in full legality! Official science tacitly assents to the destruction of such literature. Where the cosmic connections become too obvious, too strong and too undeniable, the innovator is legally murdered. The 20th century has done no better than any other in promoting truth, but it exceeds all others in its promotion of murder.

Our description of the basic instrument of cosmic electronics will begin with the Drown instrument having nine dials. Her initial instrument had seven dials, but the reason for extension to nine dials will become evident shortly. Instruments in this nine-dialed configuration were being used and made by Dr. Drown up until her death.

In all the world of electronics there is no piece of apparatus *seemingly* simpler than a Drown tuner. Each of the nine dials has ten positions, numbered from 1 to 10. Behind each position is a metal stud, leading to the back of the panel. At the back of the panel a simple loop of wire passes around each stud on its way to the next stud.

These nine ten-position dials, with a single loop of wire around each dial position behind the panel, are led out at one end to two metal plates. At the other end, the dials are connected via the rubber detecting pad to ground. In a diagnostic set-up, the patient sits with his feet on the foot-plates. The patient is the source of the bioenergetic "signals"—originating in his tissues—that are to be tuned by the instrument.

In nonmedical applications, any specimen may be placed on the footplates and tuned by the instrument. The principles of analogy will have value here in clarifying the functioning of the Drown instrument. The analogy is with radio broadcasting, but it should be constantly borne in mind that with a Drown instrument we deal with life energy and not electricity. The sketch sharpens the analogy.

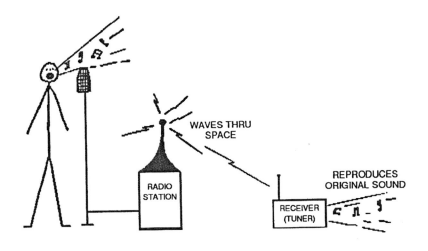

RADIO ANALOGY

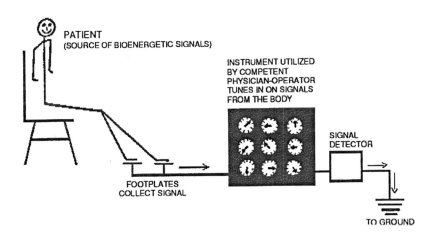

DROWN DIAGNOSIS (Simplified)

All radio and older TV receivers have an antenna with which they pick up the vibrations transmitted through the ether[1] by the various broadcasting stations. The antenna collects these vibrations, leads them into the radio set, and there they are tuned, resisted, amplified and finally made audible (and visible in TV) by the processes in the receiver.

The Drown instrument similarly begins with an antenna or aerial. The *patient* is the broadcasting source, emitting a jumble of different bioenergetic vibrations. This vibratory activity is conducted to the tuning section of the Drown instrument, which is simply an antenna that may be tuned to bioenergetic vibratory activity.

The patient sits with his feet on the end of an antenna. The antenna collects his "broadcast"—the totality of all the emanations from his body tissues and organs. From this totality, the various "transmissions" from heart, lungs, liver and other individualized structures are separated by a tuning process. These transmissions are then compared with what are known to be the normal transmissions from healthy organs and tissues.

The vibratory rate, or frequency, of any organ or other bodily structure is established by the molecular arrangement of material substance from which the organ is formed. This vibratory rate is expressed numerically. The rate of liver tissue, for example, *is always the same throughout the human and animal kingdoms*. This rate does not change unless disease or malfunction cause changes in the *molecular structure of the organ*.

All diseases similarly have their own distinctive vibratory rates, and may be thereby identified. Years of cautious labor were needed to establish these "rates" as they are called in Drown therapy. They equate, for analogy, with the frequencies assigned to radio stations by the Federal government. Tissue rates, by contrast, are assigned by the cosmos.

The rate of tissue in Drown therapy is established by placing a healthy specimen on the footplates and then tuning in on the molecular broadcast. The broadcast is broken down, in the tuning process, into its component vibrations in general accord with Fourier's Theorem. This process is termed "establishing a rate". An example would be the tissue supporting the human kidneys, which has a rate of 379. Healthy tissue of this type always has this rate and no other.

Drown diagnosis is entirely numerical in character, and is more firmly anchored in mathematics than any conventional approach to diagnosis. The Drown system, by virtue of this mathematical basis, has an inherent affinity for truth consistent with all other mathematical processes at human disposal. All Drown diagnosis, therapy and radiovision photography depends upon the use of numbers. Significantly enough, the *chemical ether* which we identified in Chapter 11, "Cosmic Blueprint", as the correct, occult name for the orgone energy, is also known to occultists as the *number ether*.

The vibratory rate of anything in creation may be established by this process. Application of the instrument is not confined to human structures or to diagnostic processes. Any element, compound or mixture may be examined in this way and its rate established. There are multitudes of resonances between the human body and other kingdoms of nature, and the ancient art of gems is one small example of how men of old perceived—and often prized—stones and other substances sympathetically resonant to various parts of the body and the emotions irradiating those parts.

An ordinary photograph of an artifact of any kind can be placed on the footplates and a vibratory rate established. A photo reproduces form, and form is produced by molecular arrangements. In the UFO field, much has been learned through this particular application of the Drown work, and it will be dealt with fully in its place. Human beings today can literally tune in on UFOs—of all types—by these methods, and personal experience taught me that this is the way that certain unseen entities of advanced mentality also can—and do—tune in on humans.

When a rate is established, it is entered in an atlas, and becomes a formal part of the technique. All bodily organs, glands, tissues and fluids, as well as all common diseases, have their rates listed in the atlas. Anything that can be placed upon the input plates of the instrument can be expressed numerically in this way. The number is derived from the emanations originating in the substance and intimately related to its form.

The vibratory rates of known elements such as iron, copper, tin and so on are numerically consistent with the known atomic weight of those elements. The atomic weight of an element is the experimentally-determined number of electrons in the orbital system of that element. The atomic weight is not an arbitrary number, and the mathematical correspondence between the Drown vibratory rate and the atomic weight should give firm indication that the Drown rates are not a arbitrary numbers, but rather, cosmic numbers.

This simple-yet-complex tunable antenna, by virtue of having nine dials each with ten positions, will tune to over two billion different combinations. Behind each dial division, inside the cabinet, is a small length of wire, or stub. Each stub is loop-connected to the following stub, and each ten-position dial is connected in series to the next. From a formal point of view, the arrangement could be considered a type of impedance rheostat, but gross errors result if we seek to establish too close a relationship with existing electrical concepts. A Drown instrument has a direct-current resistance of one ohm.

Each stub being numbered, the vibrations detected are expressed numerically, as they are extracted one at a time from the totality of any vibration passing through the instrument. The operator strokes a rubber diaphragm placed over the upper plate of the signal recording condenser. When the tuner comes into resonance with any component of the incoming vibrations, the friction of the operator's hand upon the rubber diaphragm grounds these charges. The effect is felt as a definite stickiness of the hand on the rubber pad, and heard as a sharp snapping sound.

In the case of the radio receiver, the "carrier wave" of the station being tuned strikes the antenna. On this carrier, the broadcasting station has superimposed the announcer's voice or music. Whatever its nature, this superimposed element of the signal is called the *modulation*. The radio receiver picks up the carrier wave, extracts and amplifies the modulation from the carrier, and reproduces it through a loudspeaker.

In our analogy, the electrostatic energy of the human body—which remains with it after death—is the carrier wave in the Drown conception of these processes. The vital energy (orgone) of the human being, which flows into and through and animates the human tissues, is the *modulation* or intelligence. This energy is not present in the body after death. Absence of this energy turns a living human into a corpse. This vital or orgone energy enables our organs and tissues to maintain their form. The Drown instrument tunes into the histological structure to determine whether or not the "carrier" in the various organs and glands is receiving the proper amount of intelligence or modulation, i.e., life energy.

In the Drown instrument, the longest of the nine antenna connections is coupled to the first dial. The base rates of vibration it is intended to detect are usually set up in the human structure by some lowering or impingement of function. In simple terms, if any organ is so afflicted that the passage of vital energy through it is impeded, then a low vibration will be set up that can be detected on the disease dial. Acupuncture on the contemporary scene lends additional force to the concept of energy impingement in radionic medicine. The *chi*, or vital force of acupuncture, is functionally identical to the orgone energy of Wilhelm Reich. In its own right, acupuncture is a valid alternative to modern, official medicine, with its self-defeating dependence upon synthetic chemicals and total incomprehension of the life energy.

As the disease dial is rotated from stud to stud, the diagnostician strokes the detecting pad. When the disease dial is turned on to the stud that matches the length of the antenna to the disease vibration coming from the patient, the diagnostician's finger sticks on the pad and makes a snapping sound. Visual indicators using a cathode ray tube were built and used by Dr. Drown. The arrangement proved less desirable and efficient for the expert operator than the tactile aural detector, and was in time discarded.

No energy enters the instrument in diagnosis except that of the patient. The instrument itself is connected to ground. Inside the cabinet are no electron tubes, pilot lights, transistors, batteries or connections to the commercial power. Just nine circles of loop-connected studs confront the investigator. The sheer simplicity of construction and appearance led to the automatic condemnation of the whole thing as quackery.

The Ahrimanic powers, whose votaries control human science, have good reasons for wanting to wipe this allegedly "quack" instrument off the face of the earth. Those nine dials, and what they connect to and represent and manipulate, are the break-in to cosmic electronics. That is why, in Dr. Drown's entire career, not one representative of official medicine, not one court or law officer, not one physicist investigating the work for the law or the universities, not one district attorney planning to frame her, ever went to her and asked how the instrument was supposed to work. Their high academic qualifications led them to think that they already knew.

They knew nothing about the instrument whatever, and were determined not to find out. The cosmic connections of those nine seemingly-innocuous rings of loops would have been beyond their comprehension. Those connections have awaited the attention of a younger, better, more Life-positive generation that wants its mind blown and its creative imagination fired. Those young people are on the earth now, and they mean business. Cosmic electronics is for them.

In the diagnostic process as developed by Dr. Drown, the function of every organ and gland was systematically measured. A blood chemistry, a differential blood count and urinalysis were all made *in the body*, and the accuracy of these tests was repeatedly established by comparison with regular laboratory results. The first cause of the patient's illness was established by these systematic tunings, and the result was a complete, dynamic blueprint of the individual being examined. Many years of careful work were necessary to bring radionic diagnosis to the level of efficiency that she reached.

A cross-sectional radiovision photograph could be made of any part of the patient's body, if necessary or desirable. Since subsequent nonmedical researches with orgone energy by myself—using variants of Dr. Wilhelm Reich's cloudbuster—have revealed an intimate relationship between orgone energy and terrestrial magnetism, dvd and videotape recording has a thundering future in the New Medicine. The probability is high that with minimum work, the cross-sectional photographs made by Dr. Drown and her co-workers during her lifetime, could be produced in the near future as motion pictures. There can be no doubt that it is possible to go inside a diseased organ, and *objectify it on a TV screen as it functions in the living human.* All this without need for an incision, or any of the magnets or mechanism essential to CAT scans.

In therapy, the patient is placed in a complete circuit with himself. His energy is collected on a plate of block tin placed over the solar plexus. This energy is passed through the instrument and back to the patient via his feet—the latter resting on two plates of German silver. The minute electrical current resulting from the junction, via the patient's body, of the dissimilar metals of tin and German silver, acts in the therapy hookup as a carrier for the vital energy.

The patient's energy normally would be radiated into the ethers to return to him after a world-circling journey. Under this therapeutic system, the energy is passed through the instrument instead. The tuning of the device governs the precise area of the body into which the totality of the patient's energy returns. The focusing of his total energy in this way results in increased cell division in the tissue so treated. New cells come in healthy, and the diseased condition is gradually overcome.

Over a period, and with systematic and careful monitoring by the doctor, the affected organ or tissue is largely restored. Never should it be forgotten that the energy involved in therapy is possessed inherently of form-giving power. As regeneration begins, the vibratory rate of the area under treatment begins changing back toward a normal, healthy rate of vibration. Regular checking by the doctor is necessary to keep the instrument in tune with the tissue under treatment. Many thousands of persons through the years have been able to feel directly the focusing of their energy by these methods. Similarly felt at sites far remote from the physician's office is the changing of rate patterns when the physician makes adjustments. This brings us to the special propertics of blood in this new medicine.

Early in the history of the Abrams work, it was found that a patient's blood sample was equivalent—in a bioenergetic tuning sense—to having the patient present. A patient could be treated via his blood sample. Later on, experience established that a single, dime-sized spot of a patient's blood—held in crystalline form in a chip of blotting paper—permitted the diagnostician to tune in on the patient regardless of the patient's physical location in the world. Physical separation between diagnostician and patient was of no consequence, for when he was tuned in on, he was *bioenergetically present at the input to the instrument.* This bore unequivocal testimony to the existence of a unifying bioenergetic continuum. The "etheric web" of ancient occultism had become a modern scientific reality.

Dr. Drown extended this finding technologically. She made complete diagnoses—including all blood and urine tests—provided therapy, and also made cross-sectional tissue photographs. From Europe in 1939 she made an unequivocal radiovision photograph of a stomach cancer in a Connecticut patient—confirmed by post mortem surgery conducted by the patient's own doctor, a qualified osteopathic physician and surgeon. Dr. Drown herself had patients in Europe—treated from California—until her death. The blood crystal was the visible end of an invisible line connecting crystal and patient for life—and for a short time after death.

Rutted down as they are in the Old Knowledge, it is difficult for mechanistic scientists to conceive of such relationships as being valid. Their skepticism is usually rooted in their peculiar "tunnel vision" about cosmic laws and canons, and accentuated by their lack of investigation of the New Knowledge. That a single drop of blood is indeed the bearer of these energy patterns is verified by European scientists following Dr. Rudolf Steiner's indications. This work was done independently of American pioneering in the New Medicine, but has been subject to the same sequestration and obscurantism that has attended the birth of the New Knowledge in the New World.

Physicians, scientists and all others interested in this pattern-bearing power of a single drop of blood should read two overwhelming books. They are:

> *Sensitive Crystallization Processes: A Demonstration of Formative Forces in the Blood,* by Ehrenfried Pfeiffer, 1931; republished 1975 by Anthroposophic Press, RR 4, Box 94-A1, Hudson, New York 12534.
> *Formative Forces in Crystallization,* by Ehrenfried Pfeiffer, published by the Anthroposophic Press, New York, 1936.

These books outline and describe the experimental methods by which a single drop of blood, subject to sensitive crystallization processes, demonstrably contains the pattern forms of various diseases. These disease forms or signatures are distributed according to definite law within the crystallization field. These crystallizations enable the physician to diagnose the patient's ailments by studying these blood crystallizations. The work has been clinically verified by qualified physicians in Europe over a period of many years, and is still in use there.

The pattern-bearing power of a single drop of human blood is a demonstrated scientific fact, and the so-called skeptic should inform himself concerning this process.

Throughout her life, Ruth Drown walked in reverent awe of the Creator's works. She was as close to being a holy woman as anyone could come and still remain human. Accordingly, she considered herself a trustee of the work she she initiated and took to a very high level, rather than its inventor or creator. She would quickly correct anyone claiming she was the originator of the work and never ceased asking in her daily devotions for guidance.

In the early days, she particularly asked for guidance regarding the true meaning of those nine circles made up of chains of loops. These requests were in due course answered. In 1930, a friend gave Dr. Drown a copy of MacGregor Mathers's book, *The Kabbalah Unveiled*. One of the major classics of qabalistic literature, it was for years the handbook of Meade Layne, whose early theoretical penetration of the UFO mystery has already been described. At the time Ruth Drown was given this book, she had been a student of metaphysics for some fourteen years.

Despite this background, she was overwhelmed by the mighty classic. Mathers was in the stratosphere of extended thought. In the ensuing few years, Dr. Drown kept digging into the Mathers book—seemingly without making much progress. The volume nevertheless exerted upon her a continuous attraction. During this period Drown therapy was in its infancy. Rates for the various bodily organs and diseases were still being worked out, and a corpus of diagnostic knowledge slowly acquired. Thousands of painstaking hours were involved in the birth of the New Medicine.

The nine-dialed instrument came into being, as the first advance within Drown therapy itself. Behind all this devoted labor there loomed in the mind of the woman whose responsibility it had become, *The Kabbalah Unveiled*. With its ramified cosmic mathematics, glyphs and charts, the book became something much more. It became a *presence*.

The Qabalah is the esoteric tradition of Israel. This refers not to the contemporary Mediterranean state, but to the ancient Hebrew culture in which European spiritual culture has its roots. The Qabalah is an esoteric tradition that may be traced back to the star-worshippers of Chaldea—those ancient humans who, by virtue of their different mode of consciousness, understood cosmic workings far more deeply than does contemporary science with its mechanistic superficialities. Many of us today are those men and women of old, reincarnated repeatedly since, and with all this knowledge a part of our subconscious content. A contemporary task is to get it all up into modern consciousness and see to its rebirth in modern, technical terms.

The Qabalah has been derided as a collection of medieval forgeries, berated as rabbinical law, and variously characterized as fraud, falsehood and fiddle-faddle. These epithets fall readily from enemies of truth. Theocrats often drove the Qabalah underground. Modern scientists would call it mysticism or myth and say no more. In its various forms nevertheless, the Qabalah has come to modern man as a living system of esoteric science—appropriate to the use of both artistic and technical consciousness. Instruments for its objective study and technological employment are already on the earth, rifling the whole thing into the technical times in which we live.

All theologies have a secret tradition, or qabalah, of their own. This secret tradition may consist of only those principles imparted to priests and ministers and withheld from congregations. No religion is without such a "qabalah". The chosen people of the Qabalah today are chosen after the spirit, and not after the flesh or by race. The chosen people are those who, in this cosmic age of earth, feel a willingness to approach things behind the manifested world. Their common problem is finding the necessary *methods* for such an approach.

There are millions of such people on the earth today, and millions more are coming. The younger generation of the 1960s — many by the false path of chemical illumination — caught a glimpse of these deeper realities. Many who survived the chemical binge of the Sixties pursue the New Knowledge now by healthier pathways, and their convictions of the rottenness of the old order remain. Those who maintain the thrust of their search will not be disappointed in their spiritual aspirations. Only by genuine spiritual renewal can human life be significantly changed.

In this sketch of the Qabalah and its relationship to the work of Dr. Drown, we can once again deal only with its bare bones — the same presumption to which we were forced with Dr. Steiner's work. Accordingly we deal primarily with those aspects of the Drown work that bear upon the UFO mystery. Beyond this sketch there lies a fascinating lifetime of work and personal discovery for every awakened person drawn to this New Knowledge. Bridges have been fashioned, theoretically and physically in terms of apparatus, connecting the physical-material world to those other orders of being and substance that have begun to impinge increasingly upon earth life.

The key to the Qabalah is the formulation of a concrete symbol to represent the abstract — a *glyph* or composite symbol. In Qabalism the main glyph is the Tree of Life. This beautiful glyph, which is reproduced herein, is helm, compass and chart to the qabalist as he embarks with his thinking into the invisible that stands behind the visible world. The Tree of Life is the slide rule of the mind. Operate it correctly, and the most involved problems become comprehensible. Embodied in the Tree of Life, awaiting the penetration of the user, lie all the connections between man and the universe.

Each of the symbols on the Tree of Life represents some cosmic force or factor, and the connections that force or factor has with the cosmos. The ten Sephiroth represent planes or levels of manifestation. They are not physical locations in the universe. In their descent from Sephirah 1 to Sephirah 10, they represent the descent of power to Malkuth — the physical plane — from all that lies above the physical plane. The significance of the nine dials on the Drown instrument should already be dawningly evident.

As the ten Sephiroth represent manifestation, that from which they came is termed the Great Unmanifest, the *chaos* of the ancients. In Qabalism, the Great Unmanifest is also termed the Negative Limitless Light. Kether is the starting point of creation, the Crown of Creation, *kether* in Hebrew meaning crown. Kether is the first stirrings from the Great Unmanifest, and the highest reach of finite human consciousness. Beyond Kether, the qabalist draws the veil. In a brief review of this type, we can say nothing comprehensible whatever concerning it to the newcomer. *The Mystical Qabalah*[2] by Dion Fortune is a suitable reference text, and one that Dr. Drown recommended.

Kether contains within itself all the other Sephiroth. These Sephiroth, evolving from one another, finally descend to Malkuth, the physical-material world as we perceive it with our unaided senses. Each Sephirah contains within it all that follows after it. The alert reader will immediately discern the factual and conceptual correspondence with etherian physics as outlined in Chapter Eleven, "Cosmic Blueprint."

Chokmah, for example, which is the second Sephirah, contains within itself all that follows on the descent of power — eight potencies or densities

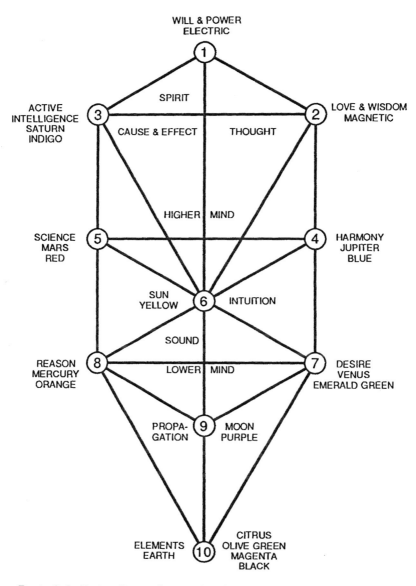

Basic Qabalistic relationships utilized in diagnostic therapeutic and photographic work of Dr. Ruth B. Drown.

in all. The paths between the Sephiroth show additional connections and relationships. Kether may be likened to the first bowl in a waterfall consisting of a chain of bowls. When it fills, it flows over into Chokmah below it, Chokmah fills and flows over into Binah and so on down. Instead of water, think now of diminishing levels of density. If you want a cosmically valid concept of God, think that *God Is Pressure*—the pressure that keeps all bowls full.

Ufologists who maintain an ostrich posture over the hostility of certain UFO entities should grasp one central principle of qabalism: there is no manifestation without differentiation into pairs of opposites. Kether differentiates into Chokmah and Binah, male and female potencies respectively, with Chokmah representing force and Binah representing form. Force and form are the two indispensable factors in manifestation.

Numbers are assigned to each Sephirah. Kether is assigned number 1. Chokmah and Binah are respectively assigned 2 and 3. On the left hand pillar of the Tree of Life, known as the Form Pillar and also as the Pillar of Severity, are found Geburah (5) and Ho (8). The right hand pillar, known as the Force Pillar, includes Chesed (4) and Netzach (7).

Between these two pillars is the central Pillar of Equilibrium or mildness. This is the qabalistic location of the Christ principle, known also to the Buddhists as the *golden mean* between the extremes of force and form. The Sephirah (6) in its location on the Tree of Life graphically and accurately depicts the central role of the Christ principle in Earth evolution.

The balanced nature of the Tree of Life illustrates the harmony of cosmic organization. The descent of power follows the numbers from Kether to Malkuth, proceeding across the Tree and down, just like a lightning flash. *Illumination* or cosmic consciousness, which may be broadly defined as instant understanding of the whole cosmic fabric, proceeds upwards from Malkuth in the same zig-zag fashion. The lightning flash often used in electrical motifs has its origin in this process of illumination.

The numbers assigned since antiquity to the various Sephiroth happen also to correspond precisely with the numbers assigned to the tuning dials on the Drown instruments. The sheer volume of these correspondences—all of them developed from years of clinical research with the instruments and objectified in radiovision photographs—excludes any possibility of accident or coincidence. From the inception of her work, Dr. Drown's instruments were the 20th century concretion of the Qabalah.

A *few* of these correspondences can now be discussed, with special emphasis upon the way the basic findings of Drown therapy bear upon the UFO problem. We shall find our way by this means to the communications methods that are at the disposal of advanced intelligences currently interested in Earth evolution—be those intelligences from above or beneath man.

NOTES TO CHAPTER FIFTEEN

1. There is little doubt in the mind of any experimenter like myself, who has dealt with both the orgone energy and electromagnetic energy, that standard radio signals are *carried by the etheric continuum* as manmade, artificial perturbations. With a Drown instrument, by contrast, we deal with the native, natural, primary perturbations of the orgone continuum. Science invented its concept of the "ether"—in distinction from the ethers identified by Rudolf Steiner—because there could not be electromagnetic waves without something for them to wave in.
2. Published by Samuel Weiser, Inc., P.O. Box 612, York Beach, Maine 03910.

Chapter Sixteen

COSMIC ELECTRONICS IN ACTION

Facts have little meaning when arrayed against prejudice.
—MANLY PALMER HALL

Beginning with Binah (3) the Sphere of Saturn, each Sephirah represents one of the seven Holy Planets of Qabalism. The numbers and planets thus assigned are:

Binah	3	Saturn
Chesed	4	Jupiter
Geburah	5	Mars
Tiphareth	6	Sun
Netzach	7	Venus
Hod	8	Mercury
Yesod	9	Moon

The divergences from the standard astronomical classifications should not be given any weight. The Qabalah penetrates far beyond the spatial positions of these bodies and the physical-material weighing and analyzing to which official astrophysics attaches such importance. The Earth is not considered one of the Holy Planets, but is deemed to be the receptacle of all that stands above it on the Tree of Life. Similarly, the Moon appears as the senior partner in the Earth-Moon binary, a role that orthodox astrophysics could neither discern nor approve. Under the concepts associated with the Tree of Life, each Sephirah is positive in polarity to the one that follows, and negative to the one preceding. These concepts—and many others—are being borne out by Space Age investigation and new physical discoveries. External facts recently determined about the human etheric double are a case in point. The human vehicle known for millennia to qabalists as the vital body, etheric double or functional body is now known as the bioplasmic body, the orgone body and the formative-force body. Twentieth century names have been given to this vehicle, and 20th century empiricism provides further verification of the ancient qabalistic truths.

The bioplasmic body is assigned to the sphere of Yesod, the Ninth Sephirah on the Tree of Life, and has been so assigned for millennia. In the 20th century scientists have been able to establish empirically that this body is positive to terrestrial light. The bioplasmic body is made tangible, i.e., negative to terrestrial light, only when it is electrically irradiated. All this work has taken place in quite recent times, but it is scientifically solid and will not be overturned.

The positive polarity of the bioplasmic body, or etheric double, relative to the physical structure was recognized by Ruth Drown three years prior to the Second World War. That is why she added, in radiovision photography, a polarity-reversing step to normal photographic processing. The film in which the etheric energy patterns have been locked is exposed to ordinary electric

179

light while in the developer. Without this polarity-reversing stratagem, the positive patterns of the etheric energy would not react with an ordinary photo emulsion.

On the basis of my experience with UFO photography, I can say unreservedly that the overwhelming majority of UFOs present in our atmosphere are *positive to light* in their normal state, and in this state *do not react* with film emulsions any more than they are seen by the normal human gaze. The ordinary human being no more sees the UFOs that are continuously present in the atmosphere than he sees his own bioplasmic double body and its emanations. The matter is as simple as that, and full technical solutions will come only when empirical work proceeds on this assumption.

Late in 1975, I found that by using Ektacolor 160 color film in a low-light, automatic Super 8mm movie camera *and employing an 18A filter to turn daylight into an artificial darkness*, UFOs may be photographed from airliner windows. They arrive on the film in *full color*, even though normal vision sees nothing outside the aircraft. Adherents of old-line optical theories will need to hustle to explain this, and numerous associated light-and-color riddles attending my efforts to solve this polarity problem.

Each Sephirah, in addition to the polarity properties indicated above, has certain qualities, purposes, properties and formative powers assigned to and under the control of its sphere. These ramified relationships establish the connections between man and the cosmos. The various organs of the human body, for example, did not develop their form and function accidentally—however comforting that concept may be to the mechanistic neurosis. The qabalistic placement of the various organs on the various Sephiroth show the origins, functions and purposes of these organs, as well as their connections to each other and to the total organism—as well as to the cosmos that created that organism. Accident is eliminated.

Understanding the workings of the various Sephiroth—the great creative rays of creation—demands a lifetime of study. Each individual human being enters incarnation with one of these great rays dominant among the forces that shape and guide him. This dominant ray is known as the Birth Ray, and is easily calculated by an occult reduction of one's birth date.

Each individual comes into the world with things he has to overcome or understand. In the case of more mature souls, they come with business to pursue on behalf of humanity as a whole. This ray is known as the Life Ray. Dr. Ruth B. Drown had Geburah for her Life Ray—the sphere to which science is assigned in qabalism. The Life Ray is calculated from the Christian, or given, name of the soul.

My Life Ray happens to be the eighth—Hod, the sphere of mercury, of concrete mind, and also of Thoth, Lord of Books. Dr. Drown first pointed out to me my own mercurial seeking after truth—running, as it were, like the liquid element of mercury into every corner where it might be found. This book is obviously an effort in the sphere of concrete mind, or reduction of the abstract. The Lord of Books has already received ten previous offerings from me. My relationship to these things is shown in my Life Ray, and it was *true the day I was born*, long before UFOs came on the scene or I ever took pen to paper as a writer. Working up the Life Ray and Birth Ray of famous people whose achievements are well-known, simply eliminates accident and substitutes design.

After the invention of her instruments, Dr. Drown spent about four years building up systematic procedures for the new means of diagnosis, and in establishing the rates of vibration of the body components and diseases. This work was a constant matching of objective activity, intuition and insight with numbers. Piece by piece, a workable and dependable numerical system of diagnosis was devised. Simultaneous study of McGregor Mathers' book, *The Kabbalah Unveiled*, resulted in the repeated presentation to Dr. Drown of consistent numerical relationships between the Tree of Life and her empirical findings in diagnosis.

Both the Tree of Life and Mathers's bookful of qabalistic mathematics gradually lost their abstruseness. Instead, they appeared as a marvelous cosmic blueprint. Connections between the Drown rates and the cosmic mathematics of the Tree of Life became undeniable. The nine innocuous loops of wire behind the panel were bringing the Great Rays down to Malkuth—the physical-mineral earth.

Dr. Drown's instruments were working in this fashion *before their inventor had divined the relationship*. Little wonder, is it, that from the first she was not only a great diagnostician, but also a gifted healer. The power of cosmic law was operating through those instruments. On these tides, many were carried to health and healing long after official medicine had resigned their cases.

Certain relationships exist between the Drown rates for various bodily organs and the numbers of the various Sephiroth. A brief resume will be given of these to dispose of the notion that these rates are fictional, or arbitrary, or otherwise meaningless.

Kether, the Crown, corresponds to the pineal gland in the human body. This is the point of entrance of the vital energy into the human being. The pineal gland rate is 98. In reducing numbers to a single digit, we find that 9 may always be omitted from the reduction without altering the final single digit. Thus, 98 expressed as $9 + 8 = 17$, which reduces again as $1+7=8$.

From the following similar reductions of the organs assigned to the first three Sephiroth, remarkable consistencies appear:

Optic Nerve	Rate 647	Reduces to 8	$8 + 8 + 8 = 24$
Pineal	Rate 98	Reduces to 8	Reduces to 6
Ocular Motor Nerve	Rate 44	Reduces to 8	

These organs and nerves correspond with Sephiroth 1, 2 and 3. Their total also reduces to 6.

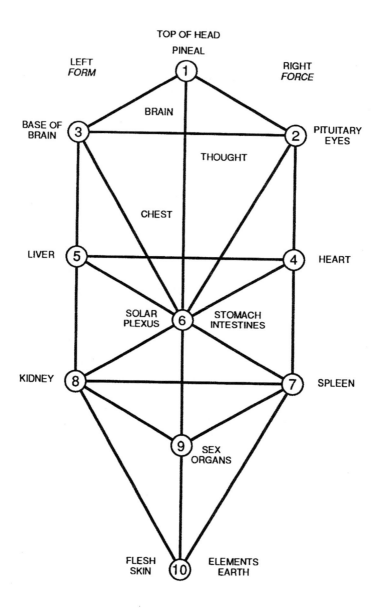

Qabalistic anatomy as utilized by Dr. Ruth B. Drown

Proceeding down the Tree of Life and correlating the Drown rates with the organs assigned to the various Sephiroth, we find the following:

Heart	Rate 256	Reduces to 4	Chesed is Sephirah 4
Liver (4 lobes)	Rate 48 481 483 87	Reduces to 5	Geburah is Sephirah 5
Stomach (Solar plexus area)	Rate 321	Reduces to 6	Tipareth is Sephirah 6
Spleen	Rate 4651	Reduces to 7	Netzach is Sephirah 7

Kidney (2 essential parts)
Glomeruli Rate   Reduces to 4
Metanephros Rate   Reduces to 4

These two rates reduce to 8 and Hod is Sephirah 8

Sex Organs
       Male Rate   954   Both   Yesod is Sephirah 9
       Female Rate   999   reduce to 9 Yesod is Sephirah 9

Skin   Rate is 10   Malkuth is Sephirah 10

In the case of the three Sephiroth at the head of the Tree of Life—Kether, Chokmah and Binah—an even more profound numerical relationship can be developed by correlating all the rates for pineal, pituitary body, small brain and the associated cranial nerves. This all lies beyond the scope of this book. The above connections may be indefinitely pursued and extended, the only limits being the knowledge and consciousness of the investigator.

The connections that are seen to exist in this new diagnostic system to the so-called psychological disturbances, must lead perforce to a complete over-haul of the mere *notions* currently held about the psychosomatic relationship. Man is a functional mind-body unit. *Let the physical integrity of a given organ—or a chain of interconnected organs—be impaired, and the quality of mind associated with that organ will have difficulty manifesting in the earth-ly personality.*

The connections between severity and liver ailments, for example, are established through Sephirah 5. Those between reasoning power and kidneys are found in Sephirah 8. The ramifications and the promise of all this are without end. The psychosomatic relationships may be systematically and numerically approached, just as soon as some brilliant young physicians devote themselves to this purpose.

At the end of her career, after some thirty years of practical diagnosis, Dr. Drown could "read" the psychological makeup of her patients with high accuracy. Patterns of ailments diagnosed with these instruments and methods reflected, and in fact were responsible for, patterns of thought and behavior conventionally termed "neurotic". These findings in turn strongly support the view of Dr. Rudolf Steiner, advanced in his medical lectures, that the spirit of man is always healthy, and so-called "mental disease" arises from inability to achieve full physical expression because of organic derangements and deficiencies originating in the physical and bioplasmic bodies.

From the point of view of ufology, the most important aspect of Dr. Drown's work is that it is possible—with instruments of this or kindred design—to tune in on any organ, gland or tissue in the human body. Recognizing and accepting this revolutionary fact is the key with which cosmic communications may be unlocked. The same key opens to our consideration *the illicit control of terrestrial minds by Ahrimanic entities from "space"*, who cannot otherwise subvert evolution.

As noted earlier, Dr. Jose Delgado in the 1960s demonstrated "control" of monkeys by injecting minute electrical currents into specific areas of the brain. Virtually any kind of emotional response could be evoked by electrical stimuli, ranging from pleasure through fright and rage. Crude but significant, these experiments required the insertion in the brain tissue of hair-like wires to carry the electric impulses. Radionics in its infancy obviates the need for wires. Brain tissue may be tuned in directly, from any point in the world—or outer space.

Dr. Drown could and did take cross-sectional photographs of brain tissue in living humans, definitively objectifying the bioenergetic differences between the sleeping and waking brain.[1] *A capability far beyond this must be attributed to the operators of the advanced space vehicles* seen, tracked and photographed in our atmosphere and beyond. Knowledgeable analysis—as outlined earlier in this book—reveals these visitants as masters of etheric propulsion. Overlooking these bioenergetic capabilities could rightly be termed the ultimate in evasion of the essential!

If the wisest men of earth choked on what Dr. Drown was able to do, imagine how ill-prepared they are to deal with other-world cosmonauts who can bioenergetically tune in right to the *medulla oblongata* of any human being—the main switchboard. Unethical entities in sophisticated possession of such capability can run humans around like so many cockroaches. Unethical entities can, and do, literally think their thoughts into and through human beings who are under such control. There is not a competent occult scientist personally known to me who is not fully aware of this baleful phenomenon, so characteristic of the modern world. Man's seemingly inexhaustible appetite for death and destruction is not an accident!

These processes represent a clear misuse of such discoveries and principles as have been outlined here. Dr. Drown would have nothing whatever to do with these potentially lethal aspects of her work, refusing even to discuss such matters. No power on this earth could induce her to do anything that was not clean, honest and upright. In the unseen, and in spaceships, all are not golden-haired gods bent on humanity's rescue—an illusion long foisted on the credulous by the ignorant. There are both earth people and space beings who misuse the power and technology of the Great Rays of creation.

On psychic control of human beings, I must speak again from practical experience—the standpoint of the participant investigator that has been adopted throughout this book. One of the most important things to happen to me as a ufologist was the medical diagnosis given to me by Dr. Drown not long after I met her. She insisted that I have this diagnosis, at no charge, so that our discussions would move from theory to practice. She was a practical woman.

At the time of this diagnosis, I was still following the exercises for inner development taught to me by Dr. Franklin Thomas. The resultant sensitivity enabled me to carry out my photographic field work. When I sat down for that diagnosis I was more sensitive bodily than at any other time in my life, before or since. One element in this diagnosis hit me with the impact of a thunderclap. I set it down now from a contemporary account.

In Dr. Drown's office were some chairs, her desk and, at the far end of the room, an instrument with nine dials in a wooden cabinet. The panel of this instrument was about twenty inches long and a foot high. On the desk top into which the instrument was built was a rubber diaphragm, measuring about four inches by two inches.

Dr. Drown waved me to a chair about four feet from the instrument and, after removing my shoes, I placed my feet on two polished metal plates. The doctor sat down in front of the instrument. Beside her was an assistant to help fill out the diagnostic sheet. The blank form was a large, rectangular chart divided into four main columns, each of which was six inches long. More than a hundred separate entries were needed to fill this chart. The column headings were "Function of Endocrine Glands", "Function of Organs", "Differential Blood Count" and "Urine Analysis".

I looked at my watch. The time was 3:05 p.m. I looked back at the formidable diagnostic chart, thinking to myself that the whole procedure would take hours. Thirty minutes later, with doctor and assistant working with remarkable mutual rapport, the chart was completed, and the first cause of my particular plexus of dysfunctions traced to bad vaccine. This came from a half-forgotten incident in my nautical career, when a filthy physician in the Levant vaccinated a ship's crew of which I was a member. This was during the postwar plague and cholera scare in the Middle East.

Before commencing the diagnosis, Dr. Drown asked me a comprehensive series of questions concerning my medical past. Any competent physician would require this same case history. Health of mother and father, causes of family deaths, sleep habits, surgery, accidents, diseases past and present, complaints and symptoms—a full standard case history. She then began making the complete functional chart of my body with the aid of the instrument.

Each time she obtained a reaction to a given setting of the dials, corresponding to the various rates and levels of function she had established, her fingers would stick on the rubber detecting diaphragm. As her fingers stuck, the rubber snapped sharply. The process was fascinating to watch, and to *feel*.

The high bodily sensitivity that I had developed was providing a dividend. As Dr. Drown turned the dials across the various studs of the instrument panel, I could feel chains of sharp and distinct resonances *inside my body*. These impacts were unequivocal. I could feel them with every alteration of the dial settings. I called this immediately to Dr. Drown's attention.

"What you describe happens with a lot of people who are, for one reason or another, especially sensitive," she said. "The more sensitive you are, the more you feel it. Some people feel nothing at all."

She made another setting.

"Where do you feel that?"

"It seems to be near my heart."

"That is correct. I am tuned now to the left ventricle."

"Have you ever had a bad fall?" she asked.

"I don't think so."

"Well, your left kidney is displaced. It's lower than it should be. Are you sure you never fell at any time?"

Then I remembered. I had fallen down a bunker hatch aboard ship in Cristobal, Panama, in 1946. In the fall, I had landed on the brassbound rim of the hatch, taking the force of the fall just below the rib cage on the left side of the back. I told Dr. Drown of this and she nodded.

Her assistant came over with a flatheaded probe, which was then plugged into the instrument. As the probe was moved over the area of the kidney, the detector pad snapped sharply. This gave the position of the kidney. The assistant moved to the right side and the procedure was repeated. The kidney on the left was distinctly lower than the one on the right. I asked Dr. Drown how she knew—without prior physical examination or the matter having come up in the case history—that the left kidney was displaced.

"The supporting tissue of the left kidney shows subnormal function, as did the kidney itself. Gravity being what it is, the chances were for all the kidney being displaced downward with the subnormal function of the supporting tissue. The probe verified that this had actually happened."

During this examination of the kidney, I felt the tuning of the organ by the instrument with particular intensity. One could not set aside such persuasive evidence. What I had forgotten of my own medical history, the instrument in the hands of Dr. Drown had unfailingly detected.

A short time later, sitting in that chair in Dr. Drown's office, I was suddenly transported back to those difficult and distressing days when I had undergone psychic experiences in attempting to communicate with UFOs. Dr. Drown checked the function of my larynx. As the instrument was tuned in on the structure of that remarkable organ, I felt the same sensation of something *locking on to me*—tuning in on me—that was always a prelude to the reception of telepathic messages.

The relationship of the larynx to the function of *hearing* is poorly understood in official medicine. Any person may satisfy oneself, by simple experimentation, that a *sympathetic movement of the larynx takes place with every sound we hear.* If the forefinger is brought into light, gentle contact with the underside of the larynx, it will be found that the organ's physical motions follow the sounds we hear.

In the mode of telepathy followed in my case, the larynx was tuned in on by the ethereans, and the sound or intelligence they wished to convey to me manifested as intelligible sound and words "inside the head" which I was later able to softly vocalize. This was a *reversal* of the speech process. The same basic process was undoubtedly followed with George Van Tassel. Others who have involved themselves psychically with UFOs have reported similarly "hearing" in this reversed fashion, inside the head. At least now I understood how it was accomplished, even if I was still ignorant of the advanced technology that utilized radionic tuning for bioenergetic communications.

This was probably what Franklin Thomas had meant when he told me that Dr. Drown had something to give me. In an instant I had grasped, purely as a result of participating freely in some of the New Knowledge as has come down to technology, the basic principles of bioenergetic communications. Dr. Drown could demonstrably tune in on *my* larynx, picking me out of the enormous sea of life by this process. If she had a crystal of my blood, she could similarly pick me out of the sea of life even if I were in Tierra del Fuego or on the moon.

She did not command the means of imposing modulation on my larynx, as did the ethereans when I had dealings with them, but as a radio technician this appeared to me as a minor aspect of this new communication method. The ear and the larynx enjoy a polar relationship in the hearing function. In Dr. Drown's work, we had the fundamentals of bioenergetic communications, using the continuum of vital energy and not connected in any way to electric power or batteries.

These functional connections all became apparent in an instant. I was at once shaken and elated. The joy of association with Dr. Drown was that such things could be discussed with her freely. Never once did I fail to learn something as a result. She evinced no surprise at all when I told her of my relating the larynx tuning to the telepathic methods of the etherean entities.

"Over and over again," she said, "for years I have stated the basic facts and no one listens. They all know without listening. *Everything is here, now. All we have to do is tune in on it.*"

She pointed to the instrument.

'This is how we tune in. And now you know for yourself that we do in fact tune in."

This revelation topped everything I had experienced in connection with UFOs and the cosmic mysteries to which they led. No wonder that with tuning of such precision, using the bioenergetic continuum, Dr. Drown could take thousands of cross-sectional radiovision photographs made since 1935. The tuning annihilated continents and oceans. When the photographs were made, the doctor could actually go back over the pictures themselves—using the same type of probe employed to check the location of my kidney—and verify the nature of the tissues and any pathologies involved.

The connections to the technology of the ethereans had been firmly established. Here among us on earth, Dr. Drown could take cross-sectional photographs inside the body of a patient, regardless of geographical separation. Had she been left to do her work in peace instead of being legally murdered, she would have produced motion pictures of the same kind. Her successors undoubtedly will do so.

Ethereans whose amazing spaceships have sped through our skies must be credited with the ability to look into and tune in on any human mind on earth. They may well occasionally study our tragicomic existence, much as we watch the inanities of our own TV plays. In their communications they have repeatedly stated that they have this capability. They watch the doings of earth people as part of their policing operations.

One must also extend to the Ahrimanic powers, standing beneath man, mastery of the same technical principles but turned to the end of corrupting and misguiding humans. If the larynx could be controlled, why not the brain in all its functions? A free human being is characterized by his personal stand—in all circumstances—against any kind of enslavement. Ethical denizens of the unseen similarly have evolved beyond the baseness of controlling other souls.

No such ethical inhibitions restrain most human beings. They are all too ready to exploit their fellow man. Avarice drags them down to any enslavement of others that the law permits. Operating from the unseen, the Ahrimanic powers are the stimulators, inspirers and often the outright controllers of human beings they wish to use in order to gain their ends of death and destruction. Such psychic control through advanced types of radionic instruments was the menace against which Franklin Thomas had warned me when I first sought his aid.

Earthmen by the thousands are already under this kind of control. Every learned occultist is aware of this phenomenon as a common presence in our age. Men and women are everywhere being dragged down beneath themselves, and a major Ahrimanic goal is the ruin of the United States. Holding the promise of reunification of all the tribes of the earth in peace and freedom, the U.S. is a great repository of ideals. The televised hearings on many U.S. controversies during the past decades have brought before the world numerous responsible individuals who well knew the difference between right and wrong. They nevertheless *acted wrongly*. One can cite no more pungent example of psychic control than this, where good men were inspired to tear the ethical and idealistic fabric of America to shreds.

Psychic control was a key element in Dr. Drown's final demise. Who drove the sweaty little district attorney to his conspiracy with the media, or drove him again and again into the public on his quackery crusade? Who inspired the prejudiced judge, who had publicly sworn to "get" Dr. Drown, to machinate himself onto the bench to hear her case? Who inspired a mighty national magazine, the existence of which depended upon the publication of lying advertisements, to smear a beloved physician for whom 10,000 patients would have been proud to publicly confess their affection and admiration? People of earth had best take note that psychic control of earth life is far, far advanced. The perpetuation of this scourge and its expansion, depends upon one central fact: *human ignorance of its reality*.

Dr. Drown was intensely interested in what I had captured from the invisible. Various photographs from our collection were placed on the footplates, and the rate of the images established. The "critters" proved to consist of calcium and fluids, the mineral *and the fluids* both being in the functional or plasmatic state.

The mineral in this plasmatic state was quite able to reflect radar echoes, in accordance with the known properties of plasmas. At the same time, the critters were ocularly invisible. This finding would resolve many of the major anomalies of ufology that were impervious to resolution under any ships-from-other-planets theory. The overwhelming majority of the photographs made by Jim Woods and myself belonged to this amazing new family of plasmatic fauna.

The rates of the ship-like discs were similarly established by Dr. Drown. By correlating these with the Tree of Life—the cosmic blueprint—it was found that a level of intelligence was involved far above and beyond that of contemporary humanity. They were entities of a high order, just as the critters were of a low order of consciousness, on about the same plane as fish. Since the same methods had recorded both types, it was more obvious than ever with Dr. Drown's assistance that these two phases of the phenomenon had been mutually confused. Any effort to approach the matter with an onlooking, nonparticipant, mechanistic consciousness was doomed to failure.

Like all great avatars, Dr. Drown regulated what she gave to others, in a way that precluded her infringing on their independence. There were certain things that she simply would not tell me or show me. Often I resented what I thought was an unjustified reservation on her part.[2] Today I see in retrospect that she gave me everything I needed, as Franklin said she would. She gave me enough to inspire me further, but not enough to involve me in her own demise. She wanted me to work out my own pathway to my own goals, as all great spiritual leaders do with those that they assist.

Once I asked Dr. Drown if she would make a radiovision photograph from a UFO photograph. She agreed. That very morning, someone had presented her with a set of UFO pictures purportedly taken by the late George Adamski from Mt. Palomar in California. In the wake of her friendship with me, she had checked many UFO photographs on her instrument, and some of the most famous had failed to respond.

She was interested herself now to see how Adamski's pictures turned out. She had heard about him and his views, but had never read his books. His pictures seemed clear and unequivocal, despite his many detractors. I was personally much excited by this first use of radiovision in a field in which I had so much interest, and to be standing beside this genius as it was all done was a rare privilege.

Dr. Drown tuned in the mother ship on the diagnostic instrument, and Adamski's famed photograph produced a long rate of eight digits. Dr. Drown set the tuning on her radiovision instrument, took a picture, and processed the film while I watched and stood beside her in the darkroom. When we took the dry 8" x 10" film out to the viewer in her office, I was flabbergasted. *The mother ship appeared on the film cross-sectioned fore and aft.*

The compartments were readily visible. So were the launching ramps leading to the underside of the craft. At the head of the ramps, in small hanger spaces, "scout" type craft could be discerned. On the after ramp, halfway to the outside and cross-sectioned, a glowing ventla could be seen, its power plant activated.

Almost overwhelmed by the significance of this technical feat, I raced home and got Adamski's book *Inside the Space Ships*. At Dr. Drown's office we could clearly see that the cross-sectional radiovision photograph duplicated in all essential detail the cross-sectional drawing Adamski had made of the spaceship he claimed to have visited. Dr. Drown had never seen Adamski's books or drawings before, and was naturally elated to have this non-medical vindication of her work, in the most advanced field of science then concerning investigators.

Years were to pass before I could apply in practical field ufology what I had learned from Dr. Drown. The overview she had given me of the new, cosmic electronics was in itself something to be dwelled upon and slowly absorbed. A whole new technology was working its way into human ken, and

I thought I might live to see the use of bioenergy—primary energy from the atmosphere—supplant electricity.

That any human organism could be tuned in on and, if vulnerable to such depredations, made a vehicle of expression for unseen entities of malefic intent, was a shocking finding. Nevertheless, it laid bare a worldwide pattern of such control of humans by the Ahrimanic powers. Evasion of the UFO mystery by scientists and governments—the irrational refusal to face what was increasingly evident as a bioenergetic revolution—was now understandable. By the late 1960s and early Seventies, scientists in secret groups and in government employ were in conscious contact with the Ahrimanic aliens, and were spread all over the world at secret bases doing the will of the enemies of mankind.

Members of certain secret societies who have infiltrated virtually all key policy posts, engineered a worldwide UFO security blackout. UFO data were stonewalled or ridiculed. These procedures were followed on a planetary scale. All governments were recruited into the service of The Liar. Sovereignty was obliterated when it came to concealing Ahrimanic activities, a chilling portent of what is planned for mankind. When New Zealander Bruce Cathie independently discovered significant mathematical information concerning UFOs, the American C.I.A. attempted to bribe or threaten him into silence right in his own country. He described his experiences to me personally in 1984.

Penetrable also now were all the mechanistic illusions projected about UFOs by the Ahrimanic powers. What investigative ventures they were unable to control, they were readily able to corrupt by this means. This thought control—again like all criminal activity—was cardinally dependent upon human ignorance and spiritual inertia. There was in progress a battle for the minds of men and women, and I had become a part of that struggle—an adventurer who had nearly gone beyond his depth.

Dr. Drown's brutal arrest, entrapment and humiliation, followed by her death by stroke at the age of 72, traumatized me to the depths of my being. For months I could barely handle business and family responsibilities. A great dynamo of the New Age had passed from the scene.

For a long time, I believed that it was this overwhelming sense of loss and grief that prevented my further application of Dr. Drown's discoveries to UFO research. Retrospect now makes it evident that it was essential for me to also understand the coeval discoveries of Dr. Wilhelm Reich, and synthesize these with what I had learned from Dr. Drown before any real progress could be made.

As in the past, chessmen were being moved on the board of life. Forces were already at work to ensure that Reich would follow Drown in my life with the precision of two railroad cars being coupled together.

NOTES TO CHAPTER SIXTEEN

1. See *Radiovision—Scientific Milestone*, published by Drown Laboratories, 1962.
2. My friend associate Robert McCullough, biological assistant to Dr. Wilhelm Reich, said that Reich acted exactly the same way over certain aspects of his inventions.

Chapter Seventeen

AVATAR EXTRAORDINARY —
WILHELM REICH

We must begin with ourselves.
—DR. RUTH B. DROWN

Craven crying after scientific heroes who will say that flying saucers are real has been a feature of establishment ufology for more than sixty years. Even statements inferential in this sense, if uttered by prominent scientists, have been repeated endlessly through the decades. Always the hope has been held out that a top scientific name would drop a blockbusting statement on UFOs—but it never happened.

The only internationally renowned scientist to have had open contact with UFOs as a direct result of his scientific work, to have applied his talents, discoveries, research and a major new invention to the problem—reporting all this fully in writing—was the late Dr. Wilhelm Reich. His findings on UFOs are the boldest and most urgent yet to emanate from any man of science. Establishment ufology has nevertheless ignored Reich. He will be dealt with here fully, truthfully and accurately—and by one who has both built and applied Dr. Reich's inventions to the UFO problem, and who offers herewith objective evidence of their relevance and potency.

Dr. Reich was a controversial figure throughout most of his professional life. Enemies of truth were relieved when he left the earth, thinking that his ideas and discoveries would die with him. The controversies that swirled around him while he was alive are but mere williwaws compared with the mind-blowing typhoons that are to come. The orgone energy that he discovered, and whose basic determinism he established, is the prime mover and power source for the ubiquitous, elusive UFOs—be they creatures or craft.

The formidable body of New Knowledge brought into human ken by Dr. Wilhelm Reich was linked to my UFO research by two of Reich's closest professional co-workers. His daughter Eva, herself a medical doctor, made a connection with me through Dr. Drown. Biologist Robert McCullough saw a functional identity between some of the photographs used to illustrate *They Live in the Sky*, and similar aeroforms whose appearance in the Arizona skies was provoked by the use of Reich's cloudbuster. Field assistant to Dr. Reich in weather control, Bob McCullough reached me through the Borderland Science Research Foundation.

Contact with these two people I count among my greatest blessings. Professional aides to Dr. Reich, their knowledge of his work and discoveries was deep and solid. By this means, I was able to cut through the miasma of falsehood and misrepresentation that has been woven around Reich's work by numerous foes of truth. When Dr. Drown died, a parallel path to the one she had blazed was thus already open to me—thanks to Eva Reich and Bob McCullough. Bringing Drown and Reich together seemed to me an obvious task, and within a few years I was tuning Reich's cloudbusters with Drown's instruments to photograph UFOs. Perhaps it was all an accident, as anxiety-ridden mechanists might assert, terrified of finding a grander design.

Dr. Eva Reich was briefly a patient of Dr. Drown. Suffering from an internal hemorrhage that conventional methods had failed to arrest, Eva had sought the aid of Dr. Drown. Without the necessity of having Eva come to California, Dr. Drown stopped the hemorrhage promptly from 3,000 miles away, utilizing her technique of grounding the parathyroid glands through the instrument, using Eva Reich's blood crystal. Such things were routine to Dr. Drown.

Dr. Eva Reich is an intelligent, perceptive woman of original mind. She saw connections immediately between the work of Dr. Drown and that of her father. She also quickly linked up my work on UFOs—to which she was introduced by Dr. Drown—with her father's contacts with space and with his far-reaching theories and findings concerning UFOs.

In a lively correspondence, she led me step by step into orgonomy. She loaned me the foundational books of Reich's work in the proper sequence, so that I felt as though I were standing at Dr. Reich's elbow. After my own grim times in previously untrodden fields of work, I appreciated fully the decades of patient work that went into the discovery of the orgone energy and its consolidation.

Without Eva Reich's thoughtful insistence that I begin at the beginning with Wilhelm Reich—and her provision of the literature—my approach to orgonomy may well have been tangential and glancing. Many persons have failed to understand Reich's work for this very reason. Dr. Reich's unique scientific accomplishments can only be fully grasped if one begins with his *Function of the Orgasm*, orgonomy's basic book.

Bob McCullough is a biologist in the broadest and best meaning of the term. With a master's degree from the Utah State Agricultural College, he was working as a research associate at the University of New Hampshire's Engineering Experiment Station in Durham after hearing as early as 1946 about the exciting new biological research being done by Dr. Wilhelm Reich. In June of 1953 he became a research associate at the Orgone Institute Research Laboratories in Rangeley, Maine. He spent two exciting and unforgettable years in varied scientific and administrative work with Dr. Reich.

Anyone who doubts that the cloudbuster manipulates orgone energy should look at Robert McCullough. He will find its signature stamped forever on the organism of this big and powerful man. Bob got a "jolt" from the device in Arizona that paralyzed his right side, and to this day he still drags his right foot from that dramatic incident. When he originally wrote me to tell me that the cloudbuster in Arizona had brought UFOs like those in my photographs into manifestation, he really started something. Close friends ever since, we are associated today in the commercial application of orgone energy weather engineering.

Study of Dr. Steiner's work, the quiet teaching of Franklin Thomas and my working friendship with Dr. Drown had all combined with my own field work and experience to functionalize my thinking before I got into Reich's work. My mental concrete had already been shattered. Reich's findings and observations, instead of being a strain to follow and tolerate—as they so often are to conventionally-trained students—became to me a flowing delight. Here were new channels for my own mercury.

Had it not been for Thomas, Steiner and Drown, and the particular sequential manner in which they had influenced me, I would not have been able to carry the discovery of the orgone energy into ufology. Recognizing this, I have sought to lead the reader along the same pathway. This is bound to convey more of the core facts and processes than technical presentations devoid of any human context. Dr. Reich's work will be more easily comprehended in this way, than if split off from connection to an immense esoteric storehouse. Reich's discoveries lend a heretofore inaccessible practical dimension to the esoteric treasures of mankind.

As with Dr. Steiner and Dr. Drown, there is an acknowledged presumption on my part in attempting to outline, in a few thousand words, the monumental work of Wilhelm Reich. An adequate biography of Dr. Reich lies far in the future because almost everything he achieved stands outside the framework of the Old Knowledge. His discoveries evoke a neurotic queasiness among the mandarins of mechanistic science. To admit validity of his major discovery is to admit simultaneously that numberless thousands of physicists have obtained scientific credentials on the basis of laws that are now demonstrably false. Few there be with such courage. A number of highly qualified men have admitted as much personally to me.

Reich brought the orgone energy to human experience as a new force in nature. He also brought certain death to the old order. He laid the bedrock of practical etherian physics and the New Technology. His cloudbuster invention, many of which I have built, designed and used, is a crushing response to skeptics. Build one big enough and you can turn the earth onto new alignments. I know the catastrophic consequences of the misuse of this force from engineering work with units of extremely modest size.

This brief treatment of Reich's work will be directed mainly to the vast contribution he made to the solution of the UFO mystery. Nobody on this earth will understand UFOs in a technical sense unless they first master Dr. Reich's discovery of the orgone energy. Lacking knowledge of the energy itself, and the functional mode of thought that comes from its understanding, human beings can only stumble and grope as they try to penetrate the mystery with the old mechanistic knowledge. The manifold and multiform UFO phenomena are essentially *orgonotic* and therefore beyond mechanistic reach.

The humiliating impasse in which official science finds itself over UFOs stems from the nonexistence within its rubric of the knowledge essential to a fruitful assault on the problem. Existing scientific reception systems continue to block this knowledge. This blocking is in turn a social objectification of neurotic blocking in individual humans. Neurotic resistance of the *individual* to radical discoveries in living functioning is reproduced on an *institutional* scale in organizations such individuals ensoul. Reich elucidates this problem. Ufology has given abundant proof of its existence in the way UFO reports are handled.

Recognizing on the basis of his brilliant career in the behavioral sciences that neurotic and irrational individuals—in or out of science—could only react by wanting to kill his lifegiving discoveries, Dr. Reich originated the independent scientific discipline known as orgonomy. By this means, he published all his findings in bulletins, journals and books for other scientists to test and replicate. He laid the foundations in psychology, biology, mathematics, physics, meteorology, sociology, astronomy and astrophysics for a viable New Technology, Life-positive and Life-giving. He wrote twenty books and untold scores of articles. He left behind *100,000 pages of unpublished manuscript*—enough material for a hundred more massive books. By 21st century science he will be accounted the greatest natural scientist of this century, and one of the greatest men of all time.

Let us now trace his steps, and member him accurately into the mores and forces that ruled his age. We will learn a great deal about UFOs by seeking to reconcile the magnitude of his achievements and benefactions with his death in a U.S. Federal prison.

Wilhelm Reich was born on a thousand-acre farm in Austro-Hungary in 1897. Educated by a private tutor, his early years centered on the farm. Life, growth and the husbanding of the living thus lay at the core of his being. Reproductive processes, to the human understanding of which he would contribute so massively, were constantly before him throughout his childhood and youth. He was an infant who *looked directly at sexual processes*, cleanly and clearly.

Interested in and concerned with the practical study of living things, he studied plant and insect life and carried those studies on through to the numerous animals on the farm. Throughout his life he maintained an abiding trust in the integrity of farmers. By the time he was 17, both his parents were dead, and he ran the farm himself while continuing his education. Invading Russian troops destroyed the Reich farm in 1915. Stripped of his home and livelihood, he joined the Austrian army. He served on the Italian front as a lieutenant until war's end.

His youthful interest in living things further manifested In a desire to make medicine his career. He entered the University of Vienna in 1918. A gifted teacher, he utilized this talent during his university days to support himself. He tutored other students, organized a sexology seminary and inevitably became interested in the work of Dr. Sigmund Freud, then centered in Vienna.

Reich became a member of the Vienna Psychoanalytic Society, a practicing psychoanalyst in 1920, and Freud himself predicted a brilliant career for the former young farmer. Graduating in medicine in 1922, Reich was chosen by Freud as first assistant physician when the latter organized the Psychoanalytic Polyclinic in Vienna that same year. By 1924 Dr. Reich was not only a teacher at the Psychoanalytic Institute, but also was breaking new ground in psychoanalytic research.

There was no way that any young physician interested in psychoanalysis could have had a finer background—or a more authoritative introduction to his chosen field. He was the protégé of the master of psychoanalysis. To be professionally associated with Freud, and a member of the great man's circle, was a rare privilege. To the end of his days, Dr. Reich acknowledged freely Freud's greatness, and emphasized that his own work—including the discovery of the orgone energy—was rooted in Freud's epochal discoveries.

While working with Freud, psychoanalytic failures, the so-called "negative therapeutic reaction", attracted Reich's early research efforts. He abandoned the traditional Freudian practice of sitting behind the patient, and sat beside the patient instead. Facing each other, looking at each other, therapist and patient now had *contact*. In view of what we have illustrated earlier in this book concerning the eyebeam or visual ray, the merits of this improvement in psychoanalytic technique will be obvious.

Resistances by the patient were a well-known reaction in psychoanalysis at that time. These were handled in essence through the *transference* phenomenon. In transference, the analyst became a substitute for the hated or traumatizing person in the patient's past. The patient's reactions were thus "transferred" from the actual person to the analyst. Dr. Reich modified this procedure from his new perspective facing the patient.

In full contact with the eyes and face of the patient, Dr. Reich detected and attacked the resistances as they appeared. He described the attitudes of the afflicted person on the couch to the patient. Reich told the patient how he was holding his face, changing his expression or shifting his gaze or his body. Reich would adopt, in caricature, the patient's attitudes and expressions as a means of bringing home to the patient the external aspects of his resistances. In medical hypnosis nowadays, the therapeutic value of mimicry is well established. Reich was using this therapeutic device nearly 70 years ago.

These procedures went against classical doctrine, and there was opposition to his methods in Vienna. Reich was nevertheless encouraged by results to persist with his research. The systematic, successive dissolution of resistances led in due course to the basic conflict upon which the patient's neurotic character structure had been erected. Dramatic changes became evident in the patient at this point.

Dissolution of the resistances led to changes in the patient's character. He became capable of genuine transference—of laying the *basic* conflict on the analyst. These findings demonstrated that the transference hitherto sought so eagerly in psychoanalysis was in most instances not a genuine transference. Rather was it a subconscious stratagem of the patient to avoid unmasking the painful, basic conflict. Reich called the new technique *character analysis*, and it is described in his book of the same name. This work is the cutoff point for Reich as far as his acceptance by formal psychoanalysis is concerned. Thereafter, goes the official legend, the brilliant innovator became "crazy".

In the accounts of these findings in Reich's own books, and in the general literature of orgonomy, the reader is struck by the needle-sharpness of Reich's observational powers. He was only in his late twenties and early thirties when he pressed the frontiers of this new healing art forward. Most evident of all is his ability to look facts right in the face and tolerate them—especially the generally evaded facts of genitality. His wholesome childhood on the farm undoubtedly contributed to this ability.

In a book directed to the UFO mystery, perhaps it may seem strange that we should find ourselves delving into the discoveries of Sigmund Freud. Let us never forget, however, that UFOs have brought out widespread irrationalism in highly educated men of science. These evasions bear an all-too-clear similarity to the reactions of psychoanalytic patients in diverting attention from painful basic conflicts. *We must know why such people cannot face the UFO problem.* We can trace the etiology of this mass evasion of the UFO problem to the neurotic mass character, of which scientists are the possessors, the same as lesser mortals. This is why we now digress into Freud.

Freud's discoveries are as crucial to ufology as they are to the rest of modern life, which increasingly incorporates them into the art of raising children. With each generation, the children are progressively less traumatized and rigidized in infancy and childhood. Generations are beginning to appear almost incomprehensible to most people born prior to the Second World War.

Freud's study of the psychoneuroses—those with a psychic basis and psychic symptoms—forced him to conclude that these neuroses were rooted in sexual impulses and inhibited memories. Hypnotic techniques used by Freud early in his work brought such memories to waking consciousness. For reasons not recountable here, Freud dropped hypnosis and developed another method of exhuming the repressed material. He called this method psychoanalysis.

Free association in psychoanalysis presses the buttons on the patient's memory bank and brings the repressed memories and experiences to the conscious level. The conscious mind then arbitrates on this material, and the neurosis is thus gradually dispersed. At least, so the general theory and practice goes. Psychoanalysis has proved a valuable tool in assisting thousands of human beings, but for Freud personally it was the unleashing of a whirlwind.

Psychoanalysis revealed the overwhelming role played in adult neurosis by infantile sexual conflicts, and the central significance of infantile sexuality to an emotionally disturbed humanity. Freud found that the unconscious emotional life of human beings is a boiling cauldron of sexuality and aggression. These findings were unequivocal. Sexual fantasies, impulses, wishes and yearnings were inseparably interwoven with childhood memories and situations.

Most of these situations and yearnings had never been fulfilled except in fantasy, and so Freud concluded that in suppression and inhibition of infantile sexuality lay the origin of the neuroses. These findings in themselves would have been sufficient to bring down on Freud the wild wrath of society and especially of organized religion, but the steady yield of psychoanalytic research took him far beyond this point. Freud broadened the concept of sexuality to encompass regions from which even the most liberal churchman could only recoil in horror.

What adult life classified as perversions or aberrations were threaded all through the childhood material. Areas of the body other than the sex organs proved to be centers of sexual sensation in the child, including the mouth, anus, breasts and urethra, but not confined to these. Actions and situations furthermore, with no perceivable outward relationship to sexuality, produced sexual stimulation and sensations. Examples would be watching fires, rural panoramas, fights, and various forms of exhibitionism. Stimulation gained in this way could produce ejaculation in the male and something resembling orgasm in the female.

By the adult standards of Freud's day, and even by today's greatly loosened conceptions, these things were perversions and aberrations. Their presence in all children to a certain age meant that *from an adult viewpoint* the child is, in Freud's words, "polymorphously perverse". The natural sexual development of the child passed through three main stages, which were, according to Freud and in evolutionary order, oral, anal-sadistic and genital or phallic.

Orderly and healthy passage through the phases is essential for the emergence of a healthy adult human being out of the child. Through his clinical research, Freud showed that disruption and interruption of this development was due mostly to interference by parents and teachers who applied the culturally accepted adult standards of sexual perversion to the evolving child. This interference entered the life of the child with crushing authority and enormous force.

The sexual life of the helpless infant became an unequal battle against endless "no-nos." These encroachments aborted or obstructed normal development to full genitality. The more powerful encroachments blocked the child's development completely at the pre-genital level. The child, and therefore the resulting adult, became in modern general parlance, "hung up". Hence the cleavage between biological impulses and cultural requirements. The mass production of sexual inhibition has been the consequence.

Freud further conceived of the adult perversions as "partial impulses", which combined to form the total sex energy of the adult. Freud called this total energy the libido. Strengthening of any one partial impulse weakened the others. Out of this, Freud developed the concept of *sublimation*, or the transfer of part of the libido from sexual to nonsexual objects and goals that were socially acceptable. Freud delineated the inevitable conflicts between sublimation and sexual satisfaction, and between free sexuality and cultural demands.

For unmasking the terrible consequences of compulsive religion in the sexual traumatization of children—and therefore in the creation of the adult neurotic—Freud was calumnized by official religion. He had awakened the volcanic forces of irrationality and destructiveness that lurk beneath the social facade of man. Every man or woman who would help heal mankind in any decisive fashion has encountered these murderous reactions. Abrams, Steiner and Drown are three we have met in this book thus far who were similarly assailed and vilified by the highly educated no less than the lowly ignorant.

Freud's elucidation of the so-called *oedipus complex* excited further hostility. International rage ripped at the brave founder of psychoanalysis. He demonstrated with the steady harvest of clinical evidence, the sexual attachment between children and their parents—an attachment that on the child's part is initially unitary, being both tender and sensual. Suppression and crushing of the sensual element early in the child's life also suppresses the tender side of the child-parent sexual relationship. Only hatred, indifference or enmity remain. Every psychiatrist's office is full of it.

By splitting the emotional life between love and sensual desire, anti-sexual upbringing results in the celebrated polarity of Eros and Sexus. This tragic schism in the emotional life produces impotence with a loved mate, together with sexual desire for a partner without love. Prostitution is a typical social phenomenon sustained by this splitting of the human emotions. Psychiatrists' offices are heavily patronized by persons whose happiness is convulsed by its effects. Civilized living provides examples of it in abundance.

In the first quarter of the 20th century, psychoanalysis had pursued these general findings against a flood of opposition and abuse. Work continued nevertheless in the therapy of the neuroses, and in the application of psycho-analytic findings to the understanding of cultural life in all its respects. Freud's basic findings on infantile sexuality were beginning to find effect in new methods of raising children—methods that would reduce adult encroachments on infantile genitality. Freud himself was convinced that a pathway would be found from psychology to biology, but he could not himself make the breakthrough.

The dynamic figure of Wilhelm Reich—Freud's protégé—now leaped upon the stage. Before his final curtain he was to change the drama of mankind. His 1925 orgasm theory was the first step on that pathway from psychology to biology that Freud said would be found. In due time, that same orgasm theory and the extensive bioenergetic research based upon it, would lead to accurate functional comprehension of spaceship propulsion, biological UFOs, typhoon and hurricane formation, control of the world's weather, and above all, to the demonstrated need for society to secure the genital rights of infants and adolescents. Reich was the first scientist to claim these rights for the innocent young. All these achievements stand among a thousand other compartmented mysteries of science and philosophy that can now be functionally connected and understood in a new way.

The world has now largely accepted Freud. In some would-be super modern quarters he is considered an old-fashioned fud. Children in western nations are increasingly raised on the bedrock of his basic findings. What Wilhelm Reich achieved rests on that same Freudian bedrock, and a century hence may well be recognized as the inevitable consequence of Freud's pioneering.

Reich's systematic study of patients who were not helped by analysis, or who later suffered a relapse, revealed a common element: the persons under study had failed to develop a satisfactory sex life. Patients aided by analysis had, by contrast, achieved a satisfactory sex life. Reich turned his energies on to the elaboration of this finding.

The idea of an *energy* was immanent in Freud's whole concept of the libido, and Reich discerned in the cases that gave a negative therapeutic reaction a stasis of libido—an energy stasis. What should move was stagnant. Sexual activity of the patient, in and of itself, did not conquer the stasis. Gratification in the sexual act was the key, and a patient lacking the capacity for such gratification could not achieve it by intensifying or expanding his sexual activities.

Reich termed the capacity for complete sexual gratification *orgastic potency*, and was at pains to differentiate it sharply from mere erective potency. To this day, the conventional psychiatrists and laypeople alike are befuddled on this issue of orgastic potency, as a study of contemporary literature, both professional and general, certainly reveals. When therapy led to orgastic potency, decisive changes took place in the individual under treatment.

Wilhelm Reich was the first scientist to undertake the systematic study of the orgasm. This primal convulsion, lying at the roots of life, had never received any scientific attention other than in the most gross and transitory way. Science in the time of Reich's early orgasm work knew far more about explosives and how to blow human beings to bits than it knew about the human orgasm. Since the energy in the organism was discharged by this process—quite aside from the sexual substances—there could be little doubt of the central role played in human sexual health by the orgasm. Up to Reich's time, nobody had asked the question, 'What is the function of the orgasm?"

Years of painstaking experimental work, including systematic galvanic recordings of the changes in human skin potential during sexual excitation and intercourse, revealed an unequivocal four-beat rhythm of tension, charge, discharge and relaxation. Reich called this the orgasm formula. The wave-form of this energy activity is the *kreiselwelle,* or spinning wave.

We have already met this waveform in Dr. Drown's instrument designs and in UFO functions. When Dr. Reich unlocked this central secret of bioen-ergetic activity, he gave psychoanalysis the most powerful forward thrust it had ever received. The libido, up to this time, had been regarded as an essen-tially psychic concept of sex energy, even though the workings of an energy were implicit in the concepts of blocking, splitting and sublimation that Freud had formulated. Now there were galvanic tracings demonstrating ener-gy activity in the human organism, during sexual orgasm, that followed a lawful but heretofore undetected pattern.

Out of this discovery there arose quickly a much more massive challenge. The bioelectric tracings were galvanic—of the order of millivolts, thousandth parts of a volt. An electrical potential of forty millivolts—forty thousandths of a volt—would not even light the tiniest penlight. Energies involved in the orgastic convulsion were obviously vast, and yet the bioelectric tracings were miniscule. The libido was demonstrably—via these tracings—a real energy. The tracings were minute indicators of a far more massive energy metabo-lism at work in the living human.

Reich also found that discharge of the excess energy in a gratifying sex-ual act stabilized the energy level in the organism. This regulation of the whole energy household was the true function of the orgasm, rather than the mere hurling out of the sexual substances. Clinical experience verified that when patients were led by analysis to the establishment of *orgastic potency as defined by Reich*, their neuroses could not be maintained.

With the dispersal of the neurosis in the individual came new social atti-tudes, including a new morality anchored within the personality. The renewed human being wishes sex only with one whom he loves. He per-ceives the irrationality of compulsive social mores, and finds it incompre-hensible that millions of people are brainwashed into such behavioral codes. Pornography is distasteful, and sexual promiscuity of no interest. Tolerance develops towards the sexually perverted. Dr. Reich termed this kind of sexu-ally healthy individual *self-regulating.*

Reich further found that the traumatizing incursions by the adult world upon infantile sexuality—all the "don't touch it" and "no-no" edicts imposed on the child—had definite physical effects on the infant's body that were car-ried into adulthood. The prohibitions and denials became somatically anchored in the musculature, even though repressed from conscious memo-ry. Through the autonomic nervous system, chronic muscular contractions were set up by these adult invasions of the infant's genital rights.

In their totality, these contractions became a permanent defense against giving or feeling on the part of the individual. The retracted pelvis, tightened buttocks and clenched teeth were typical manifestations. Dr. Reich called these muscular contractions *armor*, a singularly appropriate name. Rigidity of thought and behavior were the external expressions of the armor. This was the somatic bedrock of the neurosis, and muscular rigidity precluded orgastic potency—the discharge of the excess energy of the libido.

Reich eventually identified seven "armor rings" in the human organism. These are the ocular, oral, cervical, thoracic, diaphramatic, abdominal and pelvic rings. One or more may be implicated in a neurosis. These chronically present contractions block the normal flow of the life energy through the organism to the genital—a specialized organ for the discharge and regulation of the energy. The eye block is of special significance in ufology, for nobody can "see straight"—optically and conceptually—with an armored eye segment. Human armoring is a universal phenomenon, and it influences significantly the way we perceive things—including UFOs pervaded with life.

When repressed conflicts were resolved, in analysis, the muscular armor dissolved and orgastic potency was established. The individual's body lost its rigidity, the face become more mobile, and the patient became able to give. A notable early finding of Dr. Reich in this connection was the *terror* manifested by every patient when the end phase of analysis was reached and the muscular armor dissolved. In a body conditioned to sexual stasis and blocked energy flow, movement of the released libido energy produced *sensations of terror and a desire to clamp down*.

The same energy form that Wilhelm Reich was dealing with in these discoveries is involved intimately in everything to do with UFOs. Orgone energy propels those UFOs that are craft, and orgone energy animates the critters. Orgone energy is involved with our eyesight and with the bringing of images into consciousness. Many of these manifestations of orgone energy occur in great strength. These manifestations bring about a corresponding movement of the bioenergy of every neurotic individual with access to them—that is, of mankind in the mass. Like the patient who cannot stand the motions of his own energy, we have seen in everything to do with UFOs, *irrational clamping down*. Those who believed this clamping down was mere censorship will see that it has much deeper roots, and it is Reich alone who has led us to them.

There is much more to Dr. Wilhelm Reich that we must deal with before we proceed to his involvement with UFOs as an investigating scientist. What we have thus far laid bare of his early work is rifled right into the social problem of UFOs, and to understanding why people have reacted irrationally to this latter-day phenomenon. As we proceed, Wilhelm Reich will truly appear as a man of the future, slaving to serve a century deaf, dumb and blind to his genius and its fruits.

Chapter Eighteen

FROM ORGASM TO UFOs

No unimportant man is hated and persecuted as Reich was.
—A. S. NEILL

Dr. Wilhelm Reich's first book, *The Function of the Orgasm,* was published in 1927. Updated versions of this book are on the world market now. *Function* is Volume 1 of Reich's *The Discovery of the Orgone,* and it members this distinguished behavioral scientist into his time and into his age as a man with exceptional power to break new ground. Dr. Reich's unique personal unfoldment saw him gain an outstanding reputation in psychoanalysis at its world epicenter during the lifetime of Freud himself. That this same man was later to become personally and scientifically involved with the UFO problem—on a scale not even remotely approached by any other scientist—makes him worthy of the most careful study.

Because so much of the UFO mystery involves pathological behavior of all kinds, every person with clear judgment will recognize the importance of the behavioral element to every investigator and thinker. The original investigations and findings of Reich in both biopsychiatric and biophysical fields are sharply germane to the UFO field in all its aspects. Orgasm and UFO are closer to each other than is realized by today's narrow-spectrum UFO investigators.

Volume 2 of *The Discovery of the Orgone,* entitled *The Cancer Biopathy* (long out of print after the Federal burnings but recently reprinted) describes, for example, a number of Reich's inventions that have application to UFO investigation and which are part of the New Technology. These include the orgonoscope for visually detecting the orgone energy in the atmosphere, and the orgone field meter for measuring the extent of life fields around organisms. The orgone accumulator is also described together with its method of construction.

Those who rush to relieve Reich's work of this store of seeming goodies, all ready and available for use, will not get very far with any effort to split off these devices from the whole body of thought that gave them birth. Every civilized human of reasonable intelligence who studies Volume 1 and 2 of *The Discovery of the Orgone* will find something in these pages much more important than Dr. Reich's inventions. He will find himself, or some part of himself, leaping out of those same pages. The UFO problem will begin to appear to all who can face what they find about themselves as a problem in human functioning, with the blockages to its solution lying within ourselves. This is why Dr. Reich's discoveries in human functioning are so crucial to the UFO field, and why such detailed attention is given to them here.

Dr. Reich established in his pioneering work three major aspects of human functioning:

1. The energetic reality of the libido as an energy flowing in kreiselwelle (KRW) waveform.

2. The function of the orgasm in regulating the level of bodily energy.

3. The muscular armor as the physical agency by which the natural flow of energy is blocked or impeded in the human organism.

As part of the work that led to these findings, Dr. Reich originated the first substantial efforts to apply psychoanalytic knowledge to mass problems. He organized the workers' psychoanalytic clinics in Berlin for the Communist Party in the pre-Hitler period. His ideas and findings, which lead in the long run to true individual freedom in the sense of self-regulation and responsibility for oneself, diverged from the Red line. With Red dogma being rooted neurotic-compulsively in blocked genitality—as Reich showed in his book *The Mass Psychology of Fascism*—his separation from the Communist Party was inevitable. He was expelled in 1933.

In *The Mass Psychology of Fascism*, recently reissued by Noonday Press of New York after being burned with his other writings, Dr. Reich delineates the mass character that stands behind fascist dictatorships. Basing his work on clinical observations of German workers, he accurately defined how individuals lacking the capacity for self-regulation would follow irrational leadership—right through the gates of death. In the Second World War millions of such individuals perished, taking with them millions of innocents. A Hitler need do no more than mobilize and focus the colossal, otherwise blocked energies of the multitudes to shatter and realign all the relationships of earth life.

Hitler's advent in 1933 compelled Reich to leave Germany. Moving initially to Denmark, he was there squeezed again by Nazi pressure, crossed to Sweden and thence to Norway. His years of teaching and research at the University of Oslo's Institute of Psychology were among the most productive of his career. He was hot on the trail of the energy of the libido—the life energy—and made the breakthrough from psychology to biology and biophysics during this time.

Galvanic tracings proved that pleasure produced a small electrical charge at the skin surface, coincident with the expansion of the organism and proportionate to the pleasure. This charge disappeared in the presence of anxiety. These galvanic tracings proved to be a lawful, objective expression of basic biological functioning, and a further key to the function of the orgasm itself.

A satisfactory sexual experience caused this energy first to manifest at the skin surface, and then to discharge through orgasm. Anxiety precluded the appearance of the energy at the skin. Reich was thus empirically led to regard the genitals as a specialized skin organ capable of discharging energy and thereby regulating its level. These were new findings of tremendous import to all the life sciences.

The theorizing, intellectualizing and groping of formal psychoanalysis, together with the internecine struggles that raged between its leading figures in these years, were being left behind. Reich was plunging into biophysics. Each small finding raised further questions.

Between the miniscule electrical charges at the skin surface and the immense energies at work in the orgasm there was a contradiction. Such small energy charges could not possibly produce a convulsion of the whole human plasma system. The question forced itself again and again on Reich: *What was this life energy and where did it come from?*

Conventional thinkers maintain that we get our energy from food. While this may be partially true, advocates of this explanation remain unable to impart movement to a corpse by stuffing it with food. There is something else involved with the animation of the living. Reich's attempt to find out where the life energy came from led him, by the obviousness of the food theory, to study the breakdown of food.

Under sterile conditions with tight controls, and using a microscope magnification of 2,000–3,000X, disintegrating food was observed to produce tiny, luminous, blue-green globules that were motile and capable of culture. Dr. Reich called these pulsating energy vesicles *bions*. Emergence of these bions was systematically recorded on motion pictures, and every stage of the transformation was thereby objectified.

Germs brought near bions were killed. Dr. Reich further found that anything that can be made to swell and break down—including sand, soil and coal—will create bions. Objections naturally arose to all this, because the nonliving was appearing—seemingly—as the source of the living. Existing concepts of the origin of life were being opened to serious empirically backed questioning. Objections contained frequent references to so-called "air germs" as a contaminating source of the life-forms emerging from disintegrating mineral and plant substances. Reich's controls were complete, but he heated his materials to incandescence and still the bions appeared.

The work of British biochemist Morley-Martin in resurrecting a veritable microscopic zoo from azoic rocks, under conditions of total sterility, and the similar resurrection of living forms from the Bavarian salt deposits have already been discussed. We have come upon Reich's bions with this preparation—with our concepts of life and death changed. The bions should therefore come as no surprise, although to this day they are formally ignored or verbalized away by neurotic mandarins of science whose cosmo-conception has now been completely undermined.

The bions were discovered originally by H. Charlton Bastian of France, a contemporary of Pasteur and also something of a competitor of Pasteur's for formal scientific approval. Bastian wrote about his bions in a book called *The Beginnings of Life*, connecting his discovery with germs and disease conditions. Official science preferred the mechanistic fixity of Pasteur's conceptions, where germs exist immutably and are killed by various agencies. Pasteur won immortality and Bastian went to oblivion, although 21st century science will undoubtedly set right this historical slight.

What is important to establish here is that official science plumps always for the lifeless, the sterile and the motionless. The same basic argument ("air germs") adduced to deprecate Reich's discovery of the bions has been adduced ("reflections", "ice crystals") to deprecate UFO sightings. The mechanism of evasion is the same in both cases, and by such illustrations the investigator may become more aware not only of the reality of these evasions but also of their origin—a whole world-order threatened at its bedrock.

Reich found that the bions degenerated into so-called T-bacilli (T is an abbreviation for *todt*, German for "dead"). When injected into mice, these T-bacilli caused cancerous symptoms. Healthy bions were an antidote. The invention of the orgone accumulator—later on—proved that it was the energy charge of the bions that attacked the T-bacilli. In this early work, however, the bions appeared less as energy carriers than as quasi-bacterial agents.

A frightening experience befell Dr. Reich in the winter of 1939 when he was studying the bions. His eyes began to burn from looking into the microscope, and he developed a chronic conjunctivitis. His skin also began to tan in the depth of the Norwegian winter. The bion cultures imparted strong charges, detectable with an electroscope, to nearby metallic objects.

What force—what *radiation*—had he unlocked from these microscopic vesicles? The dangers and terrors attending the discovery of radium by the Curies in France came vaulting into Reich's mind.

Protection would surely be found in shielding, he reasoned. Shielding was a logical technical step to take under conventional concepts of radiation. Dr. Reich expected by this means to be able to confine and control the energy from the bions. Building a metal-lined box, he was alarmed and astounded to find that this box *intensified* the radiation effects within its own periphery and produced inexplicable external effects in addition. A permanent temperature differential appeared above the top of the box, for example, which appeared to negate fundamental thermodynamics.

This attempt to shield bion energy actually led to the discovery of the orgone energy accumulator. A six-sided box consisting of alternating layers, from the inside out, of metal and insulating material, the orgone accumulator provided a simple means of concentrating the energy directly from the atmosphere. This development came later, when Dr. Reich could review and begin to unravel the skein of new findings originating with his bion discovery. As an unsuccessful effort to shield bion energy, however, the six-sided box initially produced chagrin and confusion.

The energy was intensified in its manifestations by the box. Furthermore, the energy appeared to be everywhere and in everything. No defense could be contrived. No adverse effects occurred, however, and slowly Reich's apprehensions subsided. Realization seeped into him that the energy was *universally present*, and that the bions had simply brought this energy into focus—and to human attention—by local intensification.

Dr. Reich identified this energy with the galvanic energies present at the skin surface. Later on in America (1939-40) he discovered the presence of the energy in the atmosphere as a primordial, mass-free energy. He found this same energy also present in blood that was allowed to disintegrate. The connection to the single spot of dried blood, held in a blotting paper as a tuning crystal in Dr. Drown's work, will automatically commend itself to the reader at this point. The two titans, unbeknown to each other, were treading parallel paths. The presence of the orgone energy in disintegrated blood verifies from yet another angle the access given to the life energy continuum by the much-ridiculed single spot of blood.

Dr. Reich called this energy *orgone* to identify it permanently with things organic, with the orgasm and with life. He proved its existence visually, thermically, electroscopically, and later on, at the Geiger-Muller counter and by lumination in vacor tubes. *He was able to develop protozoa from the bions.*

This led him into cancer research, impelled by the prevision that cancer might have a similar origin.

Dr. Reich subsequently demonstrated that the etiology of cancer lies in sexual repression. Suffocation of the tissues and their subsequent putrefaction is due, according to Dr. Reich, to the armoring process—set afoot in infancy and culturally sustained. The cancer scourge is the consequence, raised to new heights in the 20th century by multitudinous environmental factors, including junk foods, nutritional ignorance, and carcinogenic agents reaching humanity via many routes. While a review of the cancer problem lies beyond this book, worth recording here is Dr. Ruth Drown's view of cancer. She held that it originated in the inability of the life energy to freely and fully penetrate the muscular systems of the human body. Dr. Steiner in his medical lectures similarly characterizes cancer as a disease of frustration.

Dr. Reich's bion work—despite its transcendental importance to the life sciences—provoked the hostility of group of psychiatrists at the University of Oslo. Unable to contend on a rational, scientific basis with the objective evidence that Dr. Reich was accumulating, these little men hit below the belt. They instigated a slanderous campaign against him in the Norwegian press.

A university education and qualification in psychiatry are not guarantees against irrationality today, and they were not in Norway in 1939. Dr. Reich's prior experience with destructive irrationalism on the social scene led him to ignore these attacks completely and to concentrate on his research and its bounty of new discoveries. Newspapers are not a suitable forum for either presenting or defending scientific findings.

Amid these storms, America beckoned. Dr. Reich was invited to lecture at the New School for Social Research in New York City. He accepted, and his American odyssey began as Europe disappeared under the clouds of a fearful new war.

In the U.S.A., Dr. Reich quickly found his feet and continued his pursuit of the life energy. He invented the orgonoscope for detecting the energy visually in the atmosphere, and the orgone field meter for measuring the extent of orgone energy fields. The confusing early effects of the six-sided box with which he had sought to shield the bion energy were consolidated into a major scientific discovery as the orgone energy accumulator. He proved decisively that the accumulator produced a temperature differential "out of nothing". This spectacle of a cold body warming up a hot one shakes the whole foundation of thermodynamics, since it contradicts the Second Law of Thermodynamics, one of the granite plinths of mechanistic science.

The implication of the Second Law of Thermodynamics in its widest sense, is that the entire universe is running down. Energy is being distributed through space randomly, the process is increasing, and ultimately there will be nothing but random radiation where once there was matter. The orgone accumulator demonstrates that this principle is empirically incorrect in certain specific arrangements of materials. Since no prior exception has been found to the Second Law, no one will believe that the law is wrong, although it is a statistical law and expresses only the most probable behavior.

Reich's original discovery of the orgone accumulator is only a beginning. Subsequent workers seem to have concentrated on replicating Reich's findings with the orgone accumulator, but also seem to lack direct contact with the energy. Any reasonably sensitive person who will build and play with orgone accumulators free of preconceptions will find this effect of heat appearing "out of nothing" to appear in all sorts of ways—all of them destructive of the Second Law of Thermodynamics.

In working with Dr. Reich's cloudbuster invention, and many derivatives and variants of such apparatus, for over thirty years, I can attest to the numerous magnetic and thermic anomalies created by their operation. These effects literally thrust themselves at a perceptive operator. By chance—just through participating instead of arguing—I found that when an orgone accumulator is suspended over flowing water its *underside* develops warmth. There appears to be no reason why such a differential could not be amplified to furnace levels by experimenting with both water flow and layering variations in orgone accumulator construction. As with so much else of Dr. Reich's, the problem is not data or hardware, but the threat these fundamental biophysical discoveries pose to the products of our modern day parrot-education.

As his work in the U.S. proceeded, and became ever more fruitful along bioenergetic lines, Dr. Reich saw the emotions as manifestations of a specific biological energy, blocked and impeded to produce various types of character problems. He developed techniques for freeing the orgone energy from these blocks by direct means, rather than via the roundabout psychoanalytic route. These techniques are known today as *medical orgone therapy.*

Dr. Reich trained many students and physicians in his methods for physically freeing the muscular armor, as well as in character analysis. Medical orgone therapists were required to repeat his laboratory experiments through which the orgone energy was discovered. This careful and thorough regimen included doing Dr. Reich's Experiment Twenty, which proves that protoplasmic matter can develop from concentrated free orgone energy in the accumulator. From the resulting plasmatic, bionous matter, protozoa may develop.

Medical orgone therapy also sometimes includes the use of a man-sized version of the orgone accumulator. When used therapeutically in this way, the basic six-sided box with a metal lining is increased to a volume sufficient to accommodate a human being. One side is hinged to make a door so that the patient may enter, sit on a small bench, and after closing the door, irradiate his organism with the concentrated orgone energy that the accumulator produces. Here again, one must *do*, and the doubts then take care of themselves.

The simple layered construction of the orgone energy accumulator, like the simple construction of the Drown instrument, have led to automatic conclusions on the part of complex-minded physicists that no such device can have any value. These people fail to perceive the corresponding simplicity of the voltaic cell, or of the simply layered carbon and uranium in an atomic pile. The layered structure of the electric eel, which also produces a lethal voltage "out of nothing", similarly eludes their biased gaze.

The only way to evaluate the orgone accumulator is to sit in it for extended periods, making full use of one's God-given vision. After an hour in the accumulator,[1] when your eyes have become completely attuned to this new environment, you will see luminous particles come spinning out of the walls. Their trajectory is unmistakably that of the spinning wave or KRW. As you

watch those spinning particles you will become aware that an identical trajectory is traced by every planet as it spins around the sun. The same trajectory is traced by the typhoon as it barrels across the western Pacific and curls
northeastward before dissipating. You will think on these things, and also on
the singularly cogent fact that the same waveform, expressed in the orgasms
of your parents, produced *you*. You in turn reproduce your kind by the same
energy moving in the same way.

Whatever is rational in you will then ask how it is possible for the man
who forged these mighty cosmic links for humanity to die in a Federal prison.
You will begin to wonder what forces of malignant savagery inspire the
smashing by government agents of such accumulators as were built by Dr.
Reich, devices that demonstrate directly to you that you and the cosmic
expanse *pulsate together*.

You will also now be able to synthesize into all this the propulsion systems of the ubiquitous discs in the skies. With their spinning orgonotic fields
around them, they too trace KRW forms in the ethers as they flit across our
skies. Any theory that can bind together the origins of physical life, the
weather of the world, the orbital paths described by the planets, the human
orgasm and the flying discs, can hardly be an accident or a coincidence, but
rather, the yield from organic, unrelenting study of objective facts and
processes.

Dr. Reich naturally expected that his contributions to a new physics,
rationally replacing the old physics in due and orderly course, would be
acknowledged by his contemporaries. In December of 1940, he wrote to
Albert Einstein and asked for an appointment to discuss his findings with the
great man. Professor Einstein agreed to see him, and received Dr. Reich on
13 January 1941, an interview for which Dr. Reich had carefully prepared
himself. The two men spent almost five hours together, an indicator of
Einstein's interest in Reich's findings, since the former was busy with matters leading to the atomic bomb.

Professor Einstein was willing to further investigate Reich's discoveries.
With the lights out in Einstein's study, the great mathematician had seen the
scintillations of the orgone energy through the orgonoscope. When Dr. Reich
told him about the temperature differential above the orgone accumulator,
Einstein said that if true, "it would be a bomb in physics!" Reich left the
orgonoscope with Einstein, and the latter agreed to check the temperature differential above an orgone accumulator that Reich would build for him.

Two weeks later, Dr. Reich delivered the small orgone accumulator in person to Einstein's home in Princeton. Placing the accumulator on a table in the
cellar, two thermometers were used to establish the temperature differential.
One thermometer was placed in a tube above the accumulator, the other was
suspended a yard away in the free cellar air. Professor Einstein within a few
minutes confirmed the presence of the thermic differential. Reversing the two
thermometers, the differential was again produced.

Professor Einstein was intrigued by the differential he had himself read
directly from the thermometers. He asked Dr. Reich if he could keep the
orgone accumulator for additional studies, promising to write him within a
few weeks concerning his findings. Reich agreed and the two men parted in
mutual goodwill.

On 7 February 1941, Professor Einstein wrote to Dr. Reich providing an explanation for the temperature differential that had been adduced by an unnamed assistant. Einstein's conclusion was that the anonymous aide's explanation of the thermal differential was satisfactory. By implication, the orgone accumulator did not work as Dr. Reich had claimed.

This scene by mail almost exactly acts out in the real world and with real people, the kindred events dramatized in Steiner's mystery plays. Strader, the inventor of devices of a new character and utilizing a new natural force, is impugned by an Ahrimanic know-it-all who could easily have been Einstein's assistant on this occasion. Everything of Strader's is "explained" in this same evasive fashion. Everything concerning UFOs since the Second World War has been similarly verbalized away by similarly inspired scientific know-it-alls.

Dr. Reich, like Strader in the mystery plays, was able to completely rebut the spurious explanation by the anonymous know-it-all, but it mattered not. Silence descended on the relationship from which Dr. Reich had expected so much.

Months later, Einstein returned the orgone accumulator and, after some prodding, the orgonoscope. There was no further commerce between the two men, to Reich's intense disappointment. Once more, the orgone energy had provoked the irrational in a distinguished man of science. Two years later, Alamagordo, Hiroshima and Nagasaki showed where Professor Einstein's head was—alas for poor mankind.

The orgone accumulator continues to produce heat out of nothing. In a man-sized accumulator[2] you can feel the energy coming off the walls with your hands. Your body gets warm and tingly as you sit in the box. Stay in it long enough and your skin will get red all over your body. There is no doubt about this, once you expose yourself to the phenomena.

Reich found that cancer mice kept in an orgone accumulator outlived healthy mice that were kept under normal conditions. Burns were dramatically aided by the orgone accumulator, which effected prompt elimination of pain or its significant reduction, and promoted rapid healing. Human cancer cases did not develop anemia, and many remarkable results were achieved in the reduction of tumors. Dr. Reich spelled out the story of his clinical work with cancer and the orgone accumulator—successes and failures alike—in both *The Cancer Biopathy* and in articles in the *Orgone Energy Bulletin.*

In 1947, Dr. Reich identified the Emotional Plague of Mankind (abbreviated EP), a disease of bioenergetic equilibrium that exists in the mass character by virtue of its well-nigh universal presence in individuals. Social irrationalism and destructiveness are its main characteristics. The EP is rooted in unsatisfied orgastic longings, and in the general corruption of sexual health by socially entrenched irrationalism and time-honored but life-negative institutions. The EP has manifested prolifically in UFO affairs since the earliest modern advent of UFO phenomena.

Suppression of information, ridicule of investigators, obstruction and subversion of rational work by governments themselves riddled with and addled by corruption, exemplify the EP on the UFO scene. Any investigator who gets on the true tail of the UFOs through their bioenergetic rootings and spiritual connections, will feel the full force of the EP. His telephone and his mail will be monitored. If he has evidence, his files will probably be rifled by government burglars.

While Dr. Reich was doing his giant's work, so vital to the human future, he was pilloried in 1947 by Mildred Edie Brady, a woman writer who had no factual knowledge of his work. Here again was the Emotional Plague. Dr. Ruth Drown was similarly assailed throughout her life by ignorant writers. Brady immortalized herself—in a negative way—by inferring in a *New Republic* article that Dr. Wilhelm Reich was a sex racketeer.

Dr. Reich had devoted over a quarter of a century to the scientific investigation of sexuality from the Freudian bedrock, as outlined in this book. He had published his findings. He had made monumental contributions to the therapy of the neuroses. He had opened the way to the *only* rational cure to what ails humanity—prevention of sexual crippling in babyhood, infancy and adolescence.

Pornography had been shown to have its roots in the secondary, perverse drives that develop from the blocking of the primary, natural drives. He had contributed more to accurate scientific knowledge of human sexual functioning and all its ramified socio-economic implications than any man living. Yet it was implied by Brady that he was operating a pornographic racket.

Brady's article captured the interest of the Food and Drug Administration, which at its lower levels employs hoodlum-type personnel who have always been allowed far too much latitude by their superiors. These cretins came to Dr. Reich seeking the pornographic literature they felt sure he must be disseminating. Scientific information on the orgone energy accumulator and orgone energy research was of no interest to the FDA men. In the absence of any evidence whatever of pornographic works, or against the orgone accumulator, the government investigation retreated behind the scenes. With Dr. Reich being clean, honest and upright, there was no other place for the functionaries of darkness to go but behind the scenes—there to prepare a lavish ambush,

The FDA was from this time forward out to destroy Wilhelm Reich, undoubtedly inspired and encouraged by Big Medicine. The FDA eventually used against him the identical legal stratagem employed in its suit against Dr. Ruth Drown—the interstate shipping laws. There is no little significance in the FDA electing to pursue this course after its 1951 success against Dr. Drown, which smeared her and crippled her work. The orgone accumulator offered them a chance to employ again the mislabeling provisions of the interstate shipping laws. The precedent had been established.

There is no space here to review the enormously complex and emotional nature of Dr. Reich's trial. The ordeal imposed on a private citizen by having to fight the government with all its resources is an overwhelming event in itself. The nature of Reich's work and his heavy involvement with the UFO problem made his ordeal all the more horrendous. While every city in the United States crawled with gangsters, thugs, embezzlers, thieves and crooked politicians, the government felt its proper target was an internationally renowned man of science. Only the Emotional Plague can rationalize—and legally sanctify—such monstrous irrationality.

The FDA complaint charged that Dr. Reich had entered devices into interstate commerce that were "misbranded" under the food and drug laws. The

FDA view was that if Dr. Reich sent accumulators to licensed physicians in other states so that clinical information and evidence could be gathered, or if he otherwise leased or rented them for use under a physician's supervision in another state, this constituted "shipping". The FDA said, in effect, that there was no orgone energy and that therefore the devices were misbranded.

In yet another echo of the Drown case, the FDA wanted all orgone accumulators destroyed, all of Dr. Reich's ten major books banned, and all scientific literature not in clothbound form burned. A vast experimental record in bulletin and journal form would thus be consigned to the fire. The U.S.A., to its own immortal shame, carried on in the tradition of history's book-burners. Dr. Reich was also to be banned from disseminating in any way, to anyone, information regarding orgone accumulators and kindred apparatus. This provision would legally restrain him from taking the orgonoscope or accumulator to a Professor Einstein or any other man of science. These strictures were similar in scope to those sought from the judge in the 1951 Drown case by the Federal prosecutor. All of this happened in America.

In a written response to the FDA charge, Dr. Reich asked that the matter be taken out of court completely because he held that courts were not empowered to arbitrate or regulate basic natural scientific research. The judge maintained the court had jurisdiction and a legal treadmill was set in motion that eventually led to Dr. Reich's being sent to a Federal penitentiary. He was legally outmaneuvered by the government, whose prosecutor was Dr. Reich's own former foundation attorney—one Peter Mills. Here again, the Drown case echoes. In the second assault on the Drown work, the judge assigned to her case was a man who, at a lower legal station and in a 1951 fit of frustration, had publicly sworn to "get" her. This judge had to be removed from the bench in her second trial—a sickly, unchivalrous spectacle.

Dr. Reich's actions in and reactions to these convulsive legal matters will never be understood aright without considering what he went through with the UFOs—or the manner in which the UFOs impinged upon his life and work. Glib judgment has been plenteous on the Reich trial and on his allegedly deteriorated behavior. We will gain a better perspective on all of it if we proceed with a review of his later technical and scientific discoveries.

Dr. Reich discovered the motor force in orgone energy, and successfully ran an orgone energy motor in 1948. A group of specialists known to me plans the rebuilding of the orgone motor, with a view to the permanent conquest of the fuel and energy problem. The power principle of the motor was the excitation of an orgone energy accumulator by a half volt of electricity, and the combination ran a 25 volt motor.

Qualified individuals known to me personally, saw the motor running many times. The device was essentially a biomechanical reproduction—in reverse—of the situation that exists with the millivolt galvanic charges appearing at the human skin from a primary bioenergetic power source sufficient to propel and convulse a 200 pound human being. From a half volt input (500 millivolts), Reich could run a 25 volt motor with the power developed by the orgone accumulator, or transduced by the accumulator from the primary energy continuum.

Here is the beginning of a new type of propulsion, obviously stemming from and applicable to space. So also is it the writing on the wall for fossil fuels, energy monopolies, the manipulation of people by greedy sheiks or amoral American oil barons, and the pollution of the earth. Dr. Reich had developed an essentially fuelless motor.

Starting the motor initially involved what Dr. Reich called the "Y factor", something he never disclosed. Dr. Rudolf Steiner dramatized the forthcoming advent of this motor force in his mystery plays 40 years earlier. Strader, the inventor in Steiner's prescient dramas, could in many ways have been the foreshadower of Wilhelm Reich. Dr. Steiner's spiritual scientific conceptions give access to hidden qualities of tone unsuspected by official science, and studies of this material—as well as practical experience with the cloud-buster—would indicate that starting the orgone motor involved *a specific movement of the human form.* An etheric shock effect can be produced in this way, capable of energizing the entire assemblage. Thereafter, it moves unaided and indefinitely.

In 1948, Dr. Reich also produced lumination of concentrated orgone energy in a vacor tube. This experiment had far-reaching implications in the then-gestating space age. Twenty years later, the Apollo moon missions were giving unwitting proof of his lumination discovery. The lumination of the energy in a vacuum proved that the orgone energy could exist in space. As has been mentioned in earlier chapters, the astronauts walking on the moon showed blue lumination around themselves, as they excited their personal orgone energy fields with their voice-powered radio transmitters in the vacuum of the moon's surface.

All this unwitting proof of the value and prescience of Reich's work passed over the heads of NASA physicists. The spurious nature of the blue luminescence around the moon walkers—outside their space suits that were supposed to confine them completely[3]—made the problem all the more difficult for those without understanding of the orgone energy and its characteristics. U.S. government agents had burned the experimental data on luminating orgone energy *in vacuo* ten years previously!

Research into orgone energy functions repeatedly led to the conclusion that an antagonistic relationship existed between orgone energy, which is pre-material in the atmosphere, and electromagnetic energy, which is post-material. The lumination experiments further demonstrated the ability of electrical energy to excite orgone energy. Dr. Reich decided late in 1949 to investigate further the relationship between orgone energy and electromagnetic energy in the form of atomic radiation.

On 5 January 1951, Dr. Reich put one milligram—just one-thousandth of a gram—of radium, sheathed in its lead container, into a twenty-fold orgone accumulator, i.e. an accumulator made up of twenty layerings of metal and insulating material. The package was then placed inside an orgone room, which is a metal-lined room with plywood or other insulating material adjoining the metal on all six sides. Reich left the package in the orgone room for five hours. He repeated the process daily for a week. On the last day, he left the radium and twenty-fold orgone accumulator in the orgone room for only half an hour. Conventional physics would opine that nothing could possibly happen under known laws of nuclear physics. What actually did happen blew away the foundations of that lethal and uncertain science.

The story of this event is written up fully in Dr. Reich's booklet "The Oranur Experiment—First Report 1947-51", since burned by the Food and Drug Administration. Suffice it here to quote a brief description of what occurred, by a participant eye-witness, the late Dr. Elsworth F. Baker. A distinguished author and psychiatrist, Dr. Baker remained meticulously loyal to Dr. Reich and his work until his own passing. In his final years, Dr. Baker wrought the miracle of orgonomy's resurrection, after the debacle of Wilhelm Reich's death in prison. Here is what he says of this dramatic period:[4]

"I know what happened. I was there. The count on the Geiger counter went up alarmingly and finally jammed. The building and the atmosphere around it glowed at night. One physician went into shock and nearly lost her life when she put her head in a metal cabinet in the laboratory. The mice died, and a peculiar, sickening, acrid odor pervaded the atmosphere while clouds hung over the area constantly. Reich fell ill and hovered between life and death for weeks. The whole area became uninhabitable. Although the radium was finally removed to a place eleven miles away, the process did not let up. I am confident that one day science will discover that this type of reaction accounts for the radioactive layer above our atmosphere (The Van Allen Belt) due to cosmic rays meeting the earth's energy envelope."

In the wake of this shocking, tumultuous experience, UFOs began to appear around Orgonon, Reich's 260-acre estate in Maine. Today there is little doubt among those familiar with UFOs and with orgonomy that Dr. Reich had heavily disturbed a fundamental interface of nature. By concentrating both the primary, mass-free energy of space, and the secondary, post-material energy of radioactivity together, he had signified that sleeping mankind was waking up to the cosmos.

He had aroused the interest of certain entities riding in spaceships. That they were what we now know as the Boys From Downstairs also seems beyond doubt. Before long Reich was locked in a new kind of cosmic struggle, one that few people could have comprehended prior to the writing of this book. The young, brave, supernally brilliant scientist who had taken the first intelligent look ever at the human orgasm was now in the evening of his life to do battle with UFOs on desperately unequal terms. The entities he had to deal with were advanced masters of the orgone energy he had discovered. When he defended himself in the grim days that followed, he was defending his discoveries and his life's work. His opponents were out to push him—and all his works—right off the earth.

NOTES TO CHAPTER EIGHTEEN

1. The accumulator is used here for scientific experimental purposes only and not for medical purposes.
2. The accumulator is used here for scientific experimental purposes only and not for medical purposes.
3. See the Apollo 12 photo in Appendix I.
4. From *The Journal of Orgonomy*, Vol. 1, No.'s 1 & 2, Nov., 1967.

Chapter Nineteen

COSMIC BREAKTHROUGH

Without my intention, somehow a ball of history started rolling, putting me in the center of space problems.
—WILHELM REICH, M.D.
CONTACT WITH SPACE, P. 2

When scientists amenable to negative psychic control and devoting their knowledge to death and destruction developed nuclear bombs, armored man had acquired the direct means of planetary suicide. This external physical supplement to the indirect agencies of spiritual sabotage brought the subversion of earth evolution close to achievement. The technical feat of Hiroshima was inflicted on humanity even as the world lay in ruins from the greatest of all recorded holocausts. Soon afterwards, General Douglas MacArthur said at the surrender in Tokyo Bay:

"We have had our last chance. If we will not devise some greater and more equitable system, Armageddon will be at our door. *The problem is basically theological and involves a spiritual recrudescence and improvement of human character* that will synchronize with our almost matchless advances in science, art, literature and all the material and cultural development of the past 2,000 years. *It must be of the spirit if we are to save the flesh*". (Emphasis added)

World War II and its atomic climax nevertheless conveyed little in the way of lessons to armored man. Only a few swords were beaten into plowshares before Russia and America—erstwhile grand allies—turned anew to the forge. Vying with and lying to each other, they squandered fresh billions on an atomic arms race. "Overkill" entered the language to describe the number of times over each could destroy the planet. Slaughtered and maimed millions, disrupted commerce, bombed out cities and the unprecedented mass misery left in the wake of the world conflict, faded away from memory like all bad things.

This sick posture of the human race proved yet again the two inherent qualities of armored man: structural inability to perceive the truth, and limitless ability to evade the essential. Presidents and premiers, rulers and writers, commentators, statesmen and philosophers—the official cream of this planet's intellectual populace—could offer no workable, rational alternative to more of the same. In the new nuclear dimension, the destructively inspired now sought the ultimate capability: annihilation of an entire nation—or perhaps the world—at one blow!

Amid this monstrous irrationality, culturally sanctioned and publicly financed, the steady life-positive work and discoveries of Wilhelm Reich appear like beacons in the blackness. He had identified the Emotional Plague—the biopsychic rooting of this madness that consumes mankind. He had shown how The Trap is constructed, and how to get out of The Trap. As the first scientist to speak out for the genital rights of infants and children, he had already laid the scientific and medical foundation for that very improvement of character for which General MacArthur had called for at the end of World War II.

Reich had produced inventions with the potential to turn civilization toward a new, nonpolluting and cosmically sustained technology. In the orgone energy he had discovered a universal presence—a force—with the clear potential to nullify atomic radiation as well as to power motors. Without any conscious knowledge of spiritual science whatsoever, he had by consistent, accurate observation and unceasing experiment, come upon the fundamental polarity of physical existence: the ethers above, the dense intramaterial energies below, the physical world a medium state between these two sets of forces.

The Oranur Experiment nearly cost Dr. Reich and his coworkers their lives. When it was over, the basic antagonism of orgone energy and nuclear radiation was as clearly established as any fact in physics—and none has been as frightening to its discoverer. Today it is doubtful if more than a handful of human beings really understand the dimensions of Reich's achievement. In Steiner's work, the polar relationship between the etheric formative forces and the electrical forces united with physical substance is well delineated. Until Reich, however, no one had been able to rationally assemble equipment and design experiments that would demonstrate this polarity. Would-be UFO investigators who really want to know why the fields around ether ships interrupt electrical flow should give these matters some attention.

Dr. Reich had uncovered the technical means by which the curse of atomic weaponry might be lifted from the human future. Radionics was already in existence at the time of the Oranur Experiment. Before too long, some questing young genius with the right motivation will synthesize these two technical streams. He or she will discover a primary tuning procedure by which atomic explosions may be aborted.

When passing UFOs halt electrical flow in power lines, or automobile ignitions, they are demonstrating the principle—albeit inadvertently. Oranur broke into this central principle—perhaps also inadvertently. In all its direct primitiveness Oranur had shown how the effects of nuclear radiation could be mitigated and, with the example of the UFOs halting electrical activity to guide us, perhaps nullified. Electromagnetic radiation is a perturbation of what Reich called the orgone envelope or the orgone energy continuum. Every nuclear explosion takes place against a universal but as yet barely comprehended etheric, biological interface.

"For every action there is a reaction." So there was indeed and in fact a reaction to the death-and-destruction binge of armored man in the 20th century. Those pro-evolutionary forces of the unseen who would aid mankind to fulfill its destiny in freedom had been able to work with Dr. Reich at what was an unconscious level for him. He was the only man on earth in his time with the wit and the will to mount a countering, pro-life effort against the otherwise doom-bound course of Ahrimanically-inspired mechanistic science.

There is no evidence that Dr. Reich himself consciously subscribed to such ideas, although their infrastructure looms through his book *Murder of Christ*. The historical facts nevertheless provide proof enough that Wilhelm Reich was membered into a plexus of evolutionary events as a personality of signal significance. Driving forward through the decades in pursuit of the life energy, establishing for the first time in history where the *real roots* of human liberty lie, Reich appears in heroic dimension.

On the negative side of this same period, scientists whose native ability to feel and give had been crushed in infancy, lined up in their thousands to do what was literally the work of the devil. Had these fundamentally good men been told they were doing the devil's work, naturally they would have dismissed it as mystical balderdash. Ethically and philosophically, the life-killers also have elaborate, defensive rationalizations for their misuse of knowledge. They have such rationalizations today. Socially and culturally they are supported, whatever the nation, race or ideology. The rights of the working class of the world, the good of the party, patriotism, the defense of freedom, have all been adduced as rationalizations for the destructive misuse of learning and technology.

The main connection between the misuse of technical knowledge and the unlimited funding for its promotion and prosecution is not political, but spiritual. The political merely transduces the will of the Ahrimanic powers into our world and into human formats. Through their conscious earthly collaborators in certain secret societies, who are also in control of governments, and via the psychic, radionic-type control of vulnerable key personalities outside such secret societies, the Ahrimanic powers control the world of money—communist and capitalist. In accordance with their aims, they direct the power latent in large agglutinations of money to death-dealing and destructive ends.

Few people today realize that the Bolshevik Revolution in Russia was funded by Wall Street. Bolshevik Russia is an Ahrimanic creation, as Dr. Rudolf Steiner made clear near the end of his life. Since the Russian Revolution, Wall Street and the highest officials of successive U.S. governments, have taken every necessary step to ensure that Soviet Russia does not collapse. The suppressed books of Professor Antony Sutton tell the shocking factual story of key technology transfers from the west. Soviet Russia is maintained as a Frankenstein to terrorize the western world into enormous armaments budgets. Multi-billion U.S. armaments budgets fill the banks of the conspirators, even as they submerge the USA in debt that is beyond citizen comprehension. Debt is bondage.

This sordid program, conducted almost entirely from the shadows, has blighted the twentieth century, and cruelly distorted the cultural lives of both the American and Soviet peoples. To conceive of all this as a sequence of random events, without the influence of an overarching, extrahuman, non-mortal intelligence—destructive in nature and purpose—aids that intelligence. While these clandestine machinations were proceeding, Wilhelm Reich unlocked the liberating portals of experimental etherian physics out in the open—in the light.

Misuse of this particular technical realm lies behind the psychic control of human beings. Etherian physics is also the technical *sine qua non* of UFO propulsion. Similar significance attaches to its involvement in the basic invisibility of UFOs, and to those interdimensional transductions and manipulations that create most of the phenomena we call UFOs. Dr. Reich's fertility of mind, functional thinking, energy and creative genius proceeded almost by the month to further revolutionary conclusions. If not stopped, he would bring inventions into the world one after another with the power to change the course of evolution away from destruction and toward redemption.

The door that Wilhelm Reich had opened had to be slammed shut by the Ahrimanic powers, or they would eventually be exposed—as has now happened through this book. The Boys Downstairs therefore set upon Dr. Reich. For a man consciously innocent of any knowledge of aggressive spirit forces and their containment, he put up a hell of a battle.

Those who want to study Dr. Reich's own account of events will find this in *Contact With Space*, a Record Appendix to Petitioner's Reply Brief, U.S. Court of Appeals for the First Circuit, No. 5160. Wilhelm Reich et al., Defendants-Appellants vs. the U.S.A. *Contact With Space* is the Record Appendix to Briefs for Appellants, Volume 5, Secret and Suppressed Evidence, OROP Desert Ea 1954-55. "OROP" is a contraction for orgone energy operation, and "Ea" is Dr. Reich's term for what is generally known as UFOs. Close to the core of things as usual, Dr. Reich's term is accurate and drawn from experience: "E" stands for energy and "a" is for alpha—primordial. We are indeed dealing with primary energy in the UFO field. Dr. Reich was right in this as in so much else.

The following account of Wilhelm Reich's involvement with UFOs is drawn from *Contact With Space*. Supplementary and illuminatory material has been contributed by Robert McCullough, one of Reich's research assistants at the time. A student of Steiner's work in the interim, McCullough is better able on that account to evaluate what happened at Orgonon and in Tucson during those haunted days. My own interpretations and evaluations have been also added, based upon insights, findings and experiences already described in this book.

The discovery of the orgone energy is the practical, technical break-in to the UFO problem. Appropriately enough, therefore, we find Dr. Reich's theoretical involvement with the technical principles immanent in UFOs commencing right after he discovered the orgone in the atmosphere in 1939-40. During the WWII years he worked out a mass-free energy formula, and also a pendulum formula. Both these formulae, and the mathematics from which they are derived, are published in *Contact With Space*. These formulae anticipate enormous velocities in kreiselwelle functions, and anticipate mathematically the characteristic wobbling and swinging associated with UFOs since 1947. Developed long before that, the formulae ran parallel in the time of their calculation with the first radar sightings of UFOs made by military and naval units during WWII.

Dr. Reich was a practical man. Since he could not at the time confirm these formulae with actual observations and experiments, he deposited them in the Orgone Institute archives until such time as they could be empirically confirmed. His discovery of the orgone energy had obsoleted the old physics by filling the universe with a mass-free primordial energy. He recoiled from making the kind of abstruse mathematical gambits—factually unsupported— that had emptied out that same universe, divesting it even of ether. These notions had hamstrung physical research, blocked comprehension of the orgone energy and led to systematic biasing of the scientific mind against any conception of a living cosmos.

Dr. Reich decided to wait for factual developments. In 1947 he started his work on vacua that led to blue lumination of the orgone energy. He detailed this work in the first issue of the *Orgone Energy Bulletin*, published in 1949. This was contemporaneous with Kenneth Arnold's 1947 sighting in Washington state, and the subsequent early work and writing of Major Donald Keyhoe.

In the late 1940s and early '50s, Dr. Reich occasionally heard about UFOs through newspaper reports and news broadcasts. He was so immersed in his work in biophysics at this time that he paid these reports scant attention. Considering the scope of his research and inventions, the extent of his responsibilities in teaching and running Orgonon, and his artistic diversions into writing, painting and poetry, his brushing contact with the UFO subject is understandable.

Visitors saw UFOs around Orgonon in 1951. In August of 1952, standing on his front porch, Dr. Reich *heard* something whizz by from southwest to northeast in a few seconds. He did not see the object. Although the whole subject of UFOs was still largely ridiculed, Dr. Reich himself did not consider that there was anything particularly strange in the idea of earth receiving visitors from outer space. These reverie conditions were soon banished, for in the post-Oranur period the alien invaders intruded upon Dr. Reich and his work—forcing themselves into his life with malefic intent.

This truth is far stranger than fiction. The presence of these malevolent aliens might have gone unnoticed, as it has in most of the world, had they not tackled a scientist of such phenomenal observational powers as Wilhelm Reich. His ability to single-handedly punch right through to the orgone discovery, on observation and experiment, is a matter of record. Those same observational powers rallied to his aid in 1953, when UFO attentions to Orgonon developed into an attack.

Visions of spaceships shooting destructive rays into the ground, Hollywood-style, rise to mind at the use of the word attack. The assault on Reich and his work was on a far more subtle and clever scale. Again it aimed at keeping the presence and the purposes of the attackers concealed while the destructive work was done. In the spring of 1952, Dr. Reich noticed a strange black substance settling on the rocks from which his observatory was built. The black deposits also began to appear on other rock surfaces in the vicinity.

As Bob McCullough recalls, Reich's ability to detect such subtle processes was well-nigh incredible:

"Dr. Reich's powers of observation were simply tremendous. I never found his equal among the scores of scientists I have worked with since. He was continually drawing my attention to things that had completely eluded me, even close up. His *awareness* was also astounding, as was the case with melanor, the black substance that we later connected with UFOs."

Dr. Reich noticed the black substance gathering. He made time-lapse films over a period of several weeks. These films proved beyond doubt that the blackening was an ongoing process. Other similar substances Reich detected he named brownite and orite. McCullough made preliminary analyses of these materials in 1953, his findings being published in 1955 in CORE (*Cosmic Orgone Engineering*), a publication of the Orgone Institute Press.

Melanor attacked and destroyed rocks and dried up the atmosphere. The substance created excitation of the biological energy of workers at Orgonon, who were afflicted with cyanosis, nausea, thirst and miscellaneous pains. When McCullough scraped or hammered melanor off the rocks for analysis, pressure would quickly build up in his head, and his face would flush and bum. Scraping seemed to excite the melanor, which caused a corresponding strong reaction of the biological energy of any human being nearby.

To melanor was added the compounding problem of DOR—Deadly Orgone Radiation. DOR is orgone energy that has become sequestered, and resultantly stagnant and stale. DOR clouds surrounded Orgonon and deadened the light. The landscape in the magnificent splendor of rural Maine turned bleak and somber. Trees and shrubs blackened and withered.

An enterprise dynamically devoted to life thus became surrounded by and infected with death processes. There can be little doubt today—and McCullough for one has none—that a new kind of warfare was being waged against Dr. Reich's scientific oasis. Had these processes not been rapidly detected by Dr. Reich's phenomenal observational powers, they might well have advanced to the point where this malefic engineering would have extinguished Orgonon, Wilhelm Reich and everything to which his discoveries have led us. There are very few people able to tolerate for long—biophysically and biopsychiatrically—the kind of conditions that the malevolent aliens subtly imposed on Orgonon.

Dr. Reich, McCullough and others at Orgonon became aware as events proceeded that their energy was being *drawn out of them.* In November of 1953, Dr. Reich read Keyhoe's *Flying Saucers Are Real*, and began to put two and two together. Minus any knowledge of orgone energy, Major Keyhoe had set down facts in his book that could not have failed to connect with Dr. Reich's pioneering in the borderland of etherian physics. Reich could relate the noiselessness of most UFOs to the near-noiselessness of his own orgone energy motor. The bluish lights often reported around UFOs were functionally related to the blue lamination of orgone energy in vacor tubes. The spinning rotating discs with their oft-noted swinging motions in the heavens were fully compatible with the spinning wave motion of orgone energy, and the mathematical formulae that Reich had worked out during World War II.

UFOs hung in the night sky around the laboratory. Occasionally they moved and shifted their positions. Dr. Reich took time exposures to objectify their presence. Such photographs proved that UFOs concealed themselves among the stars, a stratagem that I reported from personal experience in *They Live in the Sky*. Dr. Rcich's photographs demonstrated the ability of UFOs to materialize and dematerialize. Objects appeared during time exposures and disappeared before the exposures were terminated.

Dr. Reich and his facilities were getting a great deal of attention from UFOs. This attention was unwelcome, unwanted, disruptive, damaging and dangerous. On many nights, Reich himself was unable to sleep in the quarters at Orgonon, and drove from place to place in a station wagon, snatching fitful catnaps. McCullough has said of this baleful time: "It was as though the energy—*or something*—was following him around and you could see him becoming slowly exhausted by the strain."

What was actually going on is not too difficult to deduce, given the reality of *invisible entities* of vast technical skill in those matters to which Wilhelm Reich had broken through. There could be no more effective way to deal with this gifted, brilliant man than to *drive him crazy*, cause him to break down and to lose his mind and otherwise formidable reasoning powers. To reduce his laboratory with ray guns would only raise endless questions and investigations. Drive him crazy however and every armored nincompoop of high academic qualification would stand vindicated when saying: "I *told* you Reich was nuts."

The shutters could be drawn on etherian physics for another century.

Dr. Reich decided to fight back. His weapon was his own invention, the cloudbuster. Designed and developed to control the weather, the cloudbuster was an obvious tool to use against anything operating in the heavens and employing orgone energy functions. Dr. Reich demonstrated weather control with these devices, reported his findings fully in his journals and bulletins, and also made the information available to the U.S. government.

Others have carried on this work since Dr. Reich's death, including myself. As one who has personally worked for many years with cloudbusters, designing, building and operating them experimentally, I have no doubt whatever that they work. In 1974, because of my concern over the spurious expanding use of these devices by amateur experimenters, I personally produced and financed a 75-page survey of orgone energy weather engineering[1] and circulated it to the relevant Federal agencies. The reader will recognize that such a heavy expenditure is not made unless there is both genuine concern over the matters involved and *conviction as to the power of the device.* This is mentioned here to emphasize that in dealing with Dr. Reich's use of the cloudbuster as a weapon, I can once again write as a participant, having myself used cloudbusters to provoke the appearance of certain types of UFOs.

The original Reich cloudbuster consisted of an array of parallel, hollow, metal tubes. One end of the array is grounded into water. The tubes, or pipes, are mounted on a turntable or similar pivoting assembly so that the free ends of the pipes may be directed into the atmosphere at any elevation and on any bearing, like a battery of Oerlikon guns. Dr. Reich theorized that the hollow metal pipes "drew" orgone energy into the water, thereby permitting manipulation of the orgone energy potentials in the atmosphere. Through such manipulations, control of the weather may be exercised.

The cloudbuster gets its name simply. The device will dissipate, i.e. "bust", any discrete cloud at which it is precisely aimed. Since Dr. Reich felt that he was being drawn upon by the UFOs around Orgonon, and that the whole environment was being sucked to death by such drawing, he decided to draw upon the UFOs with his cloudbuster. There have been writers and commentators with no practical knowledge whatever of these matters who have elected to refer snidely to Dr. Reich's use of the cloudbuster in this way. He used it as a spacegun because he had no option. He was within his rights to defend his own life, his laboratory and his work against the entities who deliberately set out to harm him.

On 12 May 1954, between 9:40 p.m. and 10:45 p.m., Dr. Reich turned the cloudbuster on luminous UFOs hanging in the nearby sky. Two UFOs to the west of Orgonon were made to fade out several times by training the cloud-buster on them. This proved that Dr. Reich's invention could technically reach the strange aerial objects whose presence was associated with the overall deterioration of the environment at Orgonon. Perhaps it was possible to impair or even disable the propulsion systems of these weird craft. The scenario was that of a battle.

This experience of causing UFOs to dim out shocked Dr. Reich deeply. He feared that if he attempted further to mitigate UFO mischief upon his scientific center at Orgonon, he might precipitate an interplanetary war—something General MacArthur had inferred might lie in the near future. Accordingly, Dr. Reich took no further action against UFOs with the cloud-buster until much later that same year. Nor should it be forgotten, in understanding the terrible burdens borne by Dr. Reich at this time, that he was under simultaneous attack by the unholy alliance of Big Medicine and Big Government. Like the UFOs around Orgonon, they too wished to see Wilhelm Reich and his work extirpated.

An order dated 19 March 1954 directed Dr. Reich to halt all research activity in orgone energy, including publications and publishing. This action hampered his work at Orgonon because revenues that sustained the center came largely from accumulator rentals. Seldom if ever has scientific work of such vital moment been so mindlessly harassed in the name of the law.

There can be little doubt, with the perspective of the years since these happenings and an understanding of the New Knowledge, that the whole destructive scenario was conceived and executed by the Boys Downstairs. Psychic control of officials who are spiritually inert is about as difficult for alien technology as making a telephone call is for us. The aim of the whole venture was to bring Reich down and bury his work.

The measure of Wilhelm Reich's striking clarity of mind is his ability to keep plowing ahead despite all this engineered adversity. The advancing desolation of the area around Orgonon was associated with the incursion of these big, yellow and reddish pulsating "stars" after the Oranur Experiment. Melanor, orite and brownite appeared pursuant to the UFO onslaught. Dryness and discomfort ensued in man, animals and vegetation.

Dr. Reich began to transfer this picture to a world canvas. He began to see the connection between the withdrawal of life energy from Orgonon—the world center of thought and work on life energy—and a similar sapping of life energy *from the planet as a whole*. He could readily relate, with his functional mode of thought, the world-wide expansion of desert conditions to the local conditions at Orgonon. He began to see the drying up of the planet and the planetary problem of DOR as engineered conditions rather than natural developments.

Dr. Reich began to suspect that UFOs were undermining earth life, subtly, silently, steadily and secretly. He thus became the first scientist to anticipate from knowledge and experience the unpleasant truth about the Boys Downstairs that has been delineated in this book. All of it was based, in his case, upon observation and experiment and without any study whatever of what Dr. Steiner has said about these same matters.

The years since Dr. Reich's death have served to confirm his well-grounded suspicion of what he called the CORE men (for Cosmic ORgone

Engineering). Destructively-inspired critics have often cited Reich's suspicions as confirmation of his alleged paranoia, but the march of events permits no other conclusion than that he was one of the most sane humans alive — perceptive to the point of genius.

DOR-infestation has become characteristic of the entire planet since Dr. Reich's death. Everywhere there are humans, DOR has become concentrated and a permanent part of the environment. Breezy cities like Honolulu and San Francisco have DOR clouds anchored above them that not even trade winds or strong prevailing breezes can move. The stuff is as though tied to the ground wherever humans are found. In my own extensive world travels I have found DOR everywhere and getting worse.

More than 35 years ago, Wilhelm Reich was alert to this menace. Despite the DOR and UFO problems at Orgonon, despite financial stricture and despite the FDA harassment, he planned throughout 1954 for an expedition to Arizona. His aim was to determine, if possible, whether the cloudbuster could reverse desert development. He chose an established desert area for this experimental work. The boldness and originality of this plan, backed up with a program for its implementation, show how firmly anchored Dr. Reich was in rational work.

As plans for the Tucson, Arizona project advanced, those at Orgonon became increasingly aware that they were under surveillance. Dr. Reich records this in *Contact With Space*. Bob McCullough remembers the period indelibly:

"The whole area was infected with a very material DOR, cloying to everything, and it made things absolutely unlivable. You had to get out of it periodically by driving west, or up to some high spot, or just drive fast to keep it from dragging you down to its level. Everything was purple or purplish mauve. The white birch trees were bending over like rubber hoses, as though laden with invisible snow. This condition was all around Orgonon and down toward Farmington.

'There was a tremendous sense of something impending — of waiting for something dreadful to happen. This anticipatory waitfulness was oppressive. Something was coming and it wouldn't be good. There were periods of gremlins, also. Small objects disappeared and reappeared where no one had placed them. Pins in maps were moved or just pulled out. Unmarked aircraft repeatedly overflew Orgonon. There was a sense of harassment — of being pursued — that it is hard to fully understand."

To the occult scientist it is obvious that those at Orgonon were being psychically attacked, with the aim of obsession or psychic control. Minus any knowledge of spiritual science, it is most probable that the only thing that saved them was their high motivation and determined service to Life. Dr. Reich sought continuously to disperse the DOR conditions with the cloudbuster. He was sometimes inclined to believe that he had precipitated the problem through the Oranur Experiment, and constantly sought ways by which the adverse conditions could be eliminated. McCullough continues:

"Prior to Oranur, Dr. Reich had done work with cancer mice, and had made a little shed to house them just up toward the Observatory from the Students' Lab. The mice were gone now, but a small sample of radioactive Cobalt 60 remained in the shed throughout the Oranur and later events. I found this sample, and asked WR about it. It was not at all noxious as were radium dialed watches, for example. Rather, it was quite soft. I must assume that he took this cobalt and checked it out on the Autoscaler."

In the aftermath of Oranur three years previously, Dr. Reich had removed three milligrams of radium from Orgonon and had this material buried in heavy lead shielding, in an uninhabited area, fifteen miles away. The cobalt found in the shed by McCullough and the general DOR-UFO menace at Orgonon now prompted Dr. Reich in the latter part of September 1954 to have this material exhumed and tested.

The platinum-encased radium needles before Oranur gave more-or-less standardized counts of 16,000-17,000 counts per minute naked at 1 cm, and 7,000 counts per minute when the needle was sealed in its half-inch thick vial of lead shielding. After the involvement of these needles in the Oranur Experiment, this whole relationship was drastically reversed. Using the large and very sensitive Tracerlab Autoscaler, Dr. Reich found that the naked radium needles had lost over 90 percent of their activity! The counts had dropped by a factor of 10 from 16,000 to 1,200. Yet when these same needles were placed within their half-inch thick lead vials, the counts 1 cm outside jumped by over a factor of 10—from 7,000 to over 80,000! The empty lead shieldings themselves gave virtually zero counts. The physical sequestering of the radium had again called forth a massive atmospheric reaction.

With one needle, and the counter tube of the sensitive Autoscaler 1 cm distant from the case, the count was a shocking 163,840 CPM, against 7,000 CPM in 1951. These staggering new facts, arising out of the fundamental, polaric antagonism between etheric and intramaterial energies, served to further widen the gap between Reich's functional physics and the old conventional physics.

Bob McCullough was there and wrote the experimental protocols. He recalls the circumstances and his own reactions:

"WR was amazed that the counts were so low for the naked needles, and he kept calibrating and recalibrating the Autoscaler to make sure. In all the history of the world no one had been able to change the rate of radioactive decay in a substance by physical means.

"The rate is a statistical effect and boiling, melting, amalgamating, vaporizing etc., just does not change or affect it. And here it was drastically changed—reversed! It was a new substance.

"How did it affect us? Peculiar. If I had been a nuclear physicist instead of a biologist it would have affected me a lot more, I guess. But to us at that time, with all those events and new things pressing in on us, there were other things that seemed more important. It was a new substance, but what do you do with it? We weren't a modern AEC installation. I think if one were to run a spectrographic analysis on any ORUR substance he would get an entirely new phenomenon.

"Personally, I think that the ORANUR-exposed substances, which WR soon afterwards named ORUR, were reversed in time, and instead of decaying like normal were made 'younger'. Thus they represent earth substances at an earlier age. There's no proof on this yet, but it is something to keep in mind."

The reader who has followed earlier discussions in this book regarding the earth being much more alive in the past than it is today, will see in these events at Orgonon the first experimental evidence that when the earth was young, it was indeed much more alive—like any living organism. Concentrations of life energy at Orgonon brought these effects into the phenomenal world, permitting their expression in counts per minute. Before Dr. Reich, no one had been able to devise any satisfactory means of concentrating life energy, although it is simpler than the simplest flashlight battery.

Biophysics appears to have the characteristic of reversing classical ideas and conceptions. A mind apprised of the polar relationship between intramaterial and etheric energies would expect such reversals, since they arise out of the fundamental polarity. Classical scientific formulations in the mechanistic sense have grown out of, and are wholly based upon, material and submaterial investigation and experiment. Mechanistic science is crippled by its unipolarity. That is its crucial weakness and cardinal failing in dealing with spacecraft devised and manned by intelligences who understand the ethers fully and who have harnessed them technologically. Certain of such intelligences obviously wish to keep mankind immersed in unipolar views of the universe.

Dr. Reich put ORUR to immediate use. By placing this substance, enclosed within metal, near the BX cable connectors on his cloudbuster, he found that he could rapidly cleanse the sky of DOR. Only two to five seconds of such exposure of ORUR to the cables was necessary. Low, heavy, dark DOR lodged in the valleys and over the landscape seemed to turn blue-grey almost instantly. Using ORUR for 50 or 60 seconds with the cloudbuster caused clouds to form quickly, and in a few hours rain would ensue. The changes brought about by this new tool were dramatically obvious. Life was winning now at Orgonon as a result.

A see-saw battle followed with Dr. Reich using his 'spacegun', as he called the combination of ORUR and cloudbuster, to eradicate the DOR that UFOs continued to generate around Orgonon. To keep the region clean, he would have to use the spacegun daily. He realized that he was at war with the UFOs. By 8 October 1954 he was using the spacegun not only to keep down the DOR, but also to disable and drive off intruding UFOs.

Dr. Reich had no doubt that the spacegun directly affected the propulsion system of the invaders. When the spacegun was aimed at them, the UFOs dimmed out, disappeared, shifted position to get away from Reich's aim, and in at least one case where several UFOs were present, they all dimmed out and disappeared simultaneously, as though on common command. With his high sense of responsibility and public duty, Dr. Reich made sure that the USAF was advised of these happenings, including one September incident when a wobbling silver UFO was sighted tagging along behind two elements of USAF jets.

The Arizona desert and drought research project went forward. When McCullough reached Tucson as the head of the advance party, he made several sightings of UFOs, which he reported to Dr. Reich. At Little Orgonon, a ranch base near Tucson, the events of Maine were largely replicated. Constant UFO harassment, constant assault on the experimental work through DOR operations, and a steady fighting back by the Reich group. An

ORUR sample was flown to Arizona, and again this substance proved dramatically successful in combination with the cloudbuster.

On 6 December 1954, the zenith region at Little Orgonon was observed to be black with DOR in the morning. A Geiger counter was used to confirm the subjective impression of DOR pouring down, with counts of up to 800 CPM recorded. Drawing operations from the zenith were started, with Bob McCullough using #2 cloudbuster. His tongue had proved in the past to be an excellent indicator of DOR in the atmosphere, and now the sinister downpouring tasted like offal, strong and sour.

Then he felt a crippling sensation in his right leg. From the leg, the paralysis spread to his whole right side. His ability to move around was impaired. When he got away from the cloudbuster, his condition rapidly improved, and he emerged from this experience with his motoric reflexes normal. The next day was another story.

Early in the morning he was operating again, and this time McCullough was heavily struck by paralysis. Sick, purple-faced and crippled, he staggered into the quarters. Medical countermeasures were taken by the doctors present, and he was somewhat comforted by these actions. Nevertheless he had to leave Little Orgonon, recuperated only slowly, and was not fit for duty until the end of January 1955. He drags his right foot to this day as a reminder of this unforgettable incident.

Right before my eyes I have seen a close friend similarly struck by a UFO through a cloudbuster. The incident occurred in 1971 in southern California. The man in question had a measure of etheric sight, and had come to see me during one of the many UFO "flaps" generated that summer by my weather engineering operations. He was resting his right arm on the right hand tube of my cloudbuster. He jerked his head skywards: "Can you see that disc up there just to the right of where you're aiming this thing…" *Crack!* A bluish bolt of energy, lasting only an instant, hit the tube he was touching and traveled up his arm to his skull. He was partially paralyzed for a few minutes, the pain brought him to tears, and for that terrible instant he felt every nerve in the right side of his body irradiated by the bolt.

This incident is related not only to illustrate that what probably happened to Bob McCullough can, and has, happened to others, but also as a warning against dabbling in this work without a solid background in spiritual science. Of special importance is cleanliness of motive, for uncleanliness here opens the cloudbuster operator or other, similar experimenter, to the unseen enemies of human evolution with whom uncleanliness has a powerful, intimate correspondence. This is how "they" tune in.

The Tucson weather engineering operations of Wilhelm Reich produced a greening of the area around that flinty metropolis. The succeeding summer was the wettest in more than 20 years, with something like 12 inches of rain. The depredations of the Food and Drug Administration and its assault on Dr. Reich and his work did not permit proper and orderly completion of this bold experiment. UFO harassment was a constant, invisible and deadly cosmic counterpart to the legal attack.

Debates will rage for years over Wilhelm Reich's trial, his conviction, his death in a Federal prison in 1957, and the destruction of his scientific works by fire under Federal Court Order. Such debates are of no moment. The cosmic breakthrough made by Dr. Reich to the fundamental polarity of physical creation, and his connection of all of this to the UFO phenomenon through the efforts of the Boys Downstairs to wipe him out, is perhaps the greatest of all his legacies to his fellow men.

Let us proceed now to apply the practical scientific discoveries of Wilhelm Reich's cosmic breakthrough to some of the classic incidents, encounters and sightings in the modern epoch of ufology. In this new analysis of old happenings, let us make kindred and correlative use of what Dr. Steiner has told us, not only concerning etherian physics, but also regarding the various orders of unseen beings who are as involved in earth life as we are.

We have traveled far enough on our adventure to tackle the citadel of ufology with new weapons. The walls may not fall down at our approach. When we are done, however, we will see huge breaches in battlements that only a short time ago seemed impregnable.

NOTES TO CHAPTER NINETEEN

1. *Orgone Energy Weather Engineering, the Law and the Environmental Crisis*—privately published by Trevor J. Constable, dba Merlin Weather Engineering, San Pedro, California.

Chapter Twenty

THE NEW KNOWLEDGE AT WORK

You must look at facts because they look at you.
— WINSTON CHURCHILL

The chain of thought, work and experiment that led from my original photography of critters and discs to the later photography of the same kind using Reich's cloudbusters tuned with Drown's instruments, has opened a pathway into a borderland of nature not previously accessible. In some quarters, these findings and facts will evoke a pathological fury, and it is quite possible that an effort backed by high temporal authority will be made to sweep all of it into the ashcan. There is nevertheless a strong countering force resident in those truths that I have brought forward. This counter-force is the sweeping applicability of the basic theories and findings to the fall spectrum of UFO phenomena.

With the master key provided by the orgone energy discovery, buttressed by demonstration of the imperceptible-physical level of existence, we can for the first time really tackle the whole UFO scene and make it talk. Let us proceed to analyze some classic UFO incidents from various parts of the world—happenings completely independent of me and my work. The incidents we will deal with are in most instances typical of many similar happenings, recorded and verified over a period of several decades.

These occurrences have defied penetration or elucidation by conventional knowledge. Standard investigative procedures have generated much data, but the findings by their very nature lead not at all to any causative areas. We know what happens to the grass, the sand, the water, the weeds, the people and various forms of apparatus, but we do not know why or how. Materialistic mechanistic science is excellent at asking "What is it?", and answering that question with precision. Ask "How does it arise?"—the question that leads to the formative forces of nature—and the one-sidedness of mechanistic knowledge and cognition becomes glaringly evident.

The New Knowledge permits us to get sense out of these otherwise enigmatic collections of data and experiences that have thus far beaten the world's best minds. These instances will establish for the reader, beyond any rational doubt, that even with those elements of the New Knowledge we have broached in this book, we have been able to find our way to the end of the beginning in ufology. Some bright young people are needed to carry it all forward.

As a typical scanning of enigmatic UFO incidents, we may cite the NICAP publication *Strange Effects From UFOs*, compiled with NICAP's usual scrupulous attention to detail. The publication cites frequent electromagnetic (EM) interference phenomena and physiological effect cases. The latter include witnesses being burned, feeling heat, feeling numbness, experiencing temporary blindness and unconsciousness. Physical evidence cases include reports of damaged and scorched trees and bushes, crushed foliage, burned ground and road areas, and effects on numerous animals including cattle, horses, dogs, cats, birds and chickens.

Etherian physics permits the elucidation of most of these enigmas. EM activity, for example, arises out of the manipulation by man of the electric forces bound up with physical substance. These submaterial forces are polarically opposite to the etheric forces. There is therefore a basic antagonism between the two in the free state. EM signals in radio are suppressed by the ethers through which they pass and in which they are a perturbation made by man. This weakening of EM signals is known by science to take place proportionately to the square of the distance from the transmitting station. The point is eventually reached where there is *no signal*.

Where EM activity comes into the presence of highly concentrated etheric forces therefore, we should anticipate that the EM activity will be totally suppressed. Such fields of etheric force exist around certain spacecraft. There are examples of them objectified photographically in this book. When the ignition systems of automobiles quit in the presence of UFOs surrounded by such etheric fields we should not be surprised. On the contrary, and on the basis of the New Knowledge, we should *expect* such suppression of electrical activity. Removal or withdrawal of the high potential etheric fields should similarly result in the resumption of the electrical activity. Auto engines immediately restart, headlights burn again and full function is restored.

Many UFO photographs not made by me or my methods show that the underside of disc-type spacecraft is an area where strongly contractive etheric force is concentrated. Examples are the McMinnville, Oregon photographs, the Cluj, Rumania photographs and the photograph made in Korea by a U.S. Marine aircraft and published in this book. This portion of the spacecraft creates a void on the photographic negative, and connects directly with Dr. Wilhelm Reich's finding that concentrated orgone energy desensitizes film emulsions.

In at least one contact case (Patrolman Herbert Schinner at Ashland, Nebraska on 3 December 1967), a human being has found the underside of a landed spacecraft, a ladder attached there, and the inside of the ship itself to be strangely cool.[1] Can chance be the only reason all these things begin to weld themselves into a functional whole? We are still only at the beginning, so let us proceed.

The presence of powerful fields of etheric force must inevitably overwhelm and distort EM radio and television signals. Such signals take place as perturbations of the life, chemical and light ethers. When the concentrations of these ethers begin to run in multiples of their free, natural levels, the agitatory power of EM transmitters is no longer sufficient to create waves in the medium by which the signals are conveyed. To provide an analogy, we may say that it is easy to send chains of waves across a pond of water using a child's paddle. If it is a pond of fuel oil we find the waves are rapidly damped out. If it is a pond of molasses, we cannot make a wave in it at all. Our problem in transferring all this to the UFO scene with the resources of formal science alone is that Professor Einstein's etherless, empty universe has us thinking in terms of EM waves that have nothing to wave in. Let in the ethers, and it all starts to sing.

The blotting out of radio reception by UFO presence is so firmly established in the phenomenology of this subject as to require no further confirmation. The most common manifestation is that radios simply fail during the

UFO presence, and return to normal upon departure of the UFO. Neither receivers nor transmitters work. The reason is in all probability a local increase in the density of the etheric interface with the material world which we have been using for radio communication in the past century or so.

In the many cases where programming or other intelligence is blotted out from radios, and regular beeping signals appear instead, we might rationally and tentatively assume such patterned beeping to originate with the cyclic whirling of the fields around the UFOs. Ordinary radios may react to such fields by generating corresponding heterodynes within themselves, since the basic circuitry is there to do just that in normal operation. The human EM signals are simply swamped by the powerful fields around UFOs.

Dimming, blurring, loss of sound, and distortion of television signals have been similarly noted, and may be attributed to the same basic cause. So also can we see how power transmission via high tension lines can be interrupted. Such direct blocking of electric power has been unequivocally associated with UFOs in many cases, both in the USA and abroad.

In Brazil on the night of 17 August 1959, power flow interruptions were indicated instrumentally at the Minais Geras power station. At Uberlandia, the staff had been advised by telephone of successive interruptions of power along the distribution system, occasioned by a UFO overflying the lines. The UFO's progress could be traced by the opening of circuit breakers along the distribution system. At Uberlandia, they were forewarned and had the staff alerted to take countermeasures.

As the switches opened due to interruption of the power flow, technicians immediately snapped them shut. No use. The single, oval-shaped UFO, by virtue of its field effect on the lines, frustrated every countering stratagem that would have kept the power flowing. Only when the UFO departed did everything return to normal. Again we see the signature: high concentrations of etheric energy suppress intramaterial energy metabolism.

The operative effects of this fundamental principle thus underlie the halting of internal combustion engines, the blocking of radios and television sets, the dimming and extinguishing of household lighting, and the interruption of commercial electricity supplies often with regional effects and consequences. These same effects extend into the corruption of standard measuring instruments which are vulnerable to high concentrations of etheric force.

Most measuring instruments used in scientific investigation depend upon the *D'Arsonval movement*. In simple terms we may describe this as the twisting effect exerted upon a coil, finely balanced and precisely located between the pole pieces of a magnet, when an electric current is passed through that coil. As the coil turns on its delicate bearing a pointer moves over a scale, which is calibrated to read whatever values the meter is designed to register.

In etherian physics, magnetism is seen as a local concentration of life ether. Anything that can upset the normally fixed concentration of this etheric force between the pole pieces of a magnet will cause spurious and uncontrollable variations of the pointer reading, when that magnetic field is integrated into a meter employing the D'Arsonval movement.

When electrical measuring instruments employing the D'Arsonval movement come under the influence of the powerful etheric fields associated with UFOs, the instruments no longer function according to their design. They go "wild". This corruption of instrumented readings in UFO encounters has been frequent enough since Kenneth Arnold's time to need no further elaboration. A force is involved in UFO propulsion that is easily able to upset the integrity of electrically-actuated measuring devices, compasses, and kindred apparatus.

Correlative experience has come to me in designing, building and operating cloudbusters experimentally. The orgone energy, or chemical ether, is the force manipulated by this device. When iron or steel tubes are employed in a cloudbuster, they become impressed with strong magnetic nodes that are easily capable of spinning a hiker's compass around 180 degrees, although the pipes on receipt from the hardware dealer exhibit no such capability. Readings on a voltmeter taken near these nodes are materially influenced, although the source of voltage remains constant, e.g. a flashlight cell.

The bioenergetic field of a healthy and lively human being is able to produce similar distortions of TV pictures and sound to those reported in association with UFO manifestations. In my home I enjoy music sent by frequency modulation from a 105-kilowatt radio station, the most powerful in Southern California. Yet by simply positioning myself in a certain way when the receiver is running at normal volume, I can suppress the signal. I can cause my bioenergetic field to shield those radio waves, and it is my contention that this action is functionally identical to radio stoppage caused by UFOs. The same energy is at work.

Bioenergy in the form of so-called "static" electricity can be drawn off the human hair with a comb and, when brought into proximity with electrical measuring instruments, cause their readings to be changed. I have sometimes amused myself in this way with the many meters employed in shipboard electronic equipment. A functioning radiofrequency ammeter in a transmitter circuit, for example, reading a normal 10 amperes, can be readily lifted to 12 amperes by this stratagem. Rapid near-clapping of the hands will build up charges on the hands that can change a 24-volt D.C. supply to read 27 or 28 volts.

Mechanistic investigation of UFOs has been characterized in the past—at least in public pronouncements—by the assertion that many UFO happenings "could have" been analyzed properly "if only" sufficient instrumentation could have been brought to bear on the scene at the time. Since we know from both observation and experience now that we are dealing with etheric forces in UFOs capable of inducing huge errors in instruments, we will be wise to abandon the expectation of accurate analysis by unaided mechanistic means. Not only is a new thinking necessary, but so is a *new generation of instruments. The New Knowledge will have to be supported by a new metrology.* Right now, our investigative resources, like our thinking, are out of balance. UFO phenomena scream this fact at us every time they impinge on our world.

Close correlations between known microwave phenomena established in our laboratories, and microwave effects observed and recorded in connection with UFOs are similarly full of pitfalls. The understandable tendency is to conclude that some form of microwave propulsion is employed—unless one is awake to the polar relationship between etheric and EM energies. This awakening leads to caution in reaching conclusions.

Microwave EM phenomena observed in connection with UFOs may well arise not directly from the power plants of these craft, but perhaps as a side-effect of intense etheric forces being manipulated in such power plants. Just as submaterial forces interface upwards through the material world with the etheric forces, so must the etheric forces interface downwards through the material world in the opposite way. Microwave perturbations around UFOs may well originate in this way. Wisdom would preclude any snap judgments regarding radiation of any kind at this juncture.

Moving on to physiological effects of UFOs, one overall observation seems to be an essential preliminary: the information collected to date on the physiological effects of UFOs has not thus far been evaluated by anyone aware of the reality and basic physical determinism of life energy, bioenergy, orgone or howsoever it may be called. The qualified people who have evaluated this information are at one and the same time qualified and unqualified: qualified in what this energy is not, and unqualified in what it is. No formal educational organization deals fully with bioenergy as a physical force or crucial element in the functioning of living organisms.

Witnesses who are burned by UFOs probably experience this effect because they come into contact with plasmas created in the atmosphere by the whirling fields around the craft. There are several examples of such plasmas in the photographs in this book. There is direct correlation to Dr. Reich's work in all this. An orgone energy accumulator, which concentrates orgone energy from the atmosphere within its six sides, continuously liberates heat at its own upper surface. This heat appears, to mechanistic thought, out of nothing. In a spacecraft, we see areas of concentrated orgone underneath them, and plasmas around them—just about as direct a functional connection as anyone could demand.

Plasmas of the order of 200-400 degrees F. may well be created in the atmosphere by the whirling etheric fields around the craft. Burning of humans could also result from microwave penetration of the skin, whether such microwaves originate directly from the power plants, or are an interfacial phenomenon as previously suggested. Virtually the whole range of heat effects connected with UFOs can be at least partially elucidated now that we can, as conscious earthmen, generate warmth by concentrating orgone energy.

Such extreme instances of heat effects, as the burning of road surfaces and the scorching of foliage and trees, are the opposite end of the whole spectrum of heat effects that begins with human feeling warmth. All now appear in a new perspective. High etheric charges moving in the atmosphere produce high heat, just as the low charge in a small, stationary orgone accumulator produces a small but constant temperature differential. Thanks to Wilhelm Reich's penetrating pioneering, we can understand other effects arising from high etheric potentials.

The law of reversed or ergonomic potential, as formulated by Reich, says that a body of higher charge attracts a body of lower charge and withdraws that charge. The limit on this withdrawal of charge appears to lie in the capacity level of the more highly charged body. In ufology, when a human being comes into contact with high orgonotic (etheric) potential his own bioenergy will be withdrawn by the higher charge. Such a withdrawal of energy is bioenergetically tantamount to driving out his etheric double with anesthetic. Where it occurs locally, is *numbness*. Where it is general, there is unconsciousness. How strange it is that orthodox medicine can proceed generation after generation never asking *what actually happens* when a patient is numbed or rendered unconscious.

Blindness from UFO encounters is similarly connected with high orgonotic charges. The optic nerve, according to Dr. Reich, is highly sensitive to orgone energy. Furthermore, when a human being looks at a space vehicle that is propelled by extremely high orgonotic potentials, he connects with it bioenergetically through his eye beam or visual ray. Earlier it was explained how the visual ray may be used to detect otherwise invisible UFOs through the subtle "shock effect" felt through the visual ray when it intersects with a high potential source.

My experience has been that the eye can be worked on both ways by UFO encounters. Energy can be sucked out of the organism in this way by the high potential fields around UFOs, and in other cases the organism of the observer may become heavily charged via the eyes. Conjunctivitis is perhaps the mildest eye condition arising from UFO observation. Meriting recall here is the conjunctivitis that afflicted Dr. Reich while he was observing the bions—microscopic blue energy vesicles that he discovered on his pathway to the larger discovery of the orgone energy. The orgone energy radiating from the bions produced the conjunctivitis.

This effect demonstrates not only how conjunctivitis may come from observing even distant UFOs, but also the polar relationship between the blue and red principles in living organisms. The ancient art of healing with color is based on this and other fundamental polarities, which in imbalance constitute disease. The blue orgone energy in high concentrations provokes the red principle from within the organism. Burning, tanning and reddening of human skin in UFO encounters are rooted here, especially where such effects take place through the clothing. In using the cloudbuster, I always wear blue clothing and hat, to repel an excess of blue principle that would otherwise see me beet red at days end, right through my clothing.

In such an instance I would not be burned in the medical sense, but I would show an excess of red principle provoked from within by an excess of blue from without. An individual who sits in a man-sized orgone accumulator and closes the door, irradiates his organism with the concentration of orgone within the device. He becomes, after a period of exposure that varies with individuals and their overall health picture, strongly reddened.[2] The clinical literature of orgonomy reports this effect, not only from Dr. Reich's own cases, but also from the experience of other competent physicians.

There is thus an *interaction* between the biological energy of an observer and a UFO making use of biological energy for propulsion. Most investigators proceed in the dream-illusion that a human being is 160 pounds of meat, instead of a bioenergetically actuated laboratory of incredible complexity. Everything that occurs through this bioenergetic interaction is therefore not comprehended by classical thinkers and investigators—not even in its fringe aspects.

This brings us to animal effects. The armoring of the mechanistic investigator, his classical prejudices and the monstrous omissions in his world view, ill equip him to grasp that with animals one has a *purity of interaction* with UFO phenomena not possible with armored human beings. The chronic muscular contractions that afflict human beings, distorting or blocking perceptions and feelings, have not been imposed culturally on animals. Animals enjoy perceptions that are functionally unified and not blocked or split. Animals accordingly live their lives responsive to the downward reaching influences of the etheric forces, instead of blind to those forces and cut off from them.

Animals are just about the best UFO detectors we have, on account of their unblocked organ sensations, and their sensitivity to changes of atmospheric orgonotic potential brought about by either spaceships or highly charged biological UFOs, i.e. critters. The possibility of such critters, when functioning in our polarity and density, of foraging for themselves from normal terrestrial food sources is supported by numerous incidents—through the decades and ages—of kidnapped and savaged animals. The distinguished British writer and linguist, Mr. Gordon Creighton, M.A., F.R.G.S., has the admirable faculty of looking such facts in the face.

In his article *A New FSR Catalogue (Flying Saucer Review* Jan./Feb. 1970) Mr. Creighton raises in his scholarly way the possibility broached by me as long ago as 1958 that certain UFOs are environmental and have always been here. They are not sudden visitants from somewhere else in the universe. My demonstration of the invisibility of UFOs as their fundamental, natural state, lends additional strength to what Mr. Creighton has to say.

In making his catalog of effects of UFOs on animals, birds and smaller creatures, he writes:

" ... I have been greatly impressed by the total, utter, abject terror displayed by so many animals and birds in the presence of UFOs. Were the UFO phenomenon due to some long-existent environmental factor present here on the earth and in the earth's atmosphere, one would have thought that animals and birds would surely, in the course of the ages, developed some sort of familiarity with, or tolerance of, such an environmental factor, even if—as most people assume—what disturbs them so much is some kind of VHF[3] emission. That a VHF factor often seems involved I can well believe. But it looks to me at present as though this is far from accounting for the whole of the discomfort and terror shown by animals and birds."

Then Mr. Creighton asks the natural question arising from all this:

"Could this terror perhaps be something much more fundamental, elemental, springing possibly from the instinctive knowledge of our animals and birds that the 'UFO phenomenon'—or part of it—relates to some *power or agency that is utterly alien and inimical to the creatures of our world; a power or agency whose coming can only spell dismemberment, destruction or death for them*?

Fear of this type naturally attends the foraging activities of hungry predators in their own environments, as lesser and vulnerable species scatter to avoid being devoured. Animals who are likely to be eaten know, through what we call instinct, that their bodily existence is threatened. Their sensitivity to all happenings in the orgone energy envelope of the earth—the etheric continuum—undoubtedly extends to sensing what is contained or expressed in the body forcefield of the foraging tiger. They know he will literally tear them to bits. The same reactions have accompanied the presence of UFOs as recounted in Mr. Creighton's catalog. Charles Fort has also catalogued the inexplicable mutilation of animals in England and Kenya over a twenty year period, and Mr. Creighton has added to this.

That UFOs have in addition kidnapped aircraft and occupants is well established, notably in Donald Keyhoe's account of the infamous Kinross AFB case, where a jet with two occupants was seen to merge on radar with a UFO. After the merging of the two echoes into one, the one that remained streaked off the radar screen. Jet fighter and occupants were never seen again. In my 1958 book *They Live in the Sky*, I presented the affidavit of Mr. Eugene Metcalf of Paris, Illinois, who saw essentially the same happening with his own eyes in broad daylight.

Standing outdoors in good visibility, Mr. Metcalf saw a bell-shaped UFO taking a USAF jet fighter into itself through its underside—like a shark gulping down a herring. After ingesting the fully operational fighter plane, the bell-shaped UFO made off at high speed. Only the explanation of hostile entities is found preposterous by establishment ufology in such instances. Such an inability to look at facts is a simple manifestation of armored perception, which distorts what is otherwise clearly evident. Animals in their unarmored and naturally healthy constitution, respond to things as they are. Animal reactions and effects should help us override our prejudices, preconceptions and misconceptions—the burden of our structural bias—so that we can also begin to see things as they are.

From an occult viewpoint, the difference between the animal and human is the presence of an ego as a higher principle in man. This is not the ego of formal psychiatry or common usage. Animals share with man a physical body, an etheric double and an astral body. The animals have a group ego for a species, whereas man has an individual ego that he is in the process of perfecting. Physical armoring combined with the imperfection of the ego are absent in the animal, hence the clarity and integrity of their bioenergetic reactions—something shared by only a few humans.

Armed as we now are with a description of the qualities, motives and ethics of the Boys Downstairs, we would hardly anticipate that such beings would be kind to animals. There will come from them no Christ-like concern for younger brethren, and no reverence for life or love. One can well understand the reactions of animals to the approach of such entities, which would be dread and terror. A similar reaction would also result if earth animals were natural hunt objects, on occasion, of atmospheric fauna of various kinds that do not normally appear in the physical-mineral density. Certain invisible critters of this type may be bioenergetic parasites on animals. The resources of etherian physics permit us now to understand both the sensitivity and the purity of animal reaction to the incursions on their milieu made by UFOs.

Gordon Creighton has cataloged instances where animals have risen clear of the earth under the influence of emanations from UFOs. There is no point or purpose in seeking an EM explanation for this. High orgonotic potentials under the craft attract the lower potential of the animal's own orgonotic field. Because of the strong integration of the physical and etheric bodies of the animal, the physical body is simply lifted clear of the earth until such time as the influence of the higher potential field is removed—usually by departure of the UFO.

Such an analysis is consistent with the first known principles of etherian physics, chief of which is the law of reversed or ergonomic potential. This law permits us to understand what is fundamentally a bioenergetic effect, not replicable by any of the ramified techniques and devices developed by man in the EM side of technology. Finally, what is actually observed can be recognized as belonging to this new fabric of law. Wise humans will seek to know more of such law, so directly germane to otherwise impenetrable UFO phenomena.

The "falling leaf" or pendulum effect so often observed in UFO maneuvers is covered mathematically in Wilhelm Reich's formulae in *Contact With Space*, and now emerges as functionally related to the spinning waveform of orgone energy, the KRW (kreiselwelle waveform). Again, we are cheek by jowl with biological energy and bioenergetic effects, a realm in which our otherwise perfectly valid laws of EM energy do not apply or are reversed. The erratic and jerky motions often connected to UFOs are also penetrable by these same bioenergetic principles. As physicist James M. McCampbell remarks in his excellent 1973 book *Ufology—New Insights From Science And Common Sense*,[4] such UFO motions resemble the flights of hummingbirds.

The clear-thinking Mr. McCampbell goes right to the point when he further writes that "something fundamental is suggested" by this resemblance to hummingbird flight. He goes on to elaborate and speculate mechanistically when the truth is right there, where his own inquiring mind has led him: *in the bioenergetic power source of the hummingbird*. This biological power source is transduced by the organism of the bird into beating wings and physical-aeronautical activity, but *nothing would happen without the life energy*.

The bobbing, skipping and swinging of various types of UFO emerges from the fundamental KRW pattern of the energy used in propulsion, or, in the case of the critters, in their animation and natural life pulsation. Skipping discs such as those first witnessed by Kenneth Arnold may well propel themselves using the peak energies of the huge KRW waveforms that are naturally present in our atmosphere, which contribute to the spinning of the earth on its axis, and which etch their own presence in vast volumes of cloud—independent of wind motion. This would be a form of orgonotic pulse propulsion, skipping from one point of high charge to the next, covering vast over-the-ground distances. The critters very probably fly in this way.

Spinning is perhaps the most common single motion attributed to UFOs. Discs aloft either spin, or convey the illusion of spinning via moving light effects that are associated with them. There is therefore nothing unusual in finding associated with their propulsion a physical energy known to manifest with the spinning waveform or KRW. A disc surrounded by a spinning field of energy traces a KRW with that field as it moves through the atmosphere. The earth moves around the sun on a curving KRW pathway. Typhoons and anticyclones trace KRWs through the ethers as they spin across the surface of the earth. The sexual orgasm that produced everyone reading this book was a KRW. Of all the talented, gifted, wise and knowledgeable men who have butted their heads against this subject through the years, *only Dr. Reich is considered crazy.*

In the tall cane country of Northern Queensland in Australia, disc-type UFOs have been involved in incidents on the ground that dovetail in all respects with what we know of orgone energy—the chemical ether of etherian physics. The chances of the persons reporting these sightings and providing photographic evidence of these UFO "nests" being familiar with Dr. Reich and Dr. Steiner are so remote as to justify exclusion. The facts speak for themselves.

Stan Seers and William Lasich wrote up the Queensland UFO saga in the May-June 1969 issue of the *Flying Saucer Review*. There is no space here to reproduce the complete article, but the essentials can be rapidly summarized.

On 19 January 1966, at about 9 a.m., on a farm near Tully, Queensland, farmer George Pedley was operating a tractor near Horseshoe Lagoon. He was astonished to see a large, saucer-shaped object suddenly ascend from the water reeds 25 yards away. Rising to about sixty feet altitude, the object tilted a little to one side and vanished speedily to the southwest. In the reeds Pedley found a circular, flattened area within an otherwise undisturbed, dense reed growth. "The flattened stems were radially distributed in a noticeable anticlockwise manner." The nest was 30 feet in diameter.

Returning later that day with Albert Pennisi, the owner of the property, Pedley watched while Pennisi stripped and waded out to the nest. He found he could swim through from side to side below the flattened area without encountering any obstruction. This indicated that the whole mass of the nest was floating on the surface. Color photographs taken about 5 p.m. verified that the upper surfaces of the flattened reeds—green when originally discovered by Pedley in the morning—had turned brown, but only on their upper surfaces.

Other nests have subsequently appeared in the same area. Specially designed, magnetically triggered photographic monitoring of the area was set up, but films obtained up to the time the article was published in the *FS Review* were spirited out of the Australian mails. The package of film allegedly and purportedly arrived empty at Kodak's facility in Melbourne. This is mentioned because people in various parts of the world are slowly learning from experience that if they want to enjoy the fruits of their expensive, time-consuming and honest efforts to photograph UFOs, they should use materials they can process themselves. The most highly skilled burglars in the world are today in government service.

Before worrying about scientific "controls", UFO researchers should worry about retaining full control of their own work. The governments of this earth are rotten with corruption, and the new cosmic age portends their doom. Anything verifying incursions from other dimensions is being systematically suppressed, and many scientists, administrators and political figures are instrumentalities of this obscurantism. Because UFOs have their roots in laws and forces other than the purely mechanistic laws familiar to earthmen, nobody on this earth is really qualified to judge evidence of UFOs in the way that formal, orthodox technical matters may be judged.

The anticlockwise flattening of the Tully reeds is a *signature*. Such evidence of anticlockwise, energetic motion around these discs will be found woven into numerous incidents. Worth observing here is that a typhoon is an anticlockwise motion at the level of physical substance, and such a system *levitates millions of tons of water* as it crosses the ocean, dropping this water in turn as torrential rain. The same energy is likely to be involved in UFOs which levitate themselves. The common functioning principle is the orgone energy and its characteristics.

The high orgonotic potential associated with the Tully discs, for example, withdraws the orgonotic charge in the reeds under the UFO. The reeds are attracted upwards in the same fashion that animals are momentarily lifted clear of the earth by these high potential orgonotic fields. Physically, the reeds are torn free of the bottom. The whirling plasma around the disc browns the upper edge of the flattened reeds. The UFO nests thus appear as examples of orgonotic effects, and mechanical crushing of the reeds by a weighty object descending from above is insufficient to account for the observed effects.

In the Tully incident, which involved a discoidal object in the same physical density and polarity as ourselves, the question of what the object was doing nested in the reeds brings us once again to those familiar portals of biological matters. While it might be contended—albeit with some desperation—that it was a spacecraft undertaking repairs or resting its crew, such a view has little real grab or grip. What rings bells is that the nesting of UFOs *on water* amid the reeds is far more typical of living organisms than of machines, particularly of living organisms that fly, such as geese and ducks.

The biological implications immanent in the object taking flight upon the approach of a noisy tractor further reinforce a biological interpretation. The direct kinship to startled wildfowl taking flight needs no elaboration. Practical wisdom unfettered by any desire to prove that UFOs are exclusively machines from other planets will permit the merits of the biological explanation to be recorded. The very name "nests" bespeaks the living element. Subsequent discovery in the Tully area of more such nests suggests a nesting area—again typical of the living.

In this same area of Queensland, a discoidal UFO was observed pacing an airliner at high altitude. UFOs have similarly paced airliners, bombers and jet fighters all over the world. Such pacing is almost invariably interpreted as piloted action. Yet seabirds characteristically pace ships at sea, often accompanying them for hundreds of miles.

Bearing in mind what we have established about invisible biological UFOs in our atmosphere, we might reasonably anticipate that these critters will pace man's air ships just as seabirds pace man's sea vessels. Performance is moved into a faster framework, operations taking place in a less dense medium. If Ektachrome 160 Super 8mm movie film is exposed at high altitude in daylight through an 18A filter—the latter device being designed to block visible light and color—it will be found that UFOs of all the main types cavort around ordinary airliners. They may be so recorded *in full color despite the 18A filter*. The objects are "out there" more or less continuously, in positive, ocularly invisible polarity.

These objects—both creatures and craft—have become more widely visible in the past sixty years coincidentally with large scale human penetrations of the stratosphere and beyond. High and anomalous heat and pulsed EM have accompanied these penetrations. No wonder there is interest in airliners on the part of these denizens of the air jungle. The Queensland UFO nests seem far more like a nesting area for living creatures than a base for super-intelligent outer space beings. Detection of such a base would almost certainly lead to its abandonment, although no one can satisfactorily explain why highly intelligent entities would hide among reeds.

By allowing the phenomena to speak for themselves, we find ourselves returned, in the Tully incident, to the experience of Don Wood, Jr., and his encounter in the 1920s with a metallic, discoidal living organism atop a Nevada mesa. Final judgment on discs as craft should be reserved in view of the evidence. The facts strongly indicate that we are dealing within the discoidal UFO category with a multiplicity of aeroforms that may include both discoidal creatures and manned or remote-controlled constructs.

In both categories, our fundamental need continues to be additional technological application of etherian physics, further knowledge of the whole field, and the recognition that terrestrial science is lop-sided and must be brought to equilibrium. This means beginning with ourselves. The man of science in particular must ask himself how it is possible for him to move his physical bulk out of bed, drive himself to his office, walk to the lunchroom and perform all the other daily tasks that require *energy*. A science that cannot tell university students the difference between a living human being and a corpse will never penetrate the bioenergetic and etheric technology behind UFOs.

In the humanoid and construct area of this subject, the anticlockwise spinning field associated with discoidal UFOs is further verified by a 1959 encounter in France between a baker named Germain Tichit, and a landed UFO with humanoid occupant. This incident was originally published in the journal *Phenomenes Spatiaux*[5] No. 21 in September, 1969. The *FS Review* reproduced the account in its Jan./Feb. 1970 issue, translated by Mr. Gordon Creighton.

There is no space here to detail the incident fully, but upon sighting a large discoidal object on the ground near his bakery at 2 a.m.—from which a small humanoid dismounted—Tichit felt a veritable tornado blowing from this machine. The "warm and pungent" wind (a good description of a plasma) was felt by Tichit mainly on his left cheek, indicating in the words of the original article, "that the machine or some part of it was rotating in an anticlockwise direction". A powerful man and a former commando, Tichit attempted to close with the intruding humanoid, and the strong blast of warm air became quite an opposing factor in his passage across the grass.

The humanoid pointed a tube at Tichit. A beam of light shot out of it at the baker, making it difficult for him to breathe. While Tichit fought the combined effects of the beam and the blast of air, the intruding entity got back aboard his ship and took off.

The anticlockwise blast of warm air in this instance verifies yet again that the levitational force around these craft functions in this direction. The warmth reported by Tichit, who may be presumed innocent of any knowledge of Dr. Reich's discoveries, is consistent yet again with creation of a plasma in the atmosphere. This plasma may be itself the levitational agency, harnessing warmth ether to propulsion, or it may arise out of atmospheric resistance to a spinning etheric field of force that produces thermal effects.

Such physically heated air accounts for both the swirling and scorching in the Tully UFO nests, as well as for what Tichit experienced.

Other UFO nests have been found in other parts of the world—geographically, linguistically and ethnically remote from Tully, Queensland. These nests are nevertheless functionally connected with the Tully nests by the same basic evidence: high orgonotic charges, anticlockwise motion of force fields, and scorching. Worth noting is the concentration on the underside of certain discoidal craft of high orgonotic charges—cold, contractive, water-hungry energy. This affinity for water may account for part of the drying and scorching observed in these nest cases. Such withdrawal of water from plant substance may be taking place simultaneously with the generation of the whirling plasmas around the discs already mentioned.

In New Zealand on one occasion[6] manuka scrub from a 42-foot diameter nest near Ngatea, was found by horticultural consultant John Stuart-Menzies to have had every ounce of moisture in it instantaneously vaporized, cooking the plant material from the inside out. Since orgone energy penetrates all substances, we could anticipate high concentrations of this energy to withdraw water with volcanic force from any plant material near such concentrations.

Water seems likely to be heavily involved in our coming planetary technology. In the 1970s, an automobile using water as a fuel was invented in New Zealand by a man named Vincent, and demonstrated on New Zealand TV. A vehicle so propelled could revolutionize planetary life. Anyone who can find a safe way to the flash electrolysis of water, splitting it up into its component hydrogen and oxygen, will present mankind with a non-polluting, powerful and inexhaustible fuel. The intimate relationship between orgone energy and water would additionally suggest that water will be utilized in the new etherian technology that is just around the corner. What about those intelligences who are already around that corner—the UFO entities?

UFOs have manifested an interest in water from the inception of the phenomena in the modern period. They have been frequently seen down on the surface of the sea, lakes and dams, and cruising up and down rivers, on or immediately above the surface. There is a significant clue to the technical interest of UFO entities in water, and we have to go back to the summer of 1950 to revivify it with what we have learned of etheric energies in the interim.

On 2 July 1950, a senior executive of the Steep Rock Iron Mines at Steep Rock Lake, Ontario, Canada, was able, through a fortuitous set of circumstances, to make a protracted observation of a discoidal craft on the lake surface barely a quarter mile distant. About ten automaton-type figures of small stature moved about on the surface of the disc, while a vivid green hose drew water out of the lake and passed it into the vehicle through one of several open hatches. During these operations, a hoop-like antenna controlled by one of the humanoids kept rotating. With this device, the intruders were obviously able to detect the presence of a deer when it came down to drink at lake's edge.

The mine executive was observing through a cleft, and by ducking his head synchronously with the antenna's rotation past his hiding place, was able to keep himself from being detected along with the deer. Such a device bespeaks a mastery of biological energy—the ability to detect the difference between vegetation and animals and react to the biofield of the latter. Not much mental elastic is needed to conceive that the detector would also pick up human beings. Here we see how the New Knowledge won in the interim has enriched this particular account.

The mining company published a detailed version of this event in its house organ, The *Steep Rock Echo*, in two parts that appeared in the September and October issues of 1950. The observed removal of water from the lake can now also be re-evaluated as something other than a stop for drinking water. The latter is an obvious use, but does not account for the furtive quality of the visitation, with the entities taking technical monitoring precautions to ensure they would not be observed by human beings.

Water may be regarded as concentrated orgone energy—or the chemical ether in its ultimate physical expression. Small amounts of this energy, properly directed, can produce vast regional changes in the weather via the triggering effects set off by the cloudbuster. Orgone energy in the atmosphere of the earth is therefore a tremendous source of power that can be technically harnessed. Orgone energy weather engineering is only a primitive beginning. For entities riding spaceships, their technology may well involve using water as a fuel, as a portable source of chemical ether which they know how to liberate and concentrate.

Nor should we make the mistake of thinking that all our eerie and furtive visitants come from the same source, or stand at the same level of technical development, or use the same propulsion methods. The Steep Rock Lake callers may simply have a means of propelling themselves by splitting up water into hydrogen and oxygen. What we do here in this application of the New Knowledge is open the possibility of water as a fuel, both chemically and etherically, for those entities not technically able to transduce their supply of etheric force from the sun's emanations of life ether. Certainly it was not their wish to be observed taking on water at Steep Rock Lake in 1950.

Of wider scope and also involving interaction between earthly water and spacecraft, are those instances where spaceships have been observed to affect water surfaces in a quite specific way. Frank Edwards cites two of these cases in *Flying Saucers: Serious Business* on pages 303 and 304. Trusted, competent people see a discoidal vehicle five to six feet above the otherwise calm surface of a lake or river. Directly beneath the craft, and obviously caused by it, the water dances in thousands of tiny, sharp-pointed waves.

This is the *pointed condition* of orgone energy, always associated with excitation, and extensively dealt with in Wilhelm Reich's scientific works. The high orgonotic potential underneath the disc is a concentration of suctional energy in motion. The water below is strongly attracted under the law of ergonomic potential, and points upwards toward the source of excitation. High attracts low and withdraws its charge.

Movie actor Clint Walker happened to be one of the people mentioned by Frank Edwards as having observed this area of pointed, agitated water under a spaceship. Perhaps it is a long way from the fishing trip where Walker saw this manifestation out to the western Pacific where I have been in typhoons a number of times—there to observe the same effect on a vast scale. Anyone with unblocked vision can see this pointed effect of primary energy in a typhoon.

THE COSMIC PULSE OF LIFE

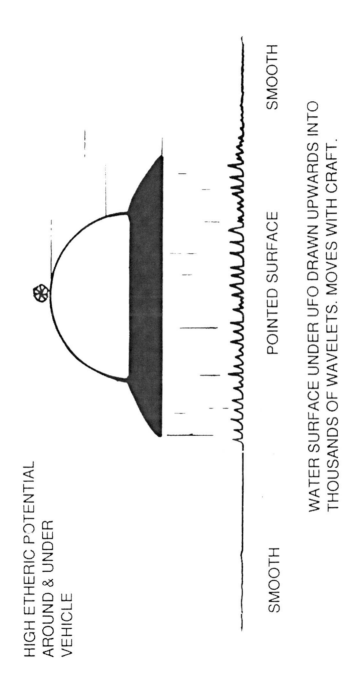

HIGH ETHERIC POTENTIAL AROUND & UNDER VEHICLE

SMOOTH

POINTED SURFACE

SMOOTH

WATER SURFACE UNDER UFO DRAWN UPWARDS INTO THOUSANDS OF WAVELETS. MOVES WITH CRAFT.

Under the influence of the extremely high orgonotic potential existing above the sea in a typhoon, the water is drawn up into characteristic sharp points that not even the wind can hold down. From these points enormous amounts of water pass over into the aeriform state. Clint Walker's UFO and the typhoon share a common functioning principle—orgone energy metabolism—and appropriate knowledge opens both happenings to further technical and intellectual enterprise.

In 1985, using advanced weather engineering apparatus on shipboard in the Pacific, I was able to reproduce this pointed condition of the sea surface in miniature. By building up a grossly abnormal high charge of etheric force above the ocean, ahead of the rapidly moving vessel, I was able to observe clearly, and to videotape, the appearance of an area of sharply pointed sea surface underneath this charged area. Thus there now exists an objective verification of what was written about the pointed condition of this energy in the original version of this book in 1975. Theoretical then, the information is factual now.

In this necessarily brief and restricted overview of the application of New Knowledge to UFO phenomena, much has obviously been left untouched and unsaid. A comprehensive book could be written today applying the New Knowledge to the gamut of UFO happenings from WWII down to the present. That is beyond the scope of this already Goliath-sized volume. What we have been able to do here is show that the way to penetrate, elucidate and understand some of the most incomprehensible UFO manifestations is to expand our knowledge and technical expertise into etherian physics. There should be little doubt that linear extensions of mechanistic method and thought will not be fruitful in this field, the roots of which lie in the ethers.

There remains now the task of rendering coherent, in some overall way, the grand design underlying UFO manifestations, especially since the wishful thought that "they" are here exclusively to bail out mankind has been torpedoed. Forces are at work in earth life—in ever increasing strength—that do not wish humanity to have any real grasp of what is going on. Our capacity to discern, digest and act upon what is happening will determine the human future. After all, earth is *our* planet. We are the ones who hold the power to decide whether or not planetary existence is going to go to the Devil.

NOTES TO CHAPTER TWENTY

1. See Eric Norman's *Gods, Demons and Space Chariots* (1970) and *Gods and Devils from Outer Space* (1973), both from Lancer Books, New York.
2. The reader is again reminded that the medical use of the orgone accumulator is Federally prohibited.
3. Abbreviation for Very High Frequency radiation, widely employed now in TV, radio communication and other applications.
4. Published by Jaymac Company, Belmont, California.
5. Published in Paris, France.
6. *FS Review*, Jan./Feb., 1969.

Chapter Twenty-one

THE BATTLE FOR THE EARTH

For we wrestle not against flesh and blood, but against principalities,
against powers, against the rulers of the darkness of this world, against
spiritual wickedness in high places.
—ST. PAUL to the Ephesians (Chap. 6:12)

The evidence, theories, findings and syntheses presented in this book are intended to delineate, in a general way, the massive struggle in which humanity is already involved—albeit unawares—for its own future in freedom. Invasion has not come upon earthmen from outer space. Mass landings in Washington, Moscow, London or Paris have not happened. Only diversions reach earthmen from the material planets, in whose material composition there is such persistent interest. While directing his gaze to outer space, man has left unguarded the gates to inner space.

Inner space, and not outer space, is the invasion route chosen by the Ahrimanic powers. A "fifth column" inside the human mind makes external force unnecessary. Washington has already been invaded from inner space. Moscow was long ago occupied by inner space invaders. Continuing ignorance of these malefic, invisible, but all-too-real forces will bring disaster not only to America, but to human evolution—to the destiny of man as a free being.

The Ahrimanic powers aim at the enslavement of mankind. If unopposed, they will overwhelm mankind and make evolution wholly Ahrimanic. That is their nature, and their function in the cosmic scheme of things. Man's future will not be salvaged for him by gaggles of golden-haired spacemen disgorging from spaceships in rescue squads. Man will win or lose the battle for the earth himself, for he is at once the goal of the battle and the battleground. The choice man makes—the extent to which he utilizes balancing forces to neutralize Ahriman's assault—will bring him victory or defeat. The stakes in this battle are not the territory, commercial advantages or political leverage of ordinary wars, but the mind and destiny of man.

Man has the right to decide whether or not earth life is going to be surrendered to Ahriman. With Ahriman's incarnation imminent in the West, man will be met rather more than half way by the adversary of the human future. UFO technology portends the kind of confrontation that lies ahead. Man faces a bewildering armory of advanced technical devices, transcendental abilities and mind-bending powers. Armed only with mechanistic thought and an unbalanced technology, minus access to the ethers, the human posture for meeting this stupendous and unavoidable event is both unstable and inadequate.

All depends in these latter days upon an expansion of human awareness. If man can be shown where the battlefield is, the nature of the terrain, and the ways in which he is already being assaulted in this inner war, then the right tactics and strategy can be brought to bear against the inimical forces. In our hard-minded, technical age, it is essential that we resist descent into mystical, ephemeral and vague concepts. Rather must we read from external,

243

objective events their true inner content—never losing our clarity of mind or our connection with everyday life and events as they unfold.

Outer space ventures by earthmen provide characteristic examples of the Ahrimanic powers at work. American programs and happenings are cited and analyzed here because the superficial facts concerning these ventures are a matter of record. What is said nevertheless applies with even greater force to the Russian space ventures and all others that have followed.

The alleged goal of Soviet space exploration was service to mankind, while Soviet mankind existed in steady-state drabness, compulsively connected to a gruesome political order from which spirit had been expunged. They lunged into space when they had no roads, and could neither feed nor house their people. A technologically backward society cannot serve mankind by going into space when its living standards are the joke of the world, for such action is manifestly irrational.

The real aim of the Soviet Union was the political extermination and absorption of the U.S.A. Spiritually inert and spiritually witless Soviet and U.S. politicos under psychic control were directed to this inhuman goal by the Ahrimanic powers. These excarnate beings of extremely high intelligence knew how to manipulate the neurotic-compulsive drives of Bolsheviks, because the subconscious is their means of access to our density and dimension. Such psychic control was also responsible for the entrainment of the U.S. leadership with the *glasnost* fraud.

The objective of the Ahrimanic powers is to pull down *all humanity*. To this end, whatever political systems and philosophies exist among men are simply manipulated. Mundane party politics, political systems and differences are therefore excluded from what is now being pointed out, because nothing definitive concerning the spiritual destiny of humanity can arise out of political studies or bickering.

The prodigious investment made by America in space exploration was trumpeted to the world as the opening of cosmic doors by inquiring man. Billions of dollars have been poured into these ventures, *but they have yet to produce a single, individual human being that we may say is outstanding in his wisdom, or in any way profound in his knowledge of cosmic processes.* The astronauts who were selected for these missions as the absolute cream of manhood, turned out to be just about as the late Chet Huntley of NBC-TV described them: "Nice guys, but dull as hell—mechanics."

Generations of astronauts have gone through the program. We have seen most of these unquestionably brave men and women descend from "cosmic exploration" to jobs as business executives, selling soft drinks, wearing out chairs in airline executive suites, popping into politics, hawking gasoline on television and in general turning completely away from their unprecedented experience as spacemen. Most have turned to things that are cosmically meaningless.

One of the exceptions is Captain Edgar Mitchell, who was transmogrified by his moon voyage, and upon his return to earth struck a trail into the new frontier of the mind. His case illustrates one phase of the battle for the earth. Other astronauts may have had parallel experiences.

Those astronauts who succumbed to various temporal lures—and remember that Ahriman is always the *tempter*—found their fame turned into fortune. Usually service officers, it was as though they had come into vast inheritances, an abundance permitting luxuries for their families and provision for their own dotage. They did not, in sum, become any kind of force for cosmic knowledge, cosmic questing or spiritual inquiry. They were not, for the most part, even inspiring men, just nice guys.

Edgar Mitchell carried out telepathic experiments on his moon voyage—on his own initiative—that showed the potential value of psychic communication. Russian workers had done far more much earlier, but Mitchell demonstrated in the front line of the American space effort the need to pursue, understand and refine these processes and develop them into practicable formats. NASA nabobs looked askance at Mitchell's doings, an individual effort to get a technical handle on the *same kind of communication used by UFOs since the modern inception of the phenomena.*

When Mitchell resigned from NASA and formed a civilian consulting firm, the goal of which was to pursue and promote investigation of the new frontier, his way became strangely hard. Assistance that he might have rationally expected to sustain his work, to underpin its growth and ensure its future, did not materialize. Other retired astronauts selling soft drinks, gasoline and airline tickets—staying away from that "psychic stuff"—had no such difficulties.

The question arising is, Why did the inner space explorer find himself with a hard way to go, while those who made no such effort were rewarded bountifully? The answer lies in the purposes and procedures of the Ahrimanic powers in subverting evolution. The prestigious Captain Edgar Mitchell, broaching the new frontier of the mind, perhaps to audiences of enthralled school children or on TV, could arouse, encourage, and reassure the inevitable consequent curiosity about the spirit. Older people, out of whom their natural connection with the spirit had been purged by mechanistic education, might awaken again, buttressed by the example of Mitchell. He was therefore a man to be restricted in every possible way. He had remained sequestered, despite his best efforts, up until 1987 when this book was translated for European readers. The astronaut selling gasoline, by contrast, got his message into every home in America.

Humans are constantly being inveigled into doing the work of the nether forces because they simply do not know such forces exist, let alone how they work into and upon earth life. Ignorance of this kind persists in the face even of the tumultuous events of the twenty-first century, and amid obvious motion of the human race toward a climactic period. Avarice is the main agency by which humans are seduced away from life-positive work and thought. Incomprehension of spiritual forces and the institutionalized denigration of the spirit in formal education, make humanity pitifully vulnerable to dehumanizing, life-negative and destructive trends.

Most of the businessmen running the financial end of the space program know that a great deal of what they are doing is unjustified, intrinsically without merit and little more than the creation of an artificial market by the Federal authorities. Yet they proceed to organize, manage and carry out these programs for the simple sake of making money. Technological spin-offs are cited as public benefits from public money expended on the programs. Medicine and surgery have been aided. We have pyroceram pots and pans. In between lie a million more inconsequential toys for humans to play with, prattle about or transmute into new sources of business profits.

As for cosmic workings, cosmic processes, cosmic rhythms, cosmic energies and UFOs, the space programs have produced a low yield of information, while burying our scientists under tons of mechanistic data. UFOs have been sighted on NASA missions, and sometimes photographed. John Glenn, Gordon Cooper, James Lovell, Buzz Aldrin, James McDivitt, Ed Gibson, Michael Collins and Alan Bean are among the astronauts publicly reported as having sighted UFOs. Some of these men, when publicly questioned, have denied any such experiences, but that may be due to security regulations and government orders. At least 15 million Americans are reported in a Gallup Poll as having seen UFOs. Scores of millions further believe that there is as yet some undetermined energy principle behind these phenomena. All these people pay taxes. Yet the current billion-dollar NASA budget contains not a dime for UFO investigation. Why?

What frightens NASA? Disquieting information? Perhaps fear exists, in high places, that someone might come across the same principles of bioenergy, cosmic electronics and cosmology broached in this book. That is a well-grounded fear. There are battalions of brighter, younger, better-trained men than me in the space program right now. I know. I have seen some of their private, avocational work. Properly supported and encouraged, they would prove out what I have merely been able to scratch the surface of. The financial despotism of the government and big business blocks that. Ahrimanic aliens put high officials under psychic control and such men say: "*I can't see spending money on UFOs.*"

Technology has made only limited advances out of a basic will to serve human progress. The driving power for technical development has come primarily in the twentieth century from the two world wars—the venting of destructive drives on a planetary scale. Without the wars, for example, it is doubtful if aviation by the 1970s would have been advanced much beyond the Fokker tri-motor. War brought today's miraculous jet age to mankind decades before man at peace could have mastered and developed the principles involved. The urge and need to bomb humans from the air created the technology behind the airliner.

Where destructive urges become institutionalized and focused into national policy, such as in a failed Russia, superior destructive activity has to be called forth to maintain what little freedom there is in the world. Destruction has been accelerating toward a climax. No thought is given, and no attention is paid, to the spiritual origin of these destructive rampages—to the way in which certain humans are inspired and tempted from the inner planes to mount emprises of slaughter that eclipse anything in prior human record. The Soviet and American peoples were both being manipulated, thesis and antithesis, so that an evil synthesis could be wrought from their conflict. With the Soviets gone, a new enemy has replaced them. Formal religion obviously does not understand the nature of evil, and therefore can provide no effective countermeasures. "Evil" is an unmarried boy and girl in the embrace of life!

Because weapons technology has expanded military action to worldwide operations in pursuit of geopolitical objectives, the occult background to political leadership, action and strategy must become far more widely known and understood. The physical power at the disposal of political leaders, which has become literally the power to dispose of the earth, makes it absolutely essential for humans to awaken to the reality of psychic powers and forces. The plain truth is that unseen, extraphysical entities have expanded their manipulations and destructive capabilities on the physical plane via humans they can control. Until recently, such suggestions would have been swamped in ridicule.

Anyone wishing an eye-witness description of Hitler going under Luciferic control, can do no better than read Group Captain F. W. Winterbotham's *The Nazi Connection*, published in the U.S.A. by Harper and Row, New York. The British master spy describes the transformation he saw, while sitting on a rostrum almost at Hitler's side, immediately before the dictator gave a speech. Esoteric students will require no further evidence of psychic control in Hitler's case. Winterbotham is a witness of unimpeachable integrity. He was later responsible in World War II for ensuring the secure delivery to Winston Churchill of the Ultra Secret decrypts of the German High Command's radio communications.

Hitler's irrational drive to assault the Soviet Union may now be seen as Lucifer making his final major effort to block the mounting strength of the Ahrimanic powers in the twentieth century. In the Luciferic tradition, Hitler was a man led out beyond himself, the whole enactment being upon the world stage.

Earlier in this book heavy emphasis was laid upon the need for Western humanity to open itself to the work of Dr. Rudolf Steiner. The Austrian philosopher possessed *exact clairvoyance*, the faculty of perceiving supersensibly with the same clarity of mind essential to scientific work. In Trevor Ravenscroft's book *The Spear of Destiny* it is stated that on his pathway to power, Adolf Hitler considered Dr. Steiner his most important and dangerous enemy. The reason? Steiner could maintain surveillance from the astral plane of the machinations of the satanistic Thule group that had recognized Hitler as the Luciferic messiah.

Nothing could be hidden from Dr. Steiner's occult powers. Although he did not concern himself with politics, Dr. Steiner had observed the Luciferic preparation for Hitler, and Hitler's preparations to become the vehicle of Lucifer. Accordingly, Steiner publicly warned of the disaster that was coming to Germany and humanity through the secret, occultly inspired aims of the Nazis.

Hitler planned the assassination of Dr. Steiner at the Munich railway station in the spring of 1922. Inside his compartment, a sawn-off, double-barreled shotgun was to be discharged into the face of this Christ-like man. Dr. Steiner knew of this planned murder through the same faculties that had opened the inner workings of the Nazis to him. His high ethical and moral standards however, would not permit him to use those powers to avoid the attack, for this would constitute black magic or misuse of those powers. He consequently appeared at the Munich railway station on time.

Before Steiner could board the train and meet the bloody fate planned for him by the Hitler gang, author Ravenscroft's mentor, Dr. Walter Johannes Stein, arrived with friends and students of Rudolf Steiner. Surrounding the great teacher, they hustled him rapidly from the Munich station to safety. Walter Johannes Stein had penetrated the Thule group around Hitler, and learned in time of the murder plot.

Without the occult background to these as yet little known happenings, no credulity could be given to *Dr. Rudolf Steiner's status in Hitler's mind as the would-be dictator's greatest enemy.* Twenty years later, Hitler was to be master of Europe and very nearly of planet earth. Mechanistic history could find no reason for elevating an obscure Austrian philosopher to primacy among Hitler's enemies. Hitler was the vehicle for the Luciferic principality of evil. Rudolf Steiner was the prophet of the Cosmic Christ in the twentieth century. No wonder Hitler wanted him off the earth.

The development of spiritual selfhood that Dr. Steiner heralded is essential if humans are to understand what is really behind UFOs—the whole fabric of new scientific principles embodied in their technology. Such an understanding is inaccessible to unaided mechanistic knowledge, and in its more compulsive elements united to sexual and bioenergetic blocking, mechanistic knowledge can be an impediment to such an extension of human cognition. Without a spiritual awakening, with each individual resolving and working upon his own further expansion of consciousness, there is no possible way to penetrate the UFO mystery. An immense expansion of consciousness is essential.

Human beings seeking to find a healthy pathway to the spirit, and determined to undertake the methodical, systematic steps on that pathway are anathema to the Ahrimanic powers. Any man or woman undertaking such a course in the full light of consciousness, and aided by the faculties developed by scientific thinking, becomes increasingly resistant to the subtle temptations of Ahriman. Eventually, such a soul escapes the clutches of the Ahrimanic powers, and therefore, cannot be tempted or seduced into life-negative, destructive, death-dealing activity. Human beings must at all costs—in the eyes of the Ahrimanic powers—be prevented from re-establishing the lawful connections between themselves and the cosmos of which they are both part and product. All of us are centuries upon centuries old. We have been Chaldean star-worshippers, Egyptian priests, alchemists, Atlantean engineers and scientists, Mayan magicians and many humbler things through the ages. The primeval cosmic wisdom lies deep in all of us. What we knew atavistically and with a different kind of consciousness in ancient times, we must now elevate and clarify in twenty-first century terms and turn into a new, life-positive technology.

The Ahrimanic powers know that if we come to understand what is behind UFOs—spiritually no less than technically—the earth will be redeemed for Christ. Therefore every effort to find the crucial, golden threads is subject to countering Ahrimanic force. A major Ahrimanic stratagem in the battle for the earth is exploitation of the neurotic deference inculcated in human beings toward the forms of "authority". Modern humans are all too willing and ready to be reassured and diverted from significant matters by generals, presidents, doctors and spokesmen for learned societies.

Where the existing neurotic mechanisms in these processes are insufficient, vulnerable human beings in positions of authority run the risk today of being placed under various forms of psychic control. Etheric tunings, via advanced radionic-type instruments, are in all probability employed to this end from the unseen. The victims think, voice and write, and otherwise place into earthly currency, the misleading falsehoods concerning UFOs that the Ahrimanic powers wish to have disseminated as part of their own continuing concealment.

High authorities in government, science and the military are in this way inspired to tell their fellow men that the whole subject of UFOs does not justify scientific study. UFO are purported to attract lunatics, fanatics and psychopaths. The media adopt juvenile and obstructive attitudes toward phenomena of the utmost import. Rational inquiry is discouraged. In those circles where humans are convinced of the reality of UFOs, the fiction is foisted that they are ships from other planets, manned by humans like us, and nothing more. What is essentially the beginning of etheric science for man is drawn back constantly to appear within the mechanistic rubric. Materializations and encounters are staged to sustain and support this flim-flam.

Political corruption in America has reached unprecedented levels. Political corruption is simply the expression, in and through the body politic, of the corruption being poured into the souls of men by the Ahrimanic powers. America has had to wait two centuries to get a series of corrupt administrations, which is not in any respect a party question. As the 21st century begins, *corruption becomes worse*. Even law enforcement officers as high as the attorney general of the United States have turned out to be crooks.

Is all of this an accident, or simple coincidence? Simple persons might believe that. Rather is it the objective expression of an accelerating inner process—corruption produced by the depredations of intelligent entities who are invisible to us. They are able to function efficiently because of the tremendous biasing power of modern, mechanistic education, which structurally blinds the most intelligent men and women to all spiritual existence.

There is an old saying that the devil looks out for his own. Worth noting here is that the convicted crooks and criminals of the Nixon Administration did not survive to live in the disgrace they deserved. On the contrary, the corrupt American publishing industry made most of them wealthy, paying them small fortunes in advance royalties on their books, filled with lies and phoney, fake contrition. These criminals and numerous other life-negative and criminal types, are the chosen people of the university lecture bureaus, who present them to America's younger generation at fees running as high as $3,000 per evening. The lesson can only be that crime pays.

In looking at the occult commerce behind the contemporary American scene, we fog our clarity of mind and thought if we permit mundane party allegiances or preferences to interfere with such analysis. In the case of Richard Nixon, his manipulation by psychic forces is painfully obvious. He came under the sway of powers whose temporal objectives include the destruction of America, which means the destruction of the opportunities abounding in America for humans to develop their spiritual selfhood in freedom.

In Mr. Nixon the people had a leader who had exhibited throughout his political life an open dislike and mistrust of Russia and the communist philosophy. When eventually elected to the world's most powerful office, after the two Kennedy murders produced dramatic re-alignments of American affairs, Nixon strangely began to manifest attitudes exactly opposite to those held by him down through the decades. Within a short time, he turned completely around. Modern history may be searched in vain for a comparable example of such a political somersault by a high official.

Suddenly, it became urgent that he go to Red China. If the "price" of the visit is the eviction of founder Taiwan from the United Nations, then pay it, and make light of the deceit. Suddenly, a *rapprochement* with the Soviet Union becomes vital. Suddenly, the President of the United States forgets that the individuals running these nations are compulsively dedicated to America's ruin. Suddenly, there is no contact with the high men of his own party, and their hundreds of man-years of experience are excluded from the management of American affairs.

The President was deceitful concerning his income tax. The verbal environment in the innermost sanctum of the White House, the political and *spiritual* center of America, becomes saturated with constant profanity. The power of words is never underestimated by persons of occult knowledge. If a man uses constant profanity he creates for himself a psychic environment that is profane. If he surrounds himself with profane men, their profane force is added to his. Soon there is created a profane center wherein flourish unseen beings with profane motives. Such an environment is derogatory to spiritual progress and perfect for enemies of spiritual progress.

Such processes as these, opening American life and fortune to control by the powers of darkness, created America's downskidding fortunes and descent from grace and respect in the world. The Watergate exposure caused the whole festering mess to erupt in full public view, but many didn't recognize salvation when it came. Consolation was sought in mere party notions, and Lyndon Johnson was "just as bad as Nixon". That is not the point. Under LBJ profanity also prevailed at the spiritual center of the nation, and the American direction was also *downward*. The relationships and facts are all there if we will but look beyond illusory superficialities.

American fortunes have been in continuous decline under profane, spiritually inert and psychically vulnerable men. Calling in Billy Graham for a prayer breakfast hardly suffices, even once a week, to repel the boarders from downstairs. Such prayers get no higher than the ceiling, and are thrown out like a sop to the millions of good American people who pray heartfully every day for their country. If LBJ is spitting expletives and cursing later in the morning of the prayer breakfast, he merely re-establishes profane conditions.

The American people will henceforth need to ensure, if their fortunes are not to come under illegal control from the inner planes, that people of unassailable moral and ethical quality alone are elected to high office. Party affiliation is of no great significance. The inner sanctum of the White House, the spiritual center of America, can only be protected in this way. If a man goes into that sanctum as a profane ignoramus of spirit, and especially if ruled by his sex drives in an immoral fashion, he will be manipulated from the inner planes like a marionette. The entities whose aim is the destruction of America can use and direct life energy far more proficiently than we use electricity.

These matters are apolitical and concern all mankind. America commands the only temporal force able to block the neurotic-compulsive aim of the

Russian leadership to impose its political rule on the world. Mr. Gorbachev is a braver man than perhaps the world realizes, for he defied the Ahrimanic powers in their own created domain. Ahrimanic hegemony over the earth is not intended to be *glasnost*, but the opposite. Tight circumscription of personal liberty and a crushing out of the spirit would characterize such a world order, of which Russia may be deemed a prototype: an elite living in grandeur, and the rest in rags. Walls, barbed wire, dogs and machine guns confine the hapless populace to the Bolshevik paradise, which we do well to remember was created by Wall Street. The confinees merit our sympathy.

Only police-state rigidity and physical force can hold such a system in place. Where material progress is set as high among the national goals as it is in Russia, with science and technology centrally directed to the promotion of material progress, we find a significant anomaly. Ninety years after the red revolution, Russia still cannot feed itself, despite decades of propaganda concerning the skill of its agronomists. In the military area, however, where the perfection of destructive devices is the goal, Russia stands second to no nation. There is in fact a superabundance of weaponry—no shortages appear there!

The enormous spiritual energies of the Russian people for good ends have never been tapped—only subverted, frustrated and blocked. The numerous nations making up Russia have populaces who are like other human being everywhere—they have been incarnated before on the earth many times. In different bodies and in different tongues and epochs they are discharging their karma and perfecting themselves as individuals, just as is the rest of humanity. All this predates communism, and will long survive any contemporary political formulation if the battle for the earth is won.

Souls with prior earthly experience in science and technology—in earlier and even prehistoric epochs—would be drawn now to incarnation in a culture enshrining science. We should expect to see evidence of this appearing in objective, everyday life and events. Russia shows abundant evidence of having more than its share of Atlantean and Egyptian scientists reincarnated among its present people. The evidence lies not only in the area of technological weapons—perversions of science—much more powerfully in the very areas where we should expect to find such ancient knowledge pressing up into modern expression: *in borderland areas that lead directly to the spirit if pursued with honesty and truth.*

Natural science everywhere has what might be termed a perimeter of acceptability. In no small way, this perimeter is influenced by the neurotic factors that produce structural blockage in human beings against approaching the roots of the living. Wilhelm Reich has laid that out clearly for humanity. There are always individuals who, through their karmic past and fortunate earthly circumstances, press freely on past the perimeter of acceptability into the borderland where they work with complete comfort and considerable achievement.

Russian scientists have pressed into this borderland both before and since the revolution. In their investigations of bioenergy and kindred phenomena they raced far ahead of the rest of the world, with the exception of Wilhelm Reich—that astounding soul who was forced by the stupidity of his contemporaries into isolation. Biological radio communications, some of the cosmic and metaphysical elements of which have been broached in this book, have occupied Soviet scientists working with EM apparatus and procedures for more than 80 years. Their achievements and findings are both ominous and enlightening.

In *Psychic Discoveries Behind the Iron Curtain*,[1] Sheila Ostrander and Lynn Schroeder have provided a panoramic overview of Soviet progress in these borderland areas of science. The bibliography to this book shows the scope of Russian inquiry. While this Russian work appears within an essentially mechanistic framework, the etheric realities leak into and through the objective findings and reports. Synthesized with the discovery of the orgone energy and its determinism, the Russian work becomes a scientific and technological bonanza.

Applicability to the UFO problem leaps out of the book everywhere. For example, V. M. Inyushin, Ph.D., Director of the Biophysics Laboratory at Kazakh State University in the former Soviet Union, has written a paper entitled *"Bioplasma: The Fifth State of Matter?"*[2] a translation by J. Paasche of San Francisco which lies before me. As a learned discussion of biofields, Kirlian photography and organismic radiation, it provides several splendid takeoff points for scientific investigation of my critters in the atmosphere.

In a 1962 book entitled *Biological Radio Communications*[3] Dr. B. B. Kazhinskiy, an electrical engineer, gives a mind-blowing summary of Soviet work in exploring mind-to-mind communication and establishing its determinism. What I have said earlier in this book about the use of the visual ray or eyebeam as a UFO detector is to be found in scientific terms and experiments made decades ago in Russia. The bio-waves accompanying every thought process have been under investigation in Russia since before the days of the vacuum tube. Experiments proved that induction of thought control was possible not only on animals, but also on humans.

The induction of images and sensations from one brain to another brain at a distance was proved possible. Demonstrated in the same chain of experimental work was the artificial postponement of the emergence of mental images and sensations. Such emergence of these images in the recipient could be postponed to a predetermined moment. This work has actually been done on earth by human beings of known identity, with proper scientific controls.

Such techniques as these will be instantly familiar to every person having adequate knowledge of UFO contact cases. The UFO entities have this knowledge in high refinement and full technical utility, and have used it on contactees.[4] The alien visitants, in this, as in so many other areas, demonstrate mastery of what remains incipient among earthmen. Knowledge is power. Knowledge of how it works leads to knowledge of how it may be countered, so that no person—seen or unseen—may play with the brain of another as is already happening to the disadvantage of mankind.

Psychic Discoveries Behind the Iron Curtain leaves little doubt of the high interest evinced by Soviet scientists in using one mind to control another. The Russian authorities had the power to clamp down on such scientific work, either through censorship and reduction of foreign contact, or through direct financial and political control of the laboratories. As the fruits of the monumental Soviet efforts in borderland areas flowed out to the world, an almost unprecedented community of interest developed with scientists outside the communist bloc. Commerce became lively, friendly and wholesome.

By the middle 1970s, however, what might have led to considerable commerce was sharply reduced by the Soviet authorities. Bioenergy and parapsychology activity became more difficult of access to outsiders. The biologist-parapsychologist Edward Naumov, guide to Ostrander and Schroeder when they made their Russian odyssey, disappeared into prison after a kangaroo trial. Official Soviet attitudes to this area markedly cooled.

All this work is intimately related to the spiritual progress of human beings. The tons of mechanistic data must inevitably be integrated into a higher synthesis—an approach to the unseen nature of humanity in its totality. Science proceeding in this way must open higher consciousness in the long run, complete with new modes of mentation. If not interfered with politically or psychically, or both, true scientific inquiry will ultimately break the Ahrimanic lock on scientific cognition. There is therefore impedance everywhere to science in areas where evidence of the ethers bursts through or is likely to do so.

In the U.S.A., the major Ahrimanic stratagem for discouragement of such scientific investigations is economic stricture and economic censorship. America is well endowed with qualified people eager to pursue borderland areas of thought and cryptoidal phenomena. Unfortunately, research that cannot be turned fairly directly into business profits or defense hardware gets little support. The great foundations largely use their tax-sheltered dollars to sustain the existing order. Active discouragement comes from official science, rooted mainly in neurotic opposition from the Old Guard with its anxieties over the maintenance of its own emotional security and the mechanistic world order. A scientific "underground" has been the result.

The underground unites informally a group of humans who have elected to reject official and government views of such phenomena as UFOs. Many are employed in government science projects. Others teach at major universities. Still more work in the aerospace industry. They know what is going on. Among them are some of the most acute New Age minds in America. They correspond. They cross-fertilize each other's work. Most of them have already privately theorized and intuitively known most of what is in this book. This underground is the free enterprise of American ideas concerning the New Knowledge—the beginning of the New Age sciences in America. Nearly twenty years ago, when I began groping with UFO phenomena, there was no such underground.

A strong voice recruited to ufology from official science is physicist Stanton Friedman, a Californian. Close to half a century of open-minded study and investigation leave him in no doubt concerning the reality of UFO phenomena. He has been active on the lecture platform for years.

Nuclear engineer Thomas E. Bearden, fortified by a strong mathematics background and with spiritual acuity sharpened by long devotion to aikido, has written several penetrating books on the UFO subject and peripheral matters. He is unafraid, from the bedrock of formal knowledge, to spring off into rational presentations on materialization and dematerialization, biological radio communications, tulpoid phenomena and a host of kindred matters.

The anthologies of author and teacher John White, *The Highest State of Consciousness, Frontiers of Consciousness* and *Future Science,* present the theories and thinking of several dozen extremely bright minds. A vast spectrum of inquiry is explored in these books.

There are, in addition, numerous young and well-qualified men and women who have resisted the biasing effects of classical education, and who, like Thomas Bearden, spring off into New Knowledge from the bedrock of the old. Veteran investigators and technologists like the late T. Galen Hieronymus—mentioned earlier in discussions of the late Ruth Drown's work—are receiving at last the intelligent curiosity of men of science that their work deserves. New interest appears in the free energy or cosmic energy technology of the late T. Henry Moray. Wilhelm Reich is discussed on every campus. Everywhere men and women are beginning to choose sides— for or against the New Knowledge—as the so-called "occult explosion" augurs a general realignment of popular thinking.

There is currently more interest in "magick" of all kinds than at any time in the past century. University students dabble in black magic and the development of paranormal powers and capacities. The old barriers and taboos against such things are being eroded. The drug binge of the Sixties led to spiritual and occult inquiry in the Seventies of a far more measured and healthy nature.

In the battle for the earth, anything that leads men to the spirit will be opposed. External events verify and reverify that the keystone that must underlie any durable world order based on true brotherly love—the spiritual-scientific understanding of the origin and constitution of all human beings— is being subtly but constantly opposed. The UFO subject has illustrated this with convincing force.

Entities riding etherically-propelled vehicles, and obviously in mastery of psychic control in all its forms, devise contact encounters with ingenuous human beings who can be used in various ways to serve certain ends. Most of these encounters produce only chatter and bewilderment. These entities often hypnotize their victims, use them for various experiments, and then give post-hypnotic suggestions of various kinds that will serve whatever end is sought. As James McCampbell has pointed out in his *Ufology*[5] on p. 113: "Frequent invitations to board their craft are issued but they are sometimes accompanied by a sinister implication or an actual threat."

In no way can such actions be evaluated as other than *unethical.* No teacher of humanity toys with the consciousness of his student. Yet what happens in all too many of these encounters is that the lives of the contactees are adversely affected. Sometimes they are psychologically disoriented or upset. Often they are treated later to additional obsessional visits by the same alleged spacemen. In their own world, the contactees are ridiculed by the media and professional skeptics.

What is the overall consequence of these depraved encounters, where an innocent human being is set upon by these weird humanoids, standing before him like the juggling fiends stood before MacBeth? The world is led to believe that material craft are involved, if convinced at all, and if not convinced, then the contactee is another "flying saucer nut". Either way, the world gets a lie, overlaid with confusion and ridicule, while the humanoids depart from view with certain etheric transductions *regarding which they tell their contact nothing.*

These are the forces opposing human redemption and the redemption of earth evolution. They are operative not only in spaceships when occasion warrants, but prolifically in tempting human beings downwards—morally and ethically. Devices, knowledge and thought leading to the spirit are subject to constant opposition. A Ruth Drown is legally murdered. A Wilhelm Reich dies in prison. Rudolf Steiner's books strangely disappear from public libraries. U.S. Government officials cannot find any money for UFO investigation, or to investigate the energy of life, but they always have at least $300 billion available for destructive devices and applications, every year. The medical profession pours millions of tons of chemicals into the national bloodstream so that half the populace is almost perpetually deadened or spiritually unintegrated.

Ahrimanic emissaries appear everywhere, unrecognized and often aided by humans who don't know that the devil is indeed alive and well—and coming to earth possibly within the lifetime of millions now living. Ahrimanic messengers inject themselves into UFO contact encounters to sow confusion and disorientation and split and shatter study groups. Many of these aliens temporarily materialize here from the lower astral plane, and in stumbling about our countryside are themselves disoriented in the terrestrial milieu, constantly asking humans the time. John A. Keel has abundantly described such incidents in his *The Mothman Prophecies*.[6]

Human beings owe it to their own personal futures not to be misled by the Ahrimanic obscurantist offensive launched through the UFO field. The experiences of contactees are not going to lead humanity out of its present troubles, or around the mounting planetary crises involving hunger, pollution, population and the threat to man's humanity posed by unrestricted proliferation of technology. Etherian physics is what UFOs are all about. That New Knowledge will open a new kind of technical development harmonious with man's own etheric being, and able to deliver him from this planet's seemingly overwhelming problems.

Certain intelligences of high moral and ethical standards have approached earthmen in vehicles created by an etheric technology. The performance and capabilities of these craft have been demonstrated in such a way that the principles can be mastered by contemporary humans. These Christ beings do not impose upon mankind a technical order in many ways contradictory to that existing on earth. Man is shown. Man has native and latent powers that he can arouse and focus. He is not presented with a power source that conflicts sharply with the destructive elements of his own mass character. He is merely shown where to begin.

If man proceeds to understand the etheric formative forces in the cosmos, earth and man, he will proceed down the road to liberation and spiritual selfhood. A new technology, including a new agriculture, would be inevitable. A new education would certainly come to replace the old, as presaged by the international Waldorf school movement founded by Rudolf Steiner. UFOs appearing to man as the Cosmic Christ's emissaries beckon to man to take the human future in full charge. Man must do it himself.

To block this wholesome development, the Ahrimanic powers have created a counter-manifestation, excising from it all connection with the etheric formative forces. UFOs thus became "ships from other planets", pursuable and penetrable with orthodox thought and orthodox methods "if only" scientists could get enough instrumentation to the right place at the right time. UFOs have failed to yield to the old ways of thought because even the "fake" UFOs dangled in front of humans by the Ahrimanic powers have to be brought into existence by the same scientific principles utilized by the Christ forces.

The battle for the earth is a spiritual struggle, but it involves enormous temporal events and processes. In this battle, we are called to redeem the earth through the creation of a new humanity—a humanity that will arise only if we can restrain ourselves from shaping the new children to the patterns of the past. Only with and through such a new humanity can a world order based on true brotherly love come into being and sustain itself. Herein lies a major evolutionary hazard for mankind.

The current political and economic drive toward world government is a stratagem of the Ahrimanic powers, to pervert and subvert authentic human renewal. Insidiously and cunningly, the idea is promoted that through an organization, human longing for freedom and peace can be fulfilled. Dr. Reich has clearly delineated for us the severe social difficulties created by the armored mass character of civilized mankind. The armored mass character precludes, for the time being, the establishment of a rational world order rooted in love. Humanity can move in this direction only slowly, and by virtue of individuals taking full responsibility for the improvement of themselves and their children.

Wise are the few present day leaders who comprehend this central, functional truth about mankind. Respecting its reality automatically moderates the wild expectations held out by the ignorant—and the devious—who promote world government today. If you have shared my adventure to this point, you will probably already be examining the "world government" scenario for those crucial signatures by which we may read this old world aright.

Who is backing, and who is behind world government in the depths? The same personalities who control existing governments without being elected. They are the inner elite of the secret societies, who have commerce with the unseen Ahrimanic beings, and who are their hands and feet on this plane. They are the financial powers behind the Bolshevik Revolution in 1917, who did their best to ensure the survival of this gruesome Frankenstein. They are responsible for the previous bondage of the Soviet population, a monstrous crime for which they are karmically accountable.

These are also the manipulators who control and mold public opinion, but not mine, and perhaps now, not yours either. The basis for successful world government cannot lie in any organization promoted by these masters of deceit, these rulers of darkness. The basis lies with YOU.

Great avatars who have walked the earth with us and shared our hopes, dreams and agonies have shown us where and how to begin, at the sacrifice of their own lives. The new civilization begins with the securing of the genital rights of babies, infants and adolescents—the only practical way to free humanity of the burden of armoring and to launch authentic human renewal.

The spirit reaches easily into an unarmored organism. Each individual—each parent—must make this beginning because it will never be found with organizations, governments, senates or soviets. Too big a job? Too long a job? You'll be dead before it's started? Perhaps. But *you'll be back* for the earth's tomorrows, and perhaps next time not as well placed as you are today.

You have the power to start pulling humanity out of its crisis, and it is the same power that stands behind UFOs. That power throbs in your heart and is behind the coursing of your blood—*the cosmic pulse of life*.

– END –

NOTES TO CHAPTER TWENTY-ONE

1. Published by Prentice-Hall, Englewood Cliffs, New Jersey, 1970.
2. Published in *Future Science*, edited by Stanley Krippner and John White, Anchor Books, New York, 1976.
3. Published by Izcatel'stvo Adademii Nauk Urkrainsey SSR, Kiev, 1962. The U.S. Defense Department saw fit to have this book translated.
4. See *Incident at Exeter* by John G. Fuller, G. P. Putnam's, New York, 1966, for a classic case.
5. *Ufology: New Insights from Science and Common Sense*, by James McCampbell. Published by Jaymac Co., Belmont, CA, 1974.
6. Published by E. P. Dutton, New York, 1975; reprinted as mass market paperback by Tor Books, 2002.

Appendix I

PHOTOGRAPHS of INVISIBLE UFOs

The reader studying these photographs should be keenly and constantly aware that reproduced here are pictures of *invisible objects in the skies of planet Earth.* They have been photographed directly from the invisible state by methods described in this book. In all save a few instances, the late Dr. James O. Woods and myself were able to capture these UFOs in scenes containing known terrestrial references.

Completely standard cameras, catalog films and filters were employed throughout, all of them obtainable from a professional photographic store. No special processing was utilized. This gallery of pictures from the unseen borderland of our own physical world, includes the classical domed shapes, luminous spheres, fast-moving balls, cigar shapes, whirling fields of plasma and living organisms that I have subsumed under the term "critters."

The Alpha Series came first. Then followed the Bravo Series subsequent to the publication of *They Live in the Sky* in 1958—now some fifty years ago! The Charlie Series first linked the orgone energy and the cloudbuster into the UFO mystery.

I was not the only person in the world to conclude, amd then prove photographically, that the mysterious UFOs were basically invisible. Samples of the work of two such pioneers are presented here in this gallery. The late Luciano Boccone and his GRCU group in Genoa, Italy, have provided prolific examples of active, unseen UFOs right down alongside us, in our own space. The Romanian engineer, Florin Georghita, has proved with his work, that he basically agrees with me—that the unseen is the source of the UFOs.

My Reverse Spectrum photos are the first examples of unseen aeroforms, both critters and craft, accompanying airliners at high altitude while remaining invisible to all on board those airliners. These photos are in full color on Super 8mm movie film, although optical scientists will be unable to explain how such images can be acquired via an 18A ultraviolet filter. An 18A filter absorbs the visible spectrum completely, yet reds and yellows and blues and whites sail right through the 18A to inscribe themselves in the film emulsion. In this book, only monochromatic reproductions are presented here for production and cost reasons.

Several photographs of UFOs from reliable sources are also presented with brief analyses based on the technical principles dealt with in this book.

Again may I emphasize in conclusion that my photographs are of *invisible objects.*

—The Author

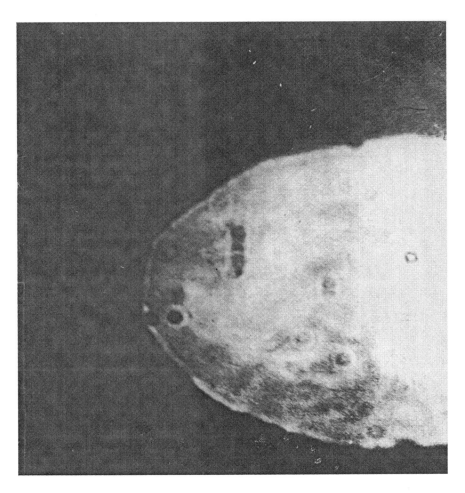

ALPHA #1

This amoeba-like invisible UFO, replete with nucleoli, vacuoles and the general appearance of a unicellular organism from the microscopic world, was photographed by the author on 25 August 1957 on the Mojave desert. The object was initially immediately over the author's head, and not directly tangible to the eye. Five successive photographs were made of the object as it moved from overhead to positions that permitted inclusion of local terrain in the pictures. Infrared film, sensitive beyond the range of human sight, was used in a Leica G 35mm camera, fitted with an 87 filter Exposure: f3.5 1/30 sec. Development: Microdol, twice normal.

259

ALPHA #2
Second photograph in series showing strange plasmatic bioform over Mojave Desert. The object appears to have expanded physically, because physical marks resembling vacuoles and nucleoli visible in Alpha #1 also appear on this object, but in smaller relative size to the main body of the object.

ALPHA #3
This is the fourth photograph in the series and bioform has now moved into a firm relationship with the desert terrain. The sky background is black, because the 87 filter employed over the lens absorbs the blue sky radiation. Once again, the physical characteristics visible in Alpha #1 and Alpha #2 are readily identifiable.

ALPHA #4

This photograph was the last in the series and provides substantial evidence that the object(s) photographed were living creatures rather than spaceships. The lower of the two objects captured here appears with obvious "eyes", a concave face and a protuberance highly suggestive of a bill or beak directly underneath the eyes.

261

Cross-sectional drawing of a ventla-type spacecraft made and published by George Van Tassel in his *Proceedings* issue of 1 December 1953. Transduction of the positive primary radiation from the sun into propulsive power takes place in the "light motor" section here in the middle of the craft. Infrared photographs of vehicles of this type made by the author directly from the invisible state are consistent in what they show with Van Tassel's description of spacecraft propulsion, obtained by telepathy over 60 years ago.

ALPHA #5

The author close to an emergent UFO on the Mojave Desert in the spring of 1958. The two-tiered object with its energy-absorbing lens on top, is only partly tangible even to infrared film, and was ocularly invisible. The whirling force-field around the object, the white lobe of radiation shown in roughly circular form to the left and slightly below the vehicle, beat against the author's own biofield. Becoming aware of this heterodyne, he called for Dr. James O. Woods to shoot to his left immediately. This remarkable photograph resulted. Photograph is believed to be the first publicly available evidence of how basic propulsion comes from spinning energy fields. Author states propulsion is through manipulations of etheric force, primal element of which was discovered by Dr. Wilhelm Reich in 1939-40, and named orgone energy. This photo should be compared with Van Tassel's cross-sectional spaceship, drawing opposite. Photo was on highspeed infrared film, with no filter, exposed at f11 and 1/50 sec. Development was 2 min. D-11.

263

ALPHA #6

UFOs have two main aspects according to the author. There are "critters"—living organisms like Alpha #1 through Alpha #4—which are normally invisible to human beings, and craft—structured, engineered vehicles propelled by etheric force. The author has photographed both types of UFOs directly from the invisible state. He interprets this as a disk on edge—only a portion of its form being tangible even to infrared—with associated field of force appearing as the whitish lobe extending earthward from the vehicle. Highspeed infrared film, no filter, shot before sunrise, near Giant Rock, California.

ALPHA #7

A companion photograph to Alpha #6, shot near noon in a Los Angeles business district while people walked to lunch just below. The whitish fields of force appear as in Alpha #5 and #6, and in this case are on the other side of the wires used as part of the terrestrial reference points. Highspeed infrared film, 87 filter, f4 1/1000 second, Leica G. Developed 5-1/2 min. in D-11.

ALPHA #8
NOW YOU SEE IT...

On 3 May 1958 this strange aeroform was recorded above the desert cabin of George Van Tassel at Giant Rock, California. A discoidal vehicle of some kind appears in the approximate middle of the photograph, on its side, and with the plane of its base roughly coincident with the vertical plane of the photograph. The plasmatic effects to the left of the disk, i.e., towards the left hand extremity of the photograph, are intense and diminish markedly towards the right hand edge of the photograph. The upper edge of the plasmatic effects is delineated quite sharply by a distinct horizontal line. Exposure was 1/1000 sec. f22 in a Leica G, highspeed infrared film, with no filter.

266

ALPHA #9
NOW YOU DON'T...
Within three seconds of the adjacent photograph, this exposure was made of the same area. Only the time necessary to advance the film lapsed between this photograph and the preceding exposure.

BRAVO #1

Fusiform, invisible UFO photographed with unfiltered high-speed infrared film near Giant Rock, California on 5 April 1958 near dawn. Note luminous corona around object. The author contends that many of these cigar-shaped UFOs are living organisms from the "critter" family of invisible denizens of our atmosphere. Leica G, f3.5,1/20 second.

268

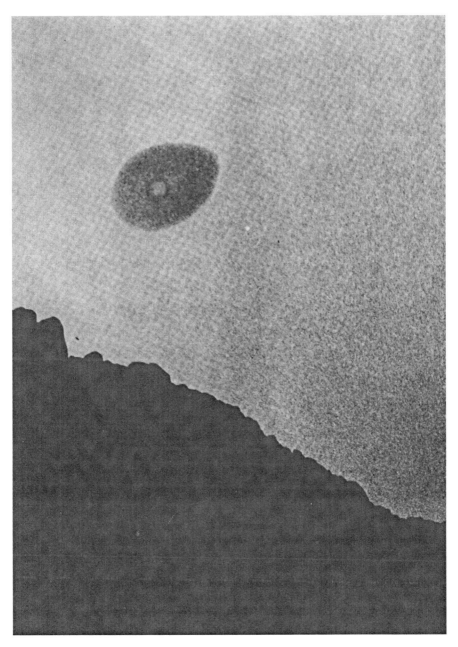

BRAVO #2

On 17 May 1958, using a Leica G without filter and highspeed infrared film, the author captured this invisible specimen of UFO life above the rocky ridge adjoining Giant Rock, California. F3.5 and 1/50 second, shot in first light of dawn.

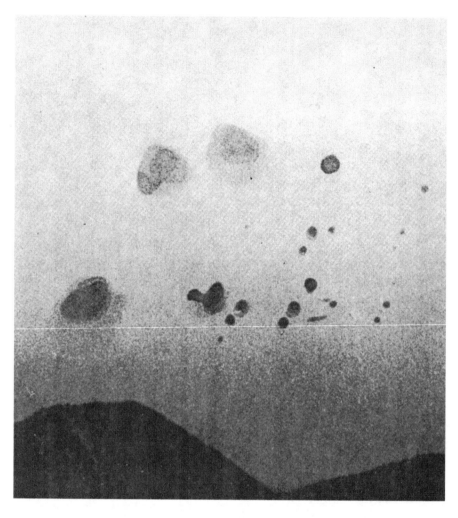

BRAVO #3

Strange, plasmatic fauna, invisible to normal human sight, cavort above the summit of Mt. Wilson, California in January 1959. The author used a Praktica FX-2 camera, exposing at f1.9 and 1/100 second, using highspeed infrared film. These organisms, according to the author, pulsate and change form when spuriously visible and account for numerous UFO reports. Compare the larger forms in this photo with those photographed by the Spanish Air Force and appearing opposite.

The Australasian Post of 8 May 1969 published this photograph of a quasi-conical UFO, observed by thousands of citizens of Madrid, Spain in 1968. Many telephoto shots were made of the object, like this one. According to the Australasian Post caption with the picture, a pursuing Spanish air force jet pilot said that on closer view, the object was revealed as "pyramid-shaped, with three bright globes at the base." Object is uncannily like the two uppermost UFOs in the author's Bravo #4 photograph, made above the summit of Mt. Wilson in California in January of 1959—nine years previously and directly from the invisible state with infrared film.

BRAVO #4

This invisible UFO—in obvious motion from left to right—was photographed by the author on 28 April 1961 shortly after sunrise, using highspeed infrared film and no filter. Site is the summit of Mt. Wilson, California, approximately one mile from the famous observatory. The author states that objects like this, which carry high orgonotic charges, reproduce on photographic prints in reverse polarity, i.e., as absorptive bodies as in this case. Such objects are photographed essentially through their charges nullifying film emulsions rather that reacting with them. Taken with a Praktica FX-2 at f11 and 1/50 second, without filter. Development was 5 min. in D-11.

BRAVO #5

Huge spaceship-type UFO looms above Giant Rock, California before sunrise on 17 May 1958, unseen by campers and visitors. This immense object, with heat obviously emanating from its underside, was invisible to normal human sight. Using highspeed infrared film, Dr. James O. Woods objectified the aeroform using techniques described in text. The lower, smaller object directly above and to the left of the flagpole, is probably another UFO. Highspeed infrared film, in a Leica G at f11 and 1/50 second, without filter and processed at 12 minutes in D-19 produced this photo.

273

BRAVO #6
CLASSIC SHAPE
Dr. James O. Woods shot this photograph of an invisible UFO near a phone pole on the Mojave desert not far from Lucerne Valley, California just after dawn on 13 September 1958. The area was heavily and repeatedly buzzed by USAF jets carrying infrared "Sidewinder" rockets. Film was 120 size Gaevert infrared, exposed at f3.5 and 1/50 second in a Minolataflex camera.

Jet →

BRAVO #7
JET

A jet fighter from George AFB near Victorville, California attempts to intercept a UFO hovering invisible above the author and James Woods at their photographic site on the Mojave desert. Repeated interceptions were attempted by USAF jets, including some low-level passes within 400 yards of the author and Woods. All the jets carried *Sidewinder* infrared homing rockets. Radar detects unseen UFOs, and they are pursued via radar. Numerous pictures unsuited to reproduction were taken of this occasion on 13 September 1958. Highspeed infrared film was used.

275

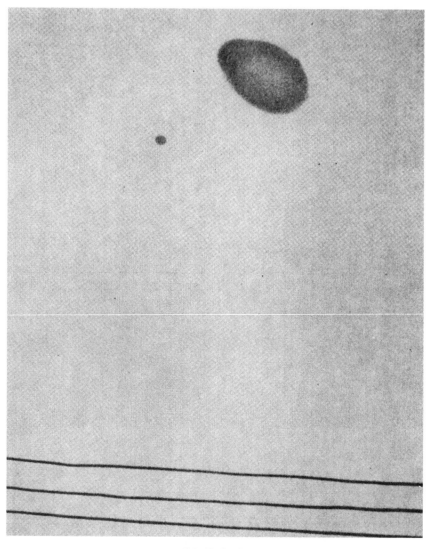

BRAVO #8

Invisible ovoid UFO of the "critter" family captured on highspeed infrared film near noon in a Los Angeles industrial district. This was one of the author's initial trials of the 18A filter, an optical device that is designed to absorb the visible spectrum while passing both its invisible ends. Exposure was at f3.5 and 1/1000 second. Use of the 18A filter in conjunction with low light Super 8mm cameras and fast color film, proved in 1975 that the "dark" or reverse spectrum of color is a verifiable reality. This was an early experimental use of the 18A in a novel, unorthodox application.

CHARLIE #1

On 11 May 1968, the author's photographic methods were synthesized with Wilhelm Reich's discoveries for the first time. A cloudbuster was used to excite the atmosphere locally, in place of the Star Exercise procedure described in the text and used previously. This UFO photo appeared on the first roll of film, adjacent to the draw tubes of the cloudbuster. As in other photos, the UFO was captured *in situ*, i.e., in the invisible state. The UFO shows obvious movement and has a corona. Foreground figure is Dr. James O. Woods. Kalloflex camera, Gaevert infrared film, f11 1/50 second, no filter. Photo made at 7:30 a.m. approximately at Thousand Palms Oasis, California.

The UFO at the top shows obvious movement and has a perceptible corona. The man in the foreground operating the cloudbuster is Dr. James O. Woods. The lower black dot reveals itself under magnification to be the head of a "critter" moving in towards the camera. It is "pushed under" in this print to maximize the upper image.

277

CHARLIE #2

The author photographed this invisible biological UFO at Thousand Palms, California, using a cloudbuster tuned to the known radionic rate of these organisms, which is 20-6. The cloudbuster was aimed south at the time, and one of its tubes may be seen in the right foreground. While UFOs like this are often similarly shaped to the classical "flying saucer," they are biological rather than electromechanical. A second UFO of the same type was captured on this same exposure at the left hand edge of the frame, but it has been cropped to accommodate the picture to reproduction mechanics. Film in this case was Gaevert 120 infrared, exposed at f8 and 1/50 second without filter. Development was 6 minutes in D-11.

CHARLIE #3

This glowing UFO was captured direct from the invisible state on 22 March 1969 by James O. Woods, using an 18A filter and high-speed infrared film. The cloudbuster shown here causes differentials in primary energy levels that underlie atmospheric movement, and its operation appears to affect other densities of existence. Photo was made at approximately 7:15 a.m., with exposure f8 and 1/100 second.

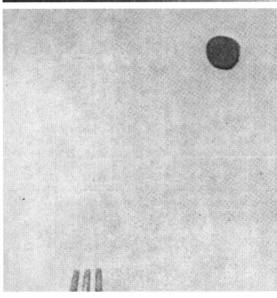

CHARLIE #4

Cloudbuster pipes in foreground evidently attracted this UFO, which closely resembles a jellyfish in general appearance. Certain jellyfish are attracted to light, and it may be that the cloudbuster provides similar attraction to these unseen aeroforms that inhabit earth's atmosphere, according to the author.

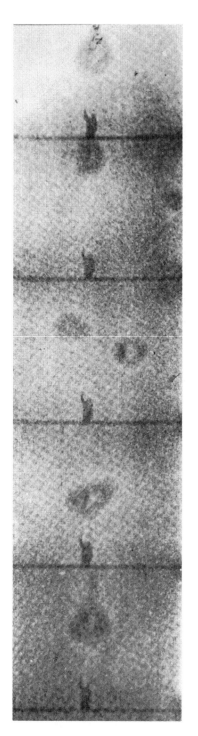

BIOFORMS IN MOVIES

Frames excerpted from infrared motion picture footage show bioforms around the author on the Mojave desert in California before sunup. The bioforms change shape, position, and also appear to divide like amoebae at tremendous speeds. This sequence was shot at 24 frames per second. The entire events shown occurred in approximately a quarter of a second. The author's view is that such materializations into the infrared occur in response to certain positions, motions and orientations of the "target human," which are poorly understood. He holds that the same factors govern UFO materializations in the earth environment.

DIANA'S CRITTER

The author's daughter, Diana, then age 11, accompanied him on a photographic and weather engineering trip to Thousand Palms, California in May, 1975. She was extremely sensitive at this age and had been trained to take photos of objects perceived etherically. While the author was busy working the weather gun (right figure) his daughter snapped some pictures with his half-frame Olympus, equipped with an 18A filter and highspeed infrared film. A critter close to the cloudbuster shows vacuoles in the best biological tradition.

WILLYS WAND—A Cloudbuster

This weather control unit, colloquially known as a cloudbuster, was built by the author from the basic inventions of the late Dr. Wilhelm Reich. As an investigator of etheric energies and forces, the author is convinced that the same primary energy manipulated in such units as this is the operative power behind UFOs of both the biological and construct type. He has used special variants of the cloudbuster to attract and photograph UFOs with techniques he developed in the 1950s.

PRIMARY ENERGY RESEARCH

The above photo shows the author in 1977 with the MK II cloudbuster, which utilized radionic tuning. Time-lapse films made during tests proved this device to be capable of dissipating clouds more rapidly than the original Willy's Wand.

The photo below shows the author (right) and his oldest and closest associate, Irwin Trent, original publisher of *The Cosmic Pulse of Life*, with the Magnum 108 cloudbuster in Banning, California, 1981. The author's innovations in research continue to the present day.

APOLLO 12 "DISCOVERS" ORGONE ENERGY

Astronaut Alan Bean walking on the moon gave off sporadic blue radiance, which NASA scientists attributed to spurious reflections from his suit. Author holds that the blue radiance is the life energy field of astronaut, irradiated by emissions from astronaut's backpack radio. The situation inadvertently replicated Dr. Wilhelm Reich's experiments published in the 1950s in which he made orgone energy luminate blue in a vacuum, work that was later suppressed by the U.S. government. Author points out that equipment dump equidistant from camera has no blue lumination, and astronaut only luminates when he talks. Author claims that orgone energy discovery is a technical break-in to new etheric technology, including spaceship propulsion.

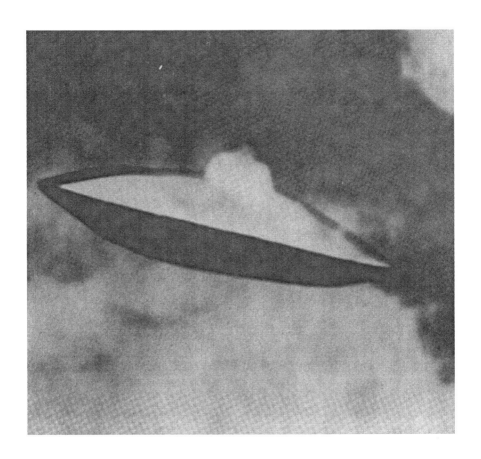

USMC AIR GROUP PHOTO OF FLYING DISC—KOREA

Photograph made by Marine Corps pilot operating over Korea illustrates technical principles dealt with in Chapter 20. The underside and left hand extremity of the disc show the total void or blackness on a photographic print associated with concentrations of contractive, cold orgone energy, the etheric force discovered in 1939-40 by Dr. Wilhelm Reich. The black area is so intense that all detail is eliminated therein, but in actual fact it is due to a high concentration of the energy that Dr. Reich showed desensitizes film emulsions. The author holds that the energy spins around the circumference of the craft. His evaluation of this case is that the camera has "stopped" the rotating field on the left side of the disc. This accounts for the otherwise anomalous "shadow" on the upper side of this vehicle, which could not appear in the form it has in broad daylight on an upper surface. In the text the author warns of so-called expert evaluation of photographs when technicians and scientists have no knowledge of the energy involved or its characteristics which, the author says, are often "totally contrary" to current ideas on energy and radiation.

Photo courtesy of the late William G. Allen, Ph.D.

The UFO in this picture was accidentally "captured" on the negative by nationally-known outdoor writer V. Lee Oertle in Utah. The two hunters only appear to be watching it. Actually, they are studying the terrain while looking for deer. The object had to be traveling at tremendous speed. It covered almost the whole arc of the sky during the 1/250 second shutter speed. Picture was taken in October, 1965 on a 9,000 foot plateau above Clear Creek Canyon, Utah. Camera: Rolleiflex 3.5, Film: Professional Plus-X at f11 and 1/250 second. Late afternoon, looking east. Photographer Oertle remembers seeing what he thought was a vapor trail in viewfinder, then forgot the incident until processing the film some time later.
Photo copyright by V. Lee Oertle, Beaver, Utah

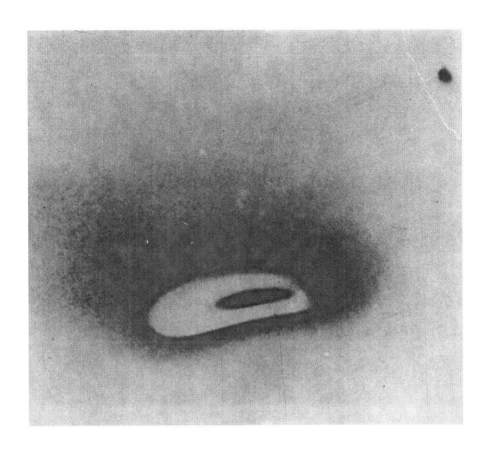

THE OERTLE UFO, ENLARGED

Author's analysis:

Starting from the center we have what is quite obviously a symmetrical form, consistent with a disc roughly at eye level. Surrounding this black symmetrical core is a white envelope, consistent with a plasma created around the disc by its whirling force-field used in propulsion. This plasma appears "bent" to the left of the disc by the obvious left-to-right motion of the vehicle across the scene. The high etheric potentials involved in the propulsion of the disc and creation of the plasma appear to be causing strong absorption of energy from the atmosphere and creating the shadowy enclosure around the disc and plasma. This "energy trail" extends backwards like the wake of a ship at sea. Small dot above and to the right of the main UFO is probably a second UFO.

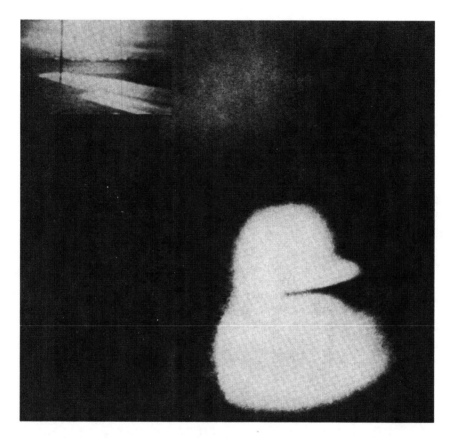

UFO OFF AIRLINER WINGTIP

November 1975

None of the sixty-odd passengers sitting on the left side of a commercial airliner flying between San Jose and Long Beach in November of 1975 saw anything outside the aircraft at 30,000 feet. Yet the author, pioneering his reverse or dark spectrum technique, was able to objectify a veritable cosmic zoo—critters and craft—and bring them back in full color on standard film.

These photographs are in a string of exposures made at 2 frames per second that show a brilliant white UFO pass close to and under the wing of the airliner at 30,000 feet. This sequence actually took place in broad daylight, at approximately 10:30 A.M. By using an 18A filter over the lens of a standard Minolta XL-400 Super 8mm movie film, the author created an artificial darkness and the UFOs near the airliner came out on the film in color, although invisible to the eye. The films were shot "blind." Ektachrome 160 was used.

In this picture, which is greatly degraded from the original color film, the wingtip of the airliner may be seen silhouetted against the body of the craft as a black, lance-like protuberance. The diffuse effect above the UFO is deep blue in color and is probably due to etheric energy luminating in connection with the propulsion of the craft.

The inset photo from the same film (upper left) shows orientation of photographer's position to airliner wing on takeoff from San Jose.

288

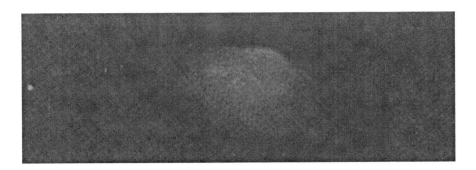

BIG RED CRITTER

This photo was originally made in color, near St. George, Minnesota by deputy sheriff Arthur Straunch on 21 October 1965 while on a hunting trip. The coloration is red on the periphery, passing through to yellow and then to white on the very brightest part of the object. The author believes that this UFO belongs to his "critter" family of UFOs—living invisible organisms native to our atmosphere that are capable of making changes in their density that render them sporadically and occasionally visible. The author photographs such critters direct from the invisible state using techniques of his own devising, described in the text.

Photo courtesy Aerial Phenomena Research Organization

LITTLE RED CRITTER

The author photographed this invisible critter from the window of an airliner at 30,000 feet between San Jose and Long Beach, California in November of 1975. He used his "dark spectrum" technique and Super 8mm Ektachrome 160 film. Four successive photos were made at 2 frames per second, in the course of which the critter "warped out," i.e., literally shrank from perceptibility while maintaining its station with airliner. This was typical density change mentioned by author in text.

Conversion to monochrome and enlargement have degraded image, but author states that in direct projection of color film there is no doubt of identity between this critter and one in Strauch photo above.

"Dark spectrum" technique simply involves placing an 18A filter over the camera lens to create an artificial darkness in full daylight. The author claims that UFOs—both critters and craft—come through onto the film and out of the artificial dark in full color. He says anyone can take such pictures.

REPLICATION OF THE INFRARED TECHNIQUE
Richard Toronto, of Vallejo, California, took this critter photograph after sundown in May 1977, near Apple Valley, California. After reading the first edition of *The Cosmic Pulse of Life*, he shot four rolls of Kodak infrared film. Two shots contained critters, although Toronto did not see anything visibly when he pressed the shutter.

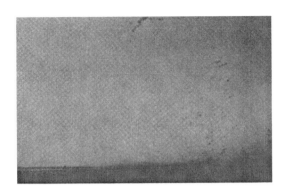

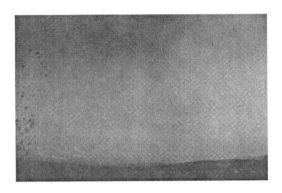

A "SCHOOL" OF ATMOSPHERIC "FISH"

In this three-frame sequence shot by Richard Toronto in Vallejo, California, on November 16, 1977, camera was moved progressively to right with each successive shot. The "heads" and "tails" appear to be the product of movement in the objects. The second frame reveals a part of the first, larger "school" of phenomena. The last frame shows a small group to the right of the main group. Kodak laboratory said that these images were probably "finger marks."

JIMMY CARTER'S CLOSE ENCOUNTER?

This photo was shot as then President Jimmy Carter took off from the signing of the Panama Canal treaty in Panama City, June 17, 1977, 2:20 PM. Professional photographer, Linda Arosemena, did not see anything in the sky and only discovered this critter after normal film development. The photo was taken with a Nikon camera equipped with a motordrive, using Kodak TRI-X black and white film, at 1/250 sec., f/16. This photo appeared in Panama's *Star & Herald*. The *Star & Herald* reported local UFO activity in the area on the previous day.

APPENDIX II

THE WORK OF LUCIANO BOCCONE
(Written 1989)

As the years passed from 1957 and my writing of *They Live in the Sky*, I received considerable mail from all over the world. The substance of my theories and my pioneering photographic work influenced more and more people. Time bore me out in my identification of UFO phenomena in general, as being inseparably bound to the etheric world and impenetrable without such knowledge. Perceptive people saw the merit and truth of my pioneer work, and wrote to me in the kindest terms.

Dealing with such a correspondence became impossible. By the late 1970s, with *The Cosmic Pulse of Life* having been published in 1976, I did not even have sufficient time to read through some of the long letters I received. One such letter, from Genoa in Italy, I did read, many times over and with burning interest.

My correspondent and soon-to-be ally and co-worker was Signor Luciano Boccone, an executive with a metals fabricating company. I reproduce Boccone's letter here in its entirety, with all its important implications for this revised and enlarged edition of *The Cosmic Pulse of Life*. Readers will thus experience the same impact from this letter as I received back in 1979:

Boccone and his group published many of their photographs in Boccone's remarkable book, *UFO—La Realta Nascosta* (*The Hidden Reality*), produced by Ivaldi Editore in Genoa. This handsome and compelling book is to my knowledge available only in the Italian language. The photographs in the book, from three independent sources, nevertheless tell a story for which words are not necessary. Three groups of workers in different parts of the world, unconnected with and unaware of each other, obtained infrared photographs of the same invisible objects in the skies of planet Earth. My pioneering work of more than 30 years ago was "on track".

Nobody in my experience was more appreciative of my pioneering work than Boccone. He understood completely the sacrifices required to pave the way for his own work. I retired from ufology by 1979, became fully engaged in weather engineering, but nevertheless found myself eager to assist and perhaps guide these enterprising Italians and their bold leader.

I began working consultatively with Boccone's group. They became my hands and feet on another continent. I could work with them as though they were Americans, because of Boccone's incredible mastery of English—right down to the subtlest nuances. Progress was spectacular. An abundant harvest of new evidence was obtained, all of it utterly destructive of the old order. At present, publication of this work would be an overdose for our decadent era, but the world will see it all someday.

The sudden, shocking death of Signor Boccone from a massive infarct in 1981, removed this radiant and dynamic man from our midst. His passing in the prime of life, and at the helm of an historic enterprise, was a tragedy of appalling impact on our lives. When humankind ceases watching the basketball games and other mindless entertainment, and takes charge of its etheric tomorrow, Luciano Boccone will be in the pantheon of the new world's heroes.

Reproduced here is a small collection of the Boccone—GRCU photographs that I am authorized to publish with this edition of *The Cosmic Pulse of Life*. This work adds an independent verification of my own labors—a verification that was unavailable when this book was originally published in 1976. Boccone and his group have moved us manifestly closer to the truth of the UFO mystery, which lies in the etheric, invisible, interpenetrating worlds around us. Through this Appendix, I am proud to have Luciano Boccone permanently associated with me.

—TJC

G.R.C.U.
Gruppo di Ricerche Clipeologiche ed Ufologiche
Arenzano

August 22, 1979

Mr. Trevor J. Constable
c/o Merlin Weather Engineering
San Pedro, California, U.S.A.

Dear Mr. Constable:

I wonder whether this letter will ever reach you, as I do not know your precise address. I hope you will receive it, anyway, and honor me with a reply, meeting our keen expectations.

I am the President of GRCU, an Italian UFO Research Group which has been operating for five years in the coastal area stretching to the west of Genoa. The scope of activities of our Group is to study both the far-off past and present UFO phenomena and to inquire into the mysterious UFO noumena that our instrumental research work reveals to us almost every day.

GRCU Arenzano has 25 members: some of us are holders of degrees in mechanical, electrical engineering, and literature; most are non-graduate engineers, scholars of archaeology, astronomy, geology, optics, physics, photography, ethnology, glottology. All of us have made ourselves acquainted with the UFO phenomena either by direct experience or through the current UFO literature. Our direct experience includes the instrumental research work that we have been performing in the infrared range of the spectrum since 1976, which thoroughly confirms your statement about the invisible state of the UFOs. Our library contains as much as 400 selected books in Italian, English and French, recounting the whole UFO story since Arnold's sighting in 1947 and earlier. Personally, I have been studying the UFO phenomena since that year, when I was 19 years old.

I am a member of CUN-Centro Ufologico Nazionale, Italy's major center for UFO studies, and a contributor to CUN's official magazine "Notiziario UFO" ("UFO News"). I also contribute to the magazine "Il Giornale dei Misteri" ("The Journal of Mysteries") which is published by a Florentine UFO Research Organization, dealing with UFO phenomena dating back to hundreds of years ago. Recently, I wrote two articles, one for "Notiziario UFO, March, April, May 1979, reporting on the first two CE4K experienced by a Genoese watchman and discussing the contribution of the narco & hypnoanalysis to the correct interpretation of such encounters, and one for "GdM", describing in detail the striking resemblance between certain bas-reliefs of "grenades" and "brooms" in the Palace of the Duke of Urbine, Italy, dating back to the XIVth century, and the spacecraft that has been reconstructed by J.F. Blumrich on the basis of Ezekiel's visions in 593-563 B.C.E. and the visible and invisible "critters" and "constructs" that can be seen or objectified today all over the world.

Together with CUN Prato and CORU La Spezia, GRCU was the third Italian UFO Research Group to start doing three years ago some concrete research work in the infrared field, that is to say, we were among the very first UFO organizations in Italy to take still pictures of UFOs directly from the invisible state, both in desert and built-up areas.

Our research methods involved since the beginning the use of standard instruments, such as Geiger counters, precision magnetic compasses and instantaneous temperature indicators, frequency, light, ultrasound and ultraviolet detectors, etc., as well as infrared, panchromatic, and color films in sophisticated gun-cameras and cameras fitted with focal distance doublers or telephoto lens, and sometimes with special filters.

In doing our instrumental research work, we follow a procedure that can be summarized in a few words. As a rule, whenever our instruments warn us of any irregular quick variation in radioactivity and/or magnetic field, air temperature, luminosity, etc. we take immediate pictures in the same direction in which such variations are detected: at low altitude in the sky or on the ground, by day as well by night. Generally, we do not see anything visibly when our instruments behave irregularly and we simultaneously press the shutter of our cameras. Incidentally, our shooting of time-lapse and flash-lamp pictures on instrumental warning may be either independent of, and/or simultaneous with, and/or subsequent to, any visible UFO sighting in our research areas.

This simple procedure has allowed us to objectify, both at low altitude and close on the ground, invisible events or noumena which prove to be strictly connected, and not only in our humble opinion, with the visible phenomena that usually go by the name of UFO manifestations.

For example, manifestations pertaining to an invisible, ultradimensional realm are objectified in most cases when our Geiger counter suddenly begins reading (and exceeding) radioactivity rates of the order of 0.20 mR/h, even if these peaks generally last only a few seconds, or when the compass needle suddenly turns to, and positions itself on, either side of the magnetic north mark, if only by a few degrees, or when some other instruments "go crazy".

A significant example of the relationship existing between the visible UFO phenomena and the invisible UFO noumena that we usually objectify under such conditions is the semitransparent, unicellular structure appearing on the picture that we snapshotted last June directly from the invisible state, when our Geiger counter was reading 0.30 mR/h per 3 seconds. This Saturn-like structure appearing on our infrared snapshot is not that different from the well-known, visible Saturn-like UFO that was photographed in 1958 by the photographer Almiro Barauna from the Brazilian ship "Almirante Saldanha".

Let me give you a few hints at our photographic equipment: Kodak B&W High-speed Infrared film is generally used in our stand-mounted Canon EF Reflex 50-mm and Voigtlaender Vito LS Special 50-mm cameras, the former fitted sometimes with a Vivitar No. 25A filter. Most of our IR pictures are however taken with no filter. Exposure varies from f1 6:1/300 second by day to f2.5:40 seconds thru 5 minutes and over by night. Last month we started to use Ektachrome (slide) infrared film. Kodacolor II, Vericolor and panchromatic Kodak Recording 2475 and Illapse and flash-lamp pictures in our Asahi Pentax, Leicaflex, Fujica, Zenith, Miranda and Minolta cameras.

Development of our IR film is 12 minutes in D-76 at 20°C. Printing of such film is regularly 20 seconds. However, in order to discern the profile of the bright nucleus, if any, of certain NL, "vapor clouds", semitransparent amoeba-like structures, etc., we usually protract this 20-second time, sometimes up to 1 minute.

About two months ago, we completed our latest UFO Study: "The Invisible UFO Evidence: July 1977-June 1979". A condensation of our previous "UFO REPORT 1977" and "UFO DOSSIER 1978" plus a selected documentation of our 'FILES 1979", this Study contains the documentary evidence of the existence of ultradimensional, biological amoeba-like lifeforms, that we have called "plasmoids" (PLAMSOrganismi Intelligenti Dello Spazio). This evidence consists of a selection of about 140 infrared, panchromatic, and color pictures of the most significant, invisible, instrumentally-detected UFO noumena that were objectified during the last two years in our area of activity.

A significant part of the "hidden reality" of the UFO problem, the events and manifestation appearing on our pictures are precisely and exactly a variety of the "invisible yet physical" world that you so vividly describe in your exceptional, outstanding book "SKY CREATURES: LIVING UFOs".

I have just finished reading it for the nth (and certainly not the last) time. Unfortunately, your original "THE COSMIC PULSE OF LIFE" is not yet on the Italian market. I fully realize that in that abridged edition you could not go practically into such detailed information on the orgone energy and the instruments that are part of the New Technology as to allow anybody to immediately put WR's [Wilhelm Reich's] inventions into practice, following simple "do-it-yourself" methods.

Your hinting at WR's major findings has however raised our curiosity to the extent that all of us are now not only anxious to learn much more on orgonomy, but also willing to go deep into the instruments of the New Knowledge.

Actually, we did not know anything here about your work until the Italian edition of Brad Steiger's book "Gods of Aquarius" was put on the market. We are therefore particularly appreciative to an old friend of ours for having made us such a precious present as your book on "Critters". We have thus learnt of our pioneering work in the field of the "invisibilium" and of the previous work performed by such men as Dr. F. Thomas, Dr. R. Steiner, Dr. R. Drown, Dr. W. Reich and his daughter Eva, who contributed decisively to your development and ability to carry the discovery of the orgone energy into ufology. We thank you very much for all this, Mr. Constable!

The purpose of this letter is twofold: first, to confirm as I said, and to support by documentary evidence if necessary, that our UFO Research Group has obtained the same results as you without knowing anything about your experiments; second, to enter into relations with you in order that, if you deem it advisable or of some interest to you, we may exchange information on the "invisible yet physical" aspect of the UFO problem. For example, in return for any information you would give us on WR's inventions that have application to UFO investigation, namely: the orgonoscope, orgone field meter, orgone energy accumulator and, particularly, the cloudbuster, we would be prepared to inform you in detail on the research work that we have performed hitherto and on the future research work that we would be enabled to do through the use of the aforesaid WR's inventions.

As I said, the photographic, instrumentally-based documentation in our hands positively confirms your photographic findings and conclusions as well as the results obtained by our close friend Florin Gheorghita in Rumania. Let me emphasize that your invisible amoeba-like lifeforms are practically identical in shape, size, density, etc. to the invisible plasmatic organisms that our Rumanian friends and ourselves have photographed in the sky and on the ground, both in Italy and Rumania. There is seemingly no difference between the invisible, unicellular, plasmatic organisms that can be seen on your infrared still pictures of "Critters", the invisible, unicellular, plasmatic structures that can be seen on Gheorghita's infrared and panchromatic still pictures of "Plasmatic Bodies", and the invisible, unicellular, ameboid structures that can be seen on the infrared, panchromatic, and color still pictures of "Plasmoids" that we have taken from our research areas on instrumental audible and visual warning.

Like your own files and Gheorghita's, our records are full of photographs showing invisible, glowing, pulsating plasmorganisms, capable of changing shape, size, density, luminosity, arrangement, position, etc. in a split second. Our pictures, for example, show invisible plasmorganisms blinking at low altitude over the Port of Genoa, pacing airliners up to the Genoese International Airport or escorting them after take-off, hovering over the Cornigliano integrated steel plant of Italsider and over many sacred places and towns, both along the Genoese coast and in the hinterland. They show invisible plasmatic bodies fleeing or cavorting in the sky at tremendous speed, or hovering or "dancing" over our research areas. Where they also land sometimes, in the form of invisible spheres of light which subsequently transmute in the fraction of a second either into invisible, glowing, seemingly-ectoplasmic, demon-like entities, or into invisible "arabesques of light" skimming over the ground, or into invisible yet corporeal, haloed human-like entities. Or where they also show up sometimes, in the form of invisible, small light balls cavorting around us or creeping up to us, seemingly going their rounds before taking off. Invisible, unbelievable flying ornithoids resembling mythological glowing griffons (the so-called "thunderbirds" or John Keel's "neopterodactyles") also appear on our pictures, flying low with stretched-out wings over the town by night. Invisible forms of energy organizing into invisible, intangible, glowing mist, fog, and amorphous masses and phantom lights taking off from the ground close to us can also be seen on our still pictures.

In short, we have got irrefutable documentary evidence of the invisible presence of such "Plasmoids" at low altitude over our mountain and marine research areas, close to us on the ground, and inside our houses, too. I repeat, we have objectified them many times directly from the invisible state on infrared, panchromatic, and color still pictures, most of which have been taken on instrumental detection only, and such plasmatic ameboid structures are more or less identical to those that have been objectified under more or less similar conditions both by you in the United States and by Gheorghita beyond the Iron Curtain.

This is the reason why, sticking to the facts and failing for the time being any exhaustive, satisfactory explanation other than their "biological nature", we agree to your interpretation, according to which most of the UFOs that have mystified men for generations are not extraterrestrial spaceships at all, but are generally invisible, ultradimensional, ultraterrestrial, biological, etheric organisms that are of our planet, that have been living with us, side by side, unnoticed, since the beginning of time. And, let me add, they have many a time been erroneously interpreted, or deceitfully explained by organized religions, as God's manifestations.

Well, notwithstanding all the documentary evidence in our hands, our efforts to have our research methods adopted by other Italian UFO Research Groups have not been rewarded with success as yet. Unfortunately, the ufologists that are prepared to accept and discuss objectively the results of our work are still a large minority today. Most of the Italian UFO investigators are still clinging to the "material" aspect of the UFOs, waiting, as it were, for a "tangible Venusian" to shake hands with, and when requested to comment on our pictures, they look at them with disbelief and carelessness. As you say, they don't know that in so doing they behave exactly like the "mechanistic scientists" whom they speak so ill of.

Last year, failing any competent counter-party in Italy with whom discussions concerning the invisible state of the UFOs might be held, and strong in the findings of our annual, instrumental sky-watch and ground research campaigns, we turned to Prof. J.A. Hynek, with a view to having the results of our work included in his UFO Classification. To this end, we wrote him a letter on July 11, 1978, proposing amendments and extensions. In support of our proposition, we enclosed a number of infrared pictures and gave a full description of the events shown there, stating where, why, and how they had been taken. In short, we proposed the extension of his Classification to all invisible yet instrumentally-detected and picture-documented UFO noumena, according to the "concrete" findings of our Group. No answer has however reached us as of the date of this letter. Perhaps he has not received, and will never receive our message, which, in all probability, has already been tampered with by... Unknown Filching Officers!

Before reading your book, the pictures taken in Cluj-Napoca by our Rumanian friends, that we visited in August 1978, and Gheorghita's opinion on his own and our photographs of ameboid and plasmatic structures were the only facts that could support our findings and our research-founded assumption, according to which unknown radiations other than alpha, beta, X, and gamma must penetrate the probe of our Geiger counter or excite the other instruments used by us. Now we learn from your book of the orgone energy and the new instruments of orgonomy! Let me say that this is a very interesting piece of news, indeed!

Now, the thing is that only a few of the books mentioned in your work can be found on the Italian market, so that we are unable to get any further immediate knowledge of the subject matter. As a result, it will be well nigh impossible for us to conduct any study in this field in the near future, i.e. during our next 3rd Campaign, which is scheduled to start in October 1979. Unless... unless you help us, Mr. Constable!

I told you of our wish to further search into the question of the invisible state of the UFOs through the application of the CB, ORAC, etc. but, as we realize the terrific consequences of the misuse of the orgonic force, we dare not do anything without precise information in our hands.

The question, for which I apologize to you in advance as I realize the trouble it will certainly cause you, is to know if you can send, or arrange for sending, us the books or manuals that give a full description of these new instruments and related CB radionic tuning equipment along with the instructions for their operation. Needless to say, we shall pay for these books on receipt. As I told you in the foregoing, we might send you, in exchange for your courtesy, the documentary evidence that we have been gathering since 1976. Your suggestion of the manner in which this should be done in order to avoid tampering would be very much appreciated.

Trusting that the U.S. Government and/or the USAF or NASA have not already put their vetoes on the divulgation of any information and data on Reich's inventions, and assuring you in advance of our gratitude, we look forward to your news with great interest.

Warmest personal regards.

Sincerely yours,

Luciano Boccone

DYNAMIC LEADER
The late Signor Luciano Boccone, of Genoa, Italy, adjusts his Reich Cloudbuster in February, 1980. All the photographs in this Appendix section were made by Boccone's GRCU research group, with original methods that they devised. They were unaware of the author's work in a similar vein 20 years before they started. Evidence obtained via infrared photography in both cases, half a world and a generation apart, is mutually corroborative.

INVISIBLE PLASMOID PROCESSION, 13 Sept. 1979. Sudden high Geiger counter readings prompted the late Boccone to shoot highspeed infrared film into an "empty" sky. He obtained FOURTEEN mutually corroborative photos of this procession of invisible, pulsating organisms passing over Genoa, Italy.

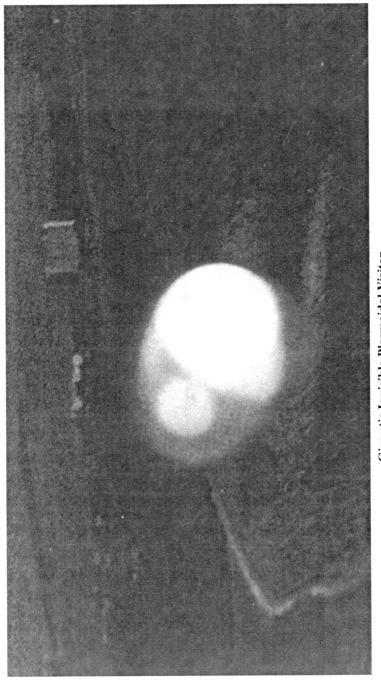

Gigantic Invisible Plasmoidal Visitor

Romanian engineer Florin Georghitza took this infrared snapshot near Cluj-Napora in Romania in September 1977. Huge size of this visitor may be judged from the adjacent road circling left of the visitor. Engineer Georghitza has made numerous UFO photos in Romania over the years, and introduced TJC to the late Luciano Boccone in Genoa.

"Falling leaf" invisible trails, objectified by infrared film and time lapse exposure, show descent of invisible objects to lights in Leghorn, Italy. Etherian physics interprets "falling leaf" pattern as entry of object into the physical world via the warmth ether, heat being seen as the fourth state of matter. Such photographs have been made by the late Luciano Boccone's group in numerous similar sequences and series, all of which are mutually corroborative. "UFOs come from and return to the invisible state," said Boccone.

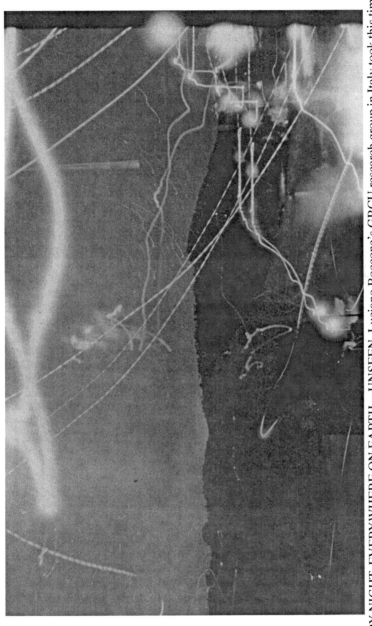

EVERY NIGHT, EVERYWHERE ON EARTH—UNSEEN. Luciano Boccone's GRCU research group in Italy took this time-lapse infrared photo of intense and invisible nocturnal activity over Arenzano, near Genoa. Invisible objects partaking of this invisible commerce etch zooming pulsatory trails in the film emulsion. Materialization and dematerialization are immanent in trails that suddenly stop. These invisible UFOs performed the recorded maneuvers during a 45-minute time exposure. They can be seen attaching themselves to street lamps at certain times.

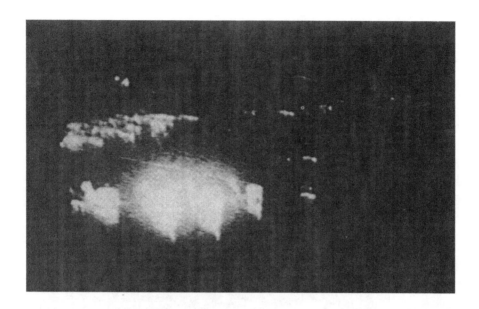

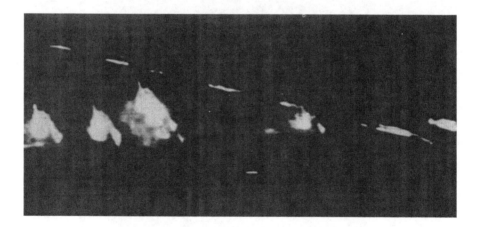

The "Neooterodactyls" was the name given by the late Luciano Boccone to these plasmoidal, griffon-like invisible creatures flying low over Arenzano, Italy. Taken on 29 October 1977, the photo was made with infrared film, using a 25A filter and a 40-second time lapse. "These blinking, turkey-like plasmoids are another example of the invisible, glowing fauna that lives unnoticed in the skies of our planet Earth," wrote Boccone.

INVISIBLE PLASMOIDS OVER BLAST FURNACE
Group of invisible UFOs floats above the steel plant of Italsider at Genoa-Cornigliano on 19 August 1977 at 10:30 pm. Site is adjacent to Cristoforo Colombo International Airport. A member of Boccone's GRCU team made this 40-second time lapse picture with unfiltered infrared film. Nothing was visible above the steel plant to the naked eye. Impression is inescapable that objects are drawing on the massive energy of the blast furnace. GRCU's photographic record documents many instances of such energy vampirism, especially attachment of UFOs to street lights and similar energetic focal points.

INVISIBLE INVASION BY NIGHT

Dozens of invisible plasmoids like these were photographed dozens of times at the GRCU research base near Genoa, Italy, on a January night in 1981. Infrared and magnetic detectors used by the researchers, as well as a Geiger counter and a Swiss orgonotester instrument, all suddenly "went crazy" with abnormally high readings. The researchers operated their cameras immediately, even though nothing was visible to the naked eye. Sheafs of mutually corroborative photos came out of the several cameras used, the above picture being made on color film. The photographic record shows these strange blue and white etheric forms falling from the zenith, and rolling down the hillside, skimming along the grass and among the cars, equipment and researchers. The author of this book maintains that this kind of unseen, etheric commerce goes on all over our world, and has done so since the beginning of time. The late Signor Boccone agreed wholeheartedly with this assessment, on the basis of a massive photographic harvest gathered by his group over many years.

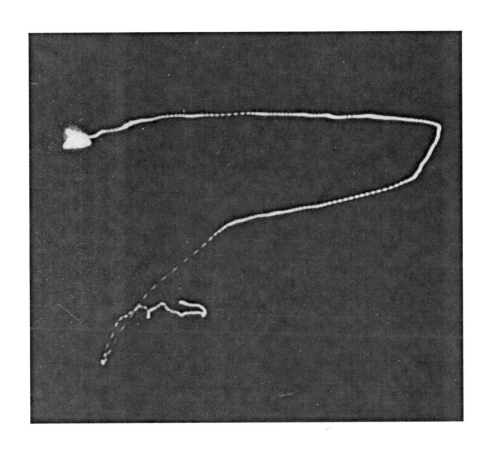

GENOA PEGLI UFO FROM GRCU

On 25 September 1979 at 10:26pm, GRCU infrared film captured this UFO for those with open minds to study. This was a 3-minute time exposure. Intermittent sections of the course traced are clearly visible, and allow speculation as to the craft's power source.

GRCU'S "FLYING GRIFFINS" — By capturing these strange, invisible flying forms on IR film near Arenzano, the late Luciano's GRCU research group provided a contribution to the author's space animals menagerie. We are not alone.

UFO NEAR A FIRE — ITALY

Snake-like plasmoidal object—invisible—is attracted to a fire near Arenzano in Italy. Luminous trail etched on IR film records the presence of an otherwise unseen visitor. A GRCU cameraman took this photograph.

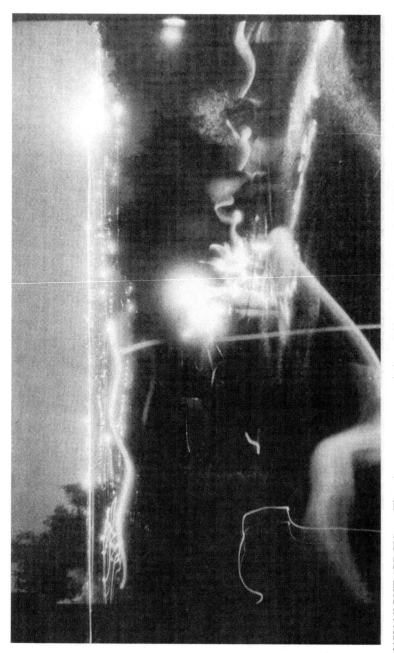

BUSY NIGHT (GRCU) — Time lapse exposure on infrared film near Arenzano in 1979, captures the tracks and maneuvers of UFO's near GRCU researchers. Particular attention is paid to the researchers' car near the exposure's center.

APPENDIX III
THE TETHER INCIDENT, 1996

In the public history of the UFO subject, there has never been anything as stupendous as the Tether Incident in February of 1996. All the carefully phrased denials and evasions of the U.S. government, through all the years from WWII to 1996, were nullified by the specially-designed camera aboard the Columbia spacecraft on NASA mission STS-75.

This superlative instrument, designed to record in the near ultraviolet spectrum—WHICH YOUR NAKED EYE DOES NOT RESPOND TO—fully justified its reputed million dollar cost. The NASA camera captured on videotape literally dozens of UFOs that surrounded the broken electrodynamic tether as it drifted 80 miles away from NASA's Columbia.

The tether appears as a 12 mile-long white bar, reversing out of the blackness of space, 300 miles above the earth. Fully unfurled and highly reflective, the tether appears with numerous UFOs, which are virtually identical with similar *living organisms* photographed in the infrared by author Constable in 1957-58. The infrared is a harmonic of the ultraviolet.

Since the Tether is 12-miles long, it is virtually usable as a means of estimating the approximate size of the associated discoids. When the camera records the UFOs going *behind the tether*, then we may conclude that the UFO is between two and three *miles* in diameter. Remember, the tether is 12 miles long. This spectacular NASA videotape also records many UFOs materializing into the ultraviolet, and many dematerializing out of the ultraviolet. That is to say, disappearing while you watch. Such happenings, faithfully reported by pilots since WWII, led government officials to infer that the pilots were drunk. Now, we can watch this NASA tape and be aware that the UFOs are capable of changing their frequency, literally. The UFOs hail from the invisible, just as Mr. Constable said, back in 1957-58. This video is in the public domain. Add it to your collection. From the videotape "EVIDENCE: The Case for NASA UFOs" Produced by Terra Entertainment 12335 Santa Monica Blvd., Los Angeles, CA 92005. Tel 310 268-1210. "EVIDENCE" is narrated by David Sereda. Available elsewhere is a 6 hour, 4 dvd set called "The Secret NASA Transmissions" containing all of the raw footage on one of the dvds. This set has been called "The Most Popular Underground Video Among Astronauts", available from The Book Tree at 1-800-700-8733 or from UFO TV, 1-800-350-4639, www.UFOTV.com.

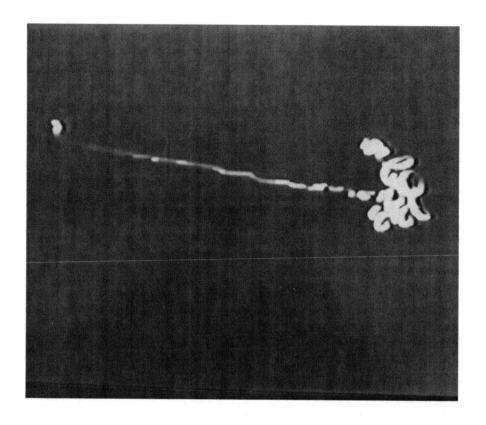

"The tether has broken... the tether has broken and it's going away from us...." So said the excited astronaut aboard the spacecraft Columbia in 1996, 300 miles above the earth. The tether was a thin, 12-mile long conductive lead, intended to absorb power from outer space, to increase the satellite's endurance and reduce its operating cost. The tether appeared, as it unfolded, like a luminous white bar against the blackness of space (see next photo). This white bar shows from 80 nautical miles astern of the Columbia, in subsequent pictures excerpted from more than a quarter hour of ultraviolet videotape made by NASA on Mission STS75.

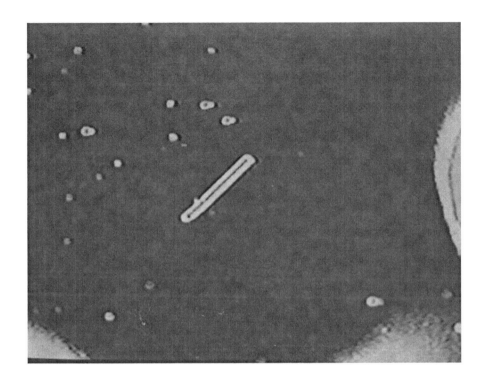

80 nautical miles away from Columbia, the 12-mile long tether appears like a luminous white bar against the blackness of outer space, 300 miles above the earth. Accompanying the tether are approximately 30 UFOs, which register on Columbia's ultraviolet-sensitive camera. Therefore they appear as "solid" as the tether, although not visible to the eyesight of the astronauts aboard Columbia. UFOs virtually *identical* to these were photographed on infrared film in 1957 by author Constable and associate Dr. James O. Woods, above the Mojave desert in California.

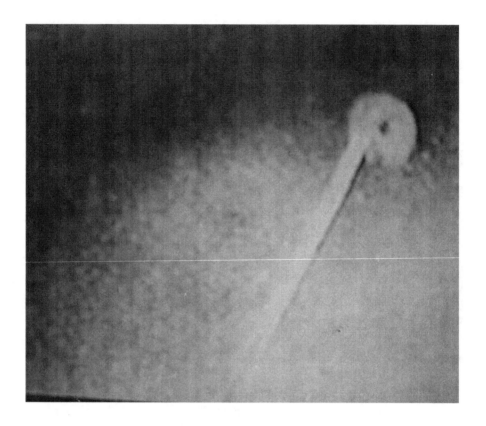

UFO that is at least 2 nautical miles in diameter, passes BEHIND the drifting electrodynamic tether that shows as a white bar going diagonally across the picture. Tether is floating 80 nautical miles from the NASA spacecraft Columbia. Note position of tether near 8 o'clock against the UFO disc.

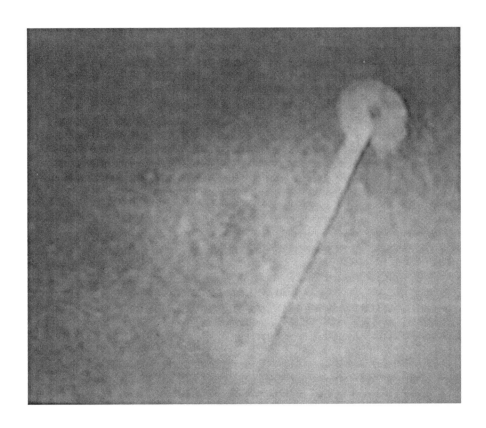

Position of tether has now moved to the right, almost to 6 o'clock on the disc. This proves that the disc is moving BEHIND the tether. The 12-mile long tether can therefore be used to estimate diameter of disc. Astronauts aboard Columbia could not visually see any of the UFOs that infested the scene. That required the ultraviolet-sensitive camera and videotape aboard Columbia.

Appendix IV

THE CRITTERS IN HISTORY

by Larry Arnold

The following excerpt is printed by permission from Larry Arnold's book *Ablaze!*, which deals with the mysteries of spontaneous human combustion—another phenomenon from which orthodoxy recoils. Larry Arnold is a respected, veteran investigator of the paranormal, and a friend of many years. His capabilities in diligent, scholarly research are evident in the following passage. I am pleased to present this contribution to "flesh out" the historical record supporting the existence of critters—which he proposes to call *Amoeba constablea*—and their interaction with the human level of life.

—TJC

For the reader who is aghast at the speculation of strange looking extraterrestrial beings, let us soothe you. One does not have to go so *far* afield to find alien life forms—alien in the sense of not being recognized by earth's biologists, rather than foreign to the biota of the planet, itself.

John P. Bessor, in July of 1947, offered a "theory" to the U.S. Air Force to explain the glowing discs seen by Kenneth Arnold earlier that year (soon to be known as flying saucers).
Claimed Bessor:

> There is a saying that Nature abhors a vacuum. If the seas of our earth are swarming with varieties of living things, both great and small, is it not logical to assume that the "sea" of our sky *abounds* with sundry forms of living things, likewise both great and small, of varied shapes, but adaptable to their celestial environment? Some may be quite invisible, others translucent, others opaque, still others capable of changing, chameleon-like, from one color to another, from one form to another, from *visibility to complete invisibility*, all in one moment [12,[1] pp. 8-9; italics added]. (Note: Bracketed numbers cite references at end.)

> Probably the phenomena represent varied species of aerial organisms [12, p. 10].

Theories aside, there is at least one organism—species, if you will—that is largely unknown to the population of this planet, and which fulfills Nature's asserted abhorrence of a vacuum; that may be involved in a series of inexplicable events.

They are called *critters*.

We first "encountered" the critters in the late 1960s in a book written a decade earlier by Trevor James Constable called *They Live in the Sky* [13].

318

This out-of-print and difficult-to-find volume purported no less than a new type of "invisible" life form living in this planet's atmosphere. Fascinating, certainly. But the book left us in that region of limbo—exposed to a new concept but without the supportive material to convince us to accept it as a reality. Years passed; and we were again confronted with evidence for these amorphous creatures in the air. This time the data were more conclusive: we were ready—some will surely say gullible, but to each his truth—to accept that earth was harboring in the vast spaces of blue sky a phenomenon that will make obsolete every current text in biology....

Yet another "story" filled with life—amazing how many one can find once a search is begun, isn't it?—comes from Crawfordsville, Indiana. Vincent Gaddis had the opportunity to read the original on-the-scene newspaper articles and to interview witnesses of a "sky monster." This self-luminous "thing" was 18-20 feet long, 8 feet wide, with neither head nor tail; it emitted "a wheezing, plaintive sound" while moving "like a fish in water" 100 to 300 feet above the ground. Gaddis condensed his notes to these facts:

> All the reports refer to this object as a living thing. A flaming, *red "eye"* was noticeable. At times it "squirmed as if in agony." Once it swooped low over a group of witnesses who said that it radiated "a *hot breath*."
>
> This visitor from the void made its appearance on two successive nights, Sept. 4-5,1891, first coming into view about midnight on both nights, and disappearing upward about 2 A.M. On the first night witnesses included the Rev. G. W. Switzer, pastor of the local Methodist Church, and his wife, who watched the sky phantom for more than an hour.
>
> On the second night *hundreds* of residents watched the monster as it moved slowly over various parts of the business district for two hours. On the third night almost every adult in the town waited for its appearance, but it never came. Many remained up all night, only to be disappointed. It appears that the monster's appearance was *strictly localized* to the sky over Crawfordsville [43, p. 33; italics added].

Those "Living" Meteorites?

On August 1,1871, while viewing the sky above Marseilles at 10:43 P.M., the French Dr. Coggia reported seeing "a large blood-red coloured meteorite, which moved *slowly* in a direction first west and next north" [21, p. 193; italics added]. In its passage it stopped three times and had five trajectory changes, the final one a "perpendicular fall to the horizon" [22, p. 454]. Mighty strange for a falling—"slowly"—rock to make a couple of right-angle turns during its descent! The astronomer A. S. Herschel suggested Dr. Coggia's "meteorite" was "a fire-balloon, and other signal lights of ordinary size" [23, p. 503]. All right. But an intelligently guided object could also perform in this manner—and one remembers Constable's mention that critters often appear "reddish."

In the same month, but the thirteenth day of 1819, "between the hours of eight and nine o'clock, was seen in the atmosphere, at Amherst, Massachusetts, a falling meteor or fire ball... of a brilliant white light resembling *burnished silver*.... Its altitude, at its first discovery, was two or three times the height of the houses; it fell *slowly* in a *perpendicular* direction, emitting great light, till it appeared to strike the earth in front of the buildings, and was instantly extinguished" [italics added]. Through other witnesses the exact point of impact was determined. "Early on the ensuing morning, was discovered... a circular form... about eight inches in diameter, and something more than one inch thickness, of a bright buff colour, with a fine nap upon it similar to that on milled cloth, which seemed to defend it from the action of the air." We've never seen a meteorite like this one in a museum! But there is a similarity with the living critter descriptions by both Wood [i.e., Don Wood, Jr.] and Constable discussed earlier: sometimes "micalike," "bright," or buff-hued.

"On removing the villous coat, a buff coloured pulpy substance of the consistence of good soft soap, of an offensive, suffocating smell appeared," the Massachusetts investigators sensed. "A few minutes exposure to the atmosphere changed the buff into a livid colour resembling *blood*. It was observed to attract moisture very readily from the air" [italics added].[2] Later the material all but "evaporated" [26, pp. 335-37].

Exceedingly strange behavior for a chunk of stony meteorite—especially as this reporter to the *American Journal of Science*, Rufus Graves, was a lecturer in chemistry at Dartmouth College and would be expected to distinguish between what he called "gelatinous" material and a rock! Yet he had "no reasonable doubt that the substance found was the residuum of the meteoric body." However, as Edward E. Free pointed out in 1910 about Graves's conclusion: "the evidence which he states is hardly satisfactory to the modern, more critical observer. It seems probable that these jellies are, in general, plasmodia of some form" [27, p. 6].

Plasmodia are "amoeboid, multinucleate masses or sheets of protoplasm characteristic of some stages of organisms," says the *Random House Dictionary*. Constable calls critters "amoeba-like life forms existing in the plasma state."

The problem with Graves's statement seems not the "residuum" (the gelatinous mass Free thought was misattributed to a meteor that fell elsewhere) but the adjective *meteoric*: if the *Amoeba constablea* is substituted—that is, that Graves and colleagues saw falling and found lying a dying or dead organism—then the episode becomes less perplexing, for an animal having independent locomotion could do what a meteor couldn't.

Around the world—in the United Kingdom, Europe, North America, and the Pacific, for examples—are found accounts of translucent or transparent gelatinous material associated with luminous falling bodies [see 25, pp. 287, 289, 294, 297-99]. It is variously called "pwdre ser, star-slough, star shoot, star shot, star-jelly or jelly, star-fall'n" [28, p. 493].

Writers have incorporated these "common and widespread" occurrences in their prose and poetry, as did William Somerville in 1740:

> Swift as the Shooting Star that gilds the night
> With rapid transient Blaze, she runs, she flies;
> Sudden she stops nor longer can endure
> The painful course, but drooping sinks away,
> And like that falling Meteor, there she lyes
> A jelly cold on earth.

Now surely those solar-furnaces called stars aren't dripping luminous jelly onto the earth; and meteors (as astronomers conceive them) don't change their aerial courses, nor fall slowly, nor leave anything but a crater or metallic fragments to attest their journeys' end.

T. McKenney Hughes, discounting popular conjectures after study, was unable to explain satisfactorily this pwdre ser:

> Nor has anybody seen it disgorged by birds.... Nor has anyone watched its growth like nostic from the ground.
> ...I was able to pack the jelly... and... send it over to Mr. Brooks, in the Botany School who reported that it was a *mass of bacteria* [28, pp. 492, 494; italics added].

Bacteria indicate life. What if, instead, these numerous reports describe organisms, atmospheric life forms which if not normally luminous, are perceived under special conditions—like an injury; or excrement, undissolved before reaching earth? Wood and his companions *saw* an injured animal, apparently of the critter type. Would it, too, if left there to die, have degenerated into a gelatinous mass, then evaporated into nothingness?

Henry More, in 1656, penned this curious passage:

> That the Starres eat... that those falling Starres, as some call them, which are found on the earth in the form of a trembling gelly, are their excrement.

Did Mr. More know something twenty-first century biologists don't? That these "Starres" were not stars but bright and luminous animals in the atmosphere? And if there's an organism, there's excrement; and if excrement, food intake has to precede such....

Dryden and Lee (1678) wrote: "The shooting stars end all in *purple* jellies" [italics added], although they proceed to mention cases of "white jelly-like matter." Lest the reader think, as do modern physicians about the archaic and "unjustified" interest in spontaneous human combustion, that all these reports of semifluid falls belong to a bygone era of superstition, scan the following article which appeared in the *Philadelphia Inquirer* for Wednesday, September 27, 1950:

Pfft—It's Gone

Flying "Saucer" Just Dissolves

Four South Philadelphia police officers had a new explanation last night for what happens to those flying saucers people
are always seeing:

They dissolve.

That's what happened last night to the airborne object just
seen about 10 P.M. by Patrolmen John Collins and Joseph
Keenan. The two officers said they were patrolling in a red car
on Vare Blvd. near 26th St. when through the windshield they
saw what appeared to be a parachute drifting slowly down from
the upper air ahead of them.

When first seen, the thing was at treetop level, they said, and
appeared to be about six feet in diameter. It settled in an open
field near 26th St. After summoning Street Sgt. Joseph Cook and
Patrolman James Casper, his driver, they went to the field to
investigate....

The four officers stood a few feet from the object, they said,
and turned their flashlights on it, whereupon it gave off a purplish glow, almost a mist, that looked as though it contained
crystals.

Collins stepped forward and tried to pick the thing up. The
part of the mass on which he laid his hands dissolved, leaving
nothing but a slight, odorless sticky residue.

Within 25 minutes, as they stood and watched, the entire substance had evaporated. It was so light, they said, that it did not
even bend the weeds on which it lighted.

Sergeant Cook notified the FBI—a little sheepishly, since, he
pointed out, he'd have nothing whatever to show them when
they arrived, except a magic circle on the ground where something purple, and quite evanescent, once had been [29, pp. 1-2].

Well, at least we can't get rebuffed by the FBI this time—they never *had*
the opportunity to analyze this residue! *The Evening Bulletin*, another
Philadelphia newspaper, headlined its story, "Big Soap Bubble or Something
Drifts Down and Goes Pfft" and added this comment by one of the officers:

"It went the way soap bubbles go," said Kennan, "crumbling
from the top and edges, the bubbles giving little pops....
Maybe it was something that came out of a factory chimney"
[30, p. 9].

Desperation: it's factory pollution—maybe. It's never been reported
before. Except for the lack of odor, the patrolmen's description of crystals-
and-bubbles easily parallels Wood's "metal-looking froth" from the creature
on the Nevada mountain. *Maybe* Philadelphia's mystery came from something that lives in the sky—excrement, perhaps. Or a dead critter itself.

Those "Living" Processions

We could continue listing similar associations. We won't, but be assured it wouldn't be difficult—just lengthy. But one aspect should be mentioned: "Why can't these unknown-to-biology life forms exist socially, that is, in groups?

Denison Olmstead chronicled [31, pp. 363-411] the supposed meteor shower of November 13, 1833, an event with strange phenomena. In Nelson County, Virginia, "animal jelly broken into fragments" fell—ostensibly from a meteor! In Rahway, New Jersey, a "fiery rain" deposited *"lumps of jelly."* A Newark, New Jersey, newspaper reported that "a mass of gelatinous matter was found, which... is supposed to have formed one of the large meteors. Its appearance resembled soft soap. It possessed little elasticity, and on the application of heat, evaporated as readily as water." At West Point, New York, Mr. Alexander C. Twining wrote how a lady there heard "a splosh" and turned to find "a round flattened mass... looking like boiled starch"; it was transparent, and soon evaporated [31, p. 396].

Olmstead concluded his paper by saying:

> Taking it as established, that such a residuum as has been mentioned was deposited by the meteors, we may infer that the matter of which the meteors were composed was both highly volatile and transparent.

We question, as would astronomers in this century, that meteors could be the "established" cause; it seems a dangerous and unfounded assumption.

An assertion perhaps less popular, but more apt to resolve the mystery, is that this material originated with heavenly organisms, possibly deposited under abnormal circumstances. As Bessor conjectured: "Possibly these leviathans of the air seek shelter in the dense atmosphere of our earth during meteor showers and other disturbances, as ships seek harbor during a tempest at sea" [12, p. 9].

Whether the substance that fell on America's Middle Atlantic states was excrement, shed "epidermis" from a collective molting by these aeroforms, or decaying organs and residue from critters killed in upper-atmospheric disturbances or by collisions with *actual* meteors, we don't know. But we're safe in stating that the "meteor" shower of November 13, 1833, consisted of a lot more than metallic rocks from space.

We would like to cite two final examples for collective—this time, cosmic—creatures that history may be quietly harboring....

A paragraph from material received by the Association for the Understanding of Man on November 18, 1966, provides the proper prefacing:

> There are strangers out there among the stars. The Earth is but an island in the sky. Where are the voyagers of those seas? Do their ships pass us by? The mind quivers, and with every ounce of its own conditioning it tends to reach out and *slap down and conventionalize* the sight of these visitors from outside the Earth.

The first and natural reaction of almost every objective observer is to classify the object first seen by conventional standards, until the observation itself becomes so acute that it is impossible to any longer do so, and an *acknowledgement* [sic] *of the unknown* is forced upon the conscious awareness [10, p. 88; italics added].

Here comes the unknown:

Across the ocean and through time, one comes to the narrative of Baden Powell, at South Mimms, Middlesex (outside London), England, on September 4, 1850. Mr. Powell, an avid amateur astronomer, was preparing to observe through his excellent telescope the planet Mercury when, as he reported to the Royal Astronomical Society:

> I observed, passing through the field of view, in a continuous stream, a great number of luminous bodies....
>
> When I first saw them I was filled with surprise, and endeavoured to account for the strange appearance by supposing that they were bodies floating in the atmosphere, such as the seeds of plants, as we are accustomed to witness them in the open country about this season; but nothing was visible to the naked eye.
>
> ...So that it was impossible I could resist the conclusion (much as I was early disposed to hesitate) that they were real celestial bodies moving in an orbit of their own, and far removed beyond the limits of our atmosphere.
>
> They continued passing, often in inconceivable numbers, from 1/2 past 9 A.M. when I first saw them, almost without intermission, till about 1/2 past 3 P.M. when they became fewer, passed at longer intervals, and then finally ceased.
>
> The bodies were all perfectly round... and they appeared *self-luminous*,... they did not change their shape or diminish in brightness.
>
> They passed with different velocities, some slowly, and others with great rapidity; and they were very various in size [32, pp. 235-37; italics added].

Self-luminous bodies that looked "just as the planets Mercury and Venus did," at differing velocity and size (from 2 to 20 seconds of arc), external to the earth and requiring six hours for the procession to pass across the telescope's viewing width—and then cease rather abruptly! An accretion of meteoroids would not behave, nor be expected to behave, in this fashion (for among other things, meteoroids are *not*—so it is claimed—self-luminous).

A Mr. Cooper, of Markree Castle, County Sligo, Ireland, also witnessed this aerial procession. A Charles B. Chalmers, Esq., F.R.A.S., reported seeing very much the same thing, only toward the end of 1849.

Mr. Powell watched the sky for another seven-and-a-half hours—nothing mysterious; members of his family, trained viewers, again peered through the scope—nothing either.

Said Powell:

I repeated my observations the following morning, and then saw one such single body pass in the same direction as those of the preceding days [32, p. 237].

If they weren't meteoroids (and certainly not comets), what were these interplanetary lights? We see two alternatives:

1. This was a convoy of intelligently controlled spacecraft obviously not of earth origin (a hypothesis sometimes mentioned by modern writers familiar with this literature).
2. It was an excursion of living organisms, pursuing their journey along the most convenient and perhaps popular thoroughfare (for there are *many* cases of this sort of "object" seen in the vicinity of the sun, Mercury, and Venus).[3]

Perhaps that solitary "single body" seen the next day by Powell was the rear guard stationed to prevent stragglers, or hurry along sightseers diverted by fascinating scenery, or was itself a straggler.
Then came the 1883 sighting and report of Professor Jose Bonilla as reported in the scientific periodical *L'Astronomie* in 1885:

Passage Across the Solar Disk of a Swarm
of Corpuscles, Seen at the Observatory
of Zacatecas, Mexico

At the Observatory of Zacatecas, situated 2,502 meters above sea level, I instituted the daily observation of the state of the solar surface, drawing, by direct means and by projection, the spots, faculae, and granulations, as well as the protuberances of the solar chromosphere, by means of the spectroscope.
To this effect, I adapted to the equatorial telescopes of 0m, 16 cm. aperture, a projection apparatus which received on a sheet of paper an image of the sun 0m, 25 cm. in diameter, the field of the telescope only projecting itself on a surface little bigger than 0m, 26 cm. When the solar disk offered some interest, I took photographs of 0m, 067 cm. diameter, by means of instant plates of silver bromide gelatin.
The dome of the observatory has small windows with opaque black curtains of a sort so that nothing penetrated across the objective save the image of the sun. This disposition permitted the noting [always with precision and clarity] of the faculae and of the least details of the spots as well as of the granulations, thanks to the transparence of the atmosphere at the altitude at which the observatory is situated under the tropical sky [22°46'34"9 North latitude].
On the twelfth of August, 1883, at 8 o'clock in the morning, I began to draw the solar spots, when I suddenly noticed a small luminous body which penetrated the field of the telescope, became visible on the paper which I used to reproduce the spots,

and crossed the disk of the sun projecting itself as an almost circular shadow.

I had not recovered from my surprise when the same phenomenon occurred again, and with such a frequency that in the period of two hours, I was able to count about 283 bodies traversing the disk of the sun.

Little by little the clouds interfered with observation, which could not be resumed until the moment the sun crossed the meridian, and then for only 40 minutes. During this interval I counted the passage of 48 more bodies. The paths followed by these bodies indicated a movement direct from the West to the East and more or less inclined to the North or South of the disk. After some minutes of observation, I noted that these bodies, which seemed black and somber, some perfectly round and others more or less elongated, while projecting their images on the solar disk, offered luminous images when leaving the edge and crossing the field of the telescope.

The intervals of the passage were variable. Sometimes one or two passed, taking only a third, or a half of a second, and up to a second to traverse the solar disk, and then one or two minutes went by before others appeared; sometimes 15 or 20 passed almost at the same time, so that it was difficult to count them. I was able to fix the trajectory of several of these bodies on the solar disk, noting the entries and exits on the paper which served me to draw the spots—this paper, as well as the equatorial telescope, followed by means of clockwork the diurnal apparent movement of the sun across the celestial vault. The figure below is a reduced copy of the drawing that I made of the solar disk that day [of 250mm. diameter] with the trajectories of the bodies of the solar spots.

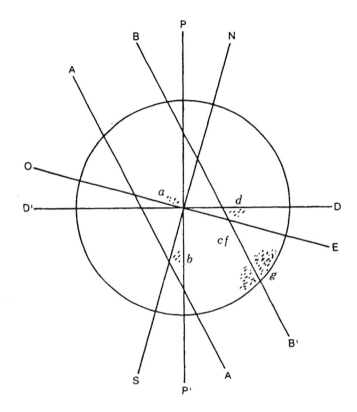

Lanes followed by the corpuscles across the solar disc.

PP¹ Declination circles
DD¹ Parallel circles
EO Solar equator
NS Polar diameter of the sun,
      apparent solar diameter-
      1899¹.

AA¹, BB¹ zone of the trajectory
      of the corpuscles on the
      solar disc
Intermediary lines were followed
by several corpuscles

Often taking photographs of the sun, when its disk presents notable spots and faculae, I also tried to photograph this rare and interesting phenomenon of the passage of these bodies on the solar disk.

To this end, I replaced, in the same equatorial, the 0m, 16 objective, by another of equal power, but with a chemical focus, to which I adapted the ocular piece and the photographic chamber. After several tries to put the bodies in focus, I succeeded in taking various photographs, of which I have sent *L'Astronomie* the most interesting. While I was taking the photographs, an aide counted the bodies with the equatorial's finder. The photograph was taken on wet collodion of .01 seconds. This rapidity did not allow me time to prepare baths conveniently; thus, the negative is a little stained. The image of the sun is not in focus, but that of the body, which offered more interest to me at that moment.

While in the projection and at first sight, all the bodies seemed round or spherical, one notices in the various photos that the bodies are not spherical but are for the most part irregular.

I have said that, in the projection of the field of the telescope, these bodies appeared luminous and showed brilliant trains; but in crossing the solar disk, they appeared opaque. When observing with attention the photograph and the negative, one notes a body surrounded with a nebulosity and with obscure trains which, in the field of the telescope and outside the disk, seem brilliant. This causes me to believe that these brilliant trains in the passage of the bodies on the disk absorbed the actinic light of the sun or diminished its photographic power.

In the afternoon, the clouds prevented observation.

I took measures then and established a plan of observation in case the phenomenon occurred the following day.

On the thirteenth of August, the two first hours of the day offered me a cloudy sky up to eight A.M. Then the clouds dispersed a bit and I was able to observe. Soon the same phenomenon appeared again, and during the forty-five minutes of observation that the state of the sky permitted, we counted 116 bodies traversing the solar disk.

Soon after the observation of the twelfth, I had telegraphed the observatories of Mexico and Puebla to ask them to observe this phenomenon, but it was invisible to these observatories. To verify in an indirect fashion the approximate distance of this swarm of bodies, I carefully adjusted the finder of the telescope, the equatorial, and a Foucault reflecting telescope of 0m, 10 diameter, directing them on the solar disk and on the bodies. I had in the night the occasion to direct them equally toward the planets and the moon (which had been for two days in its first quarter) without changing the focus, and the moon alone seemed almost in focus.

This circumstance, joined with the invisibility of the phenomenon at Mexico and Puebla or elsewhere, makes me believe that the bodies were rather close to the earth, at a distance less than that of the moon, and that their considerable parallax was the cause that at Mexico and Puebla they were projected outside of the solar disk.

> Jose A. y Bonilla,
> Director of the Observatory
> of Zacatecas, Mexico[4]

Finally, there was the 1913 meteor procession, made famous (or infamous, depending on one's perspective) by Professor C. A. Chant of the University of Toronto. "There were obviously many peculiar aspects," wrote William Corliss, "to this spectacular event."

The description that triggered the controversy began with Chant in... the *Journal of the Royal Astronomical Society of Canada* (he was editor at the time)...:

> At about 9:05 on the evening in question [February 9] there suddenly appeared in the northwestern sky a *fiery red* body which quickly grew large as it came nearer, and which was then seen to be followed by a long tail.... It resembled a rocket; but, unlike the rocket, the body showed no indication of dropping to the south. On the contrary it moved forward on a *perfectly horizontal* path with peculiar, majestic, dignified *deliberation*; and continuing in its course, without the least apparent sinking towards the earth, it moved on to the southwest where it simply disappeared in the distance....
>
> Before the *astonishment* aroused by this first meteor had subsided, other bodies were seen... emerging from precisely the same place... at the same deliberate pace, in twos or threes or fours....
>
> Several report that near the middle of the great procession was a fine large star without a tail, and that a similar body brought up the rear [33, p. 145; italics added].[5]

Chant found similar accounts from Saskatchewan to Bermuda. W. H. Denning found shipboard sightings made in the South Atlantic (*Journal of the Royal Astronomical Society of Canada* vol. 9, p. 287; vol. 10, p. 294).

William Henry Pickering, the younger of the noted father-son American astronomy family, compiled the sightings as follows:

> A procession of fire balls and meteors *all moving very slowly*... was first seen near Mortlach, 65 miles west of Regina, Saskatchewan, lat. 50° .5N long. 106°W.... and last from the steamer Newlands, lat. 3° 20'S., long. 32° 30'W. The distance between the first and last stations is 5659 *miles*. Computations indicate that the meteors traversed several thousand miles more [34, p. 632; italics added].

And he added, of these "yellow or reddish" or "slightly violet" lights that looked like "bright stars" (two of which "looked like large arc lights... of diameter equal to the moon."):

> This remarkable phenomenon was *in no sense* a meteoric shower. It was a different kind of event *altogether* [34, p. 632; italics added].

If this eminent astronomer rejects a meteor shower, then what was everybody seeing? "Fireballs," as Pickering says? But why would fireballs maintain such a uniform altitude; and what generated such a procession of them? The account, and the colors especially, reminds one of the 1833 procession of Constable's critters. What do other experts say?

"To show the exaggeration possible in an apparently well-authenticated story," C. C. Wylie refers to the 1913 display while casting aspersion on "flying saucers" and other atmospheric enigmas. Writing in the prestigious journal *Science*, Wylie asserts that "only one real fireball... not very large" appeared over North America and "disintegrated at a height of twenty-five miles near Hamilton, Ontario." Wylie simply writes off all the sightings collected by his predecessors, and concludes:

> The popular story is impossible, of course; and it is evident that an excellent but unpredicted shower of shooting stars has been "blown up" into a marvelous procession of fireballs [35, p. 127].

Wylie is a meteor specialist. Meteors can't do what Chant and Denning claimed; Pickering is discounted or ignored. Prejudice reigns.

We think of Charles Fort's introduction to his first book on oddities, *The Book of the Damned*:

> A PROCESSION of the damned.
> By the damned, I mean the excluded.
> We shall have a procession of data that Science
> has excluded....
> All would be well.
> All would be heavenly—
> If the damned would only stay damned [9, pp. 18,
> 32].

A procession of objects—this world's longest parade—for over 5,600 miles, in the sky! Scientists say, "Damned!" It is damned. Photos of pulsating, moving life forms in the atmosphere. "Damned," say the exobiologists. Astronomers look to outer space, to other stellar bodies, for exobiology.[6]

There can be no biology on earth which is *exo*—that is, "outside," "outer," "external"—the known Linnean categories of creatures. All are categorized, segmentized, and most likely segmented on some dissecting table. Half-mile-round critters in the sky: "Damn!" And damned.

Dr. Carl Sagan, the exobiologist exemplar, tells us:

> There might be a kind of biological law decreeing that there are many paths to intelligence [37, p. 89].
> A bona fide example of extraterrestrial life, even in a very simple form, would revolutionize biology... it would be truly immense [38].

Is half a mile "immense" enough? Or does "a very simple [life] form" *have* to be extraneous to earth before biology is revolutionized? We side with T. J. Constable when he writes:

> The helplessness and sometimes the irrational opposition of official science in the face of these pressing questions should not deter any thinking person from admitting the need for further investigation. Progress depends on the raising of such questions.... That is why it will take *young* people—free of neurotic dependence on the mechanistic world-conception—to press these matters forward [1, p. 54; see also 2, p. 81].

C. C. Wylie's predisposed notions reign—but not for long. Along comes a comment by Alexander D. Melbane, in a later issue of *Science*. Melbane attacks Wylie's account, graciously:

> It is perhaps not made sufficiently clear that this description of the phenomenon *differs considerably* from that which has previously appeared in the astronomical literature.... Wylie's description of the event as local in character is likewise a revision of the previously accepted version and is not easy to reconcile with the data....
> *A fully satisfactory explanation of this spectacular occurrence of 1913 has never been achieved...* it should be recognized that the *recorded* evidence is difficult, if not impossible, to reconcile with Professor Wylie's description [36, pp. 725-26; italics added].

Wylie counters the counterattack in the same issue, on the grounds that the original observations are impossible because (1) fireballs can't survive for 5,000 miles; (2) allegedly no one noted the horizon obscuring their trajectory; (3) if seen in Canada and Bermuda, people in the Middle Atlantic states would also have to see them—and Wylie asserts they were invisible there.

Well if an event doesn't fit into one's limitations, and one chooses not to do one's homework (so to speak), it becomes easy (if not simplistic) to say the observations are simply wrong!...

Melbane went out and did some homework. Guess what he found? The same thing that John O'Keefe did [39] when he went looking: "several dozen accounts in the files of newspapers in Minnesota, Wisconsin, Michigan, New York, Pennsylvania, and New Jersey." Now, unquestionably, the continuous integrity of this 5,000-mile-plus procession is confirmed. What else did O'Keefe discover? Just that by reconstructing various parabolic and hyperbolic meteoric paths, it became impossible to make the theoretical models fit the observed facts—which reaffirms (again) Pickering's 1922 statement. To reconcile data with theory, there is only one astronomically acceptable alternative left: these objects were "satellites of the earth" [39, pp. 4-8]. O'Keefe pursued this reasoning, referring to the event as the Cyrillid shower:

> Assuming that the Cyrillids were earth satellites, we might expect some of them to make more than one trip around. The next revolution... would have carried them over the Middle West, above the populated regions [39, p. 6].

Though thoroughly scanning the holdings of the Library of Congress and other archives, he and colleagues were unable "to locate a single article" indicating a second orbit. The lack of supportive observations forced O'Keefe to conclude that

> the Cyrillids were visible only on and near a great circle; and they make it very *unlikely* that any substantial part of the Cyrillid shower survived for another circuit of the earth [39, p. 6; italics added].

But we find this to be a rather strange determination as well. If this were a parade of luminous planetary satellites, then:

1. What made them so uniformly luminous for so long?
2. Why were they, if orbiting the earth (as would be indicated by the great circle arc along which they were viewed), never before reported by astronomers?
3. If their orbit was decaying, who can explain why "their flight was nearly horizontal" [39, p. 4-8]?
4. What force constricted this swarm of previously undetected satellites to a "very thin" entry corridor "about 100 miles across"?
5. What celestial mechanism caused *all* the satellites to enter earth's lower atmosphere on the *same* orbit and to *all* burn up together somewhere over the South Atlantic or Pacific Oceans?[7]
6. Why did the satellites pair up "in twos or threes or fours"?
7. How does one explain the few exceptionally bright lights amid a host of "fiery red" bodies when they should all contain the same lithoidal material?
8. If these were a familiar type of celestial object, why did the experienced astronomer Chant refer almost mystically to these bodies as "majestic," "peculiar," and "dignified"?

O'Keefe's scholarship is superb but, as these questions seem unresolvable, his premise is weak. Consider Fort's thought on these alien processions:

However, light or dark, they have been seen and reported so often that the only important reason for their exclusion is—that they don't fit in [9, p. 225].

And O'Keefe's postulate is *destroyed* if one connects the procession of February 9, 1913 with the baffling event of the next day's afternoon, when clusters of dark objects above Toronto were seen by many. "They passed from west to east in three groups," reported the *Toronto Star*, "and then returned in more scattered formations, seven or eight in all...."

Possible Trajectories and Theories for the
1913 Aerial Procession

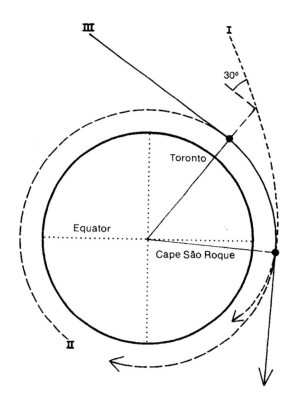

I: Typical parabolic or hyperbolic meteor path—would produce a minimum angle of 30 degrees when viewed at Toronto, rather than the horizontal flight witnessed. Rejected by almost all scholars.

II: Satellite(s) of earth—would fulfill the observational data; raises, and leaves unanswered, many dilemmas, however (see text).

III: Alien life form movement—whether organisms or their mechanical devices, the records can remain intact while problems of celestial mechanics inherent in I and II are overcome.

NOTES TO APPENDIX IV

1. Numbers in brackets refer to works listed in the References section, found at the end of Appendix IV.

2. Again an important clue to the fact that this object was organic rather than mineral. Constable theorizes, upon experimental evidence, that critters function on orgone energy (so designated by Dr. Wilhelm Reich). Concentrations of orgone energy are known to attract aqueous vapor. The implication should be obvious.

3. That an assortment of amino acids, the building blocks of life as known on earth, has been detected in deep space should dampen objections that preclude the development and existence of life forms in the "hostile" environment between planets and solar systems.

4. Editor's note (of *L'Astronomie*)—The observation of Bonilla is very interesting, but it is not easy to explain. The dates of the twelfth and thirteenth of August make one think of the meteors of those dates (Perseids), but it would be singular if none were seen at Mexico or Puebla. Were they birds? We tend to believe that it is a case of birds, or insects, or high dust—in any case, of corpuscles appearing in our atmosphere.

5. One supporting example: Walter H. Stevenson, of Fenlon Falls, northeast of Toronto, drew a picture of what he witnessed: seven bright "stars" trailed by a reddish glow, then a *bright* object as brilliant as Venus, followed by several reddish clusters, and finally a "shower of red meteors."

6. Strangely, even when they seem to find what they're looking for, they discount it (officially, at least). At the time of this writing, the first *Viking* lander on Mars was transmitting data from the red planet. This project—ostensibly to detect Martian life—drained the American coffers by $1 billion. What did this expenditure uncover? For one thing, on July 25, 1976:

> New close-up pictures of the Chryse Plain on Mars revealed what appeared to be a rock bearing the Roman letters "B" and "G" and the Arabic numeral "2." Scientists said the characters were optical illusions, caused by shadowing.
>
> "People are great at imagining things," said Dr. Carl Sagan. Dr. Bruce Murray, JPL director, agreed.
>
> But [James] Martin, an aeronautical engineer [and Viking project manager], said he wasn't so sure.
>
> "I could see the 'B' very well, but I guess I've got to accept the scientific word for it," he said. "If I were a newsman I would wonder *how a shadow could reach all the way around.*"
>
> He said he was somewhat hesitant to rule out possibilities.
>
> "There are no Martians there," Martin said. "But who knows, who has been there at some time?" [44, p. 1; italics added].

The alphabet rock was mentioned on the late evening (July 25) ABC-TV news and appeared (as above) within an AP release from the Jet Propulsion Laboratory on July 26. Then silence. Martin seems to condescend to the consensus of authority.

Viking is American; the figures are English. Why aren't these "shadows" amorphous or Chinese, if natural? Bacteria are a safe (secure) find—but "Microbes can't write!" Hmm—

7. O'Keefe first asserts the Cyrillids were "*individually* in orbit for many revolutions before they were seen" [39, p. 6]; with non-uniform bodies experiencing different drags, their entries would occur at varying times and be "spread over a number of revolutions [39, p. 6]. This refutes the observations. He later tries the theory of one large body which begins to melt (not fracture) in the upper atmosphere, whereupon the faster-moving particles eventually lap the parent body in orbit, at which time all particles *simultaneously* become visible over Canada [39, p. 7]. This sounds contrived; but then O'Keefe volunteers a statement that negates both his concepts about a lithoid origin: "Numerical integrations have failed to show *any way* in which the members of such a shower could disappear in one *revolution*" [39, p. 6; italics added].

References

1. Trevor James Constable, "UFOs Are Living Creatures," in *UFO Report* (Brooklyn) 2, no. 4 (Fall 1975).
2. Trevor James Constable, "The Case for the 'Critters,' " in *Other Worlds, Other Universes*, edited by Brad Steiger and John White. (Doubleday & Company, Garden City, N.Y., 1975).
3. "Twenty Foot Leap from Blazing Home," *London Daily Telegraph and Morning Post*, December 27, 1938.
4. Eric Frank Russell, "Invisible Death," *Fate* (U.K edition), March 1955.
5. "Shocking Ballina Christmas Tragedy," *Western People*, December 31, 1938.
6. 'Woman's Death," Croydon Advertiser, December 30, 1938.
7. Joachim, "First Level Dimensional Awareness," *See of Tranquility* (Allentown, PA), June 14, 1975.
8. Robert C. Meslin, private communication, May 29, 1976.
9. Charles Fort, *The Book of the Damned* (Ace Books, New York, 1972; The Book Tree, San Diego, 2006).
10. "Project Starlight International, Part II," *Journal of the Association for the Understanding of Man* (Austin, Texas) 2, no. 3 (Spring 1974).
11. Ray Stanford, "Firsthand Contacts with Extraterrestrial Life," Membership Conference of the Association for the Understanding of Man, August 23, 1974.
12. John P. Bessor, "Are the Saucers Space Animals?," *Fate* 8, no. 12 (December 1955).
13. Trevor James Constable, *They Live in the Sky* (New Age Publishing Co., Los Angeles, 1959).
14. Trevor James Constable, private communication, September 4, 1975.
15. John White, private communication, April, 1975.
16. Donald E. Keyhoe, *Aliens from Space: The Real Story of Unidentified Flying Objects* (Doubleday & Company, Garden City, N.Y., 1973).
17 *Introductory Space Science* (Air Force text issued in 1970 and later withdrawn from the U.S. Air Force Academy), vol. 2, chap. 33.
18. David Williamson, assistant administrator for special projects, NASA, Washington, D.C., private communication, February 21, 1975.
19. Hugh Ruttledge, *Attack on Everest* (R. M. McBride and Co., New York, 1935).
20. Walter N. Webb, "An Analysis of the Fish Model," *Pursuit* (Columbia, N.J.) 8, no. 3 (July 1975).
21. Coggia, "Extraordinary Meteor Seen at Marseilles," *Chemical News*, October 20, 1871.
22. Anonymous, *Nature*, October 5, 1871.
23. A. S. Herschel, "The Marseilles Meteorite," *Nature*, October 26, 1871.

24. "Shower of Red Matter Like Blood and Muscle," *American Journal of Science*, 1st ser. 41 (1841).

25. Ivan T. Sanderson, *Investigating the Unexplained* (Prentice-Hall, Englewood Cliffs, N.J., 1972).

26. Rufus Graves, "Account of a Gelatinous Meteor," *American Journal of Science*, 1st ser. 2 (1820).

27. Edward E. Free, "Pwdre Ser," *Nature*, November 3, 1910.

28. T. McKenny Hughes, "Pwdre Ser," *Nature*, June 23, 1910.

29. "Pfft—It's Gone: Flying 'Saucer' Just Dissolves," *Philadelphia Inquirer*, September 27, 1950.

30. "Big Soap Bubble or Something Drifts Down and Goes Pfft," *Philadelphia Evening Bulletin*, September 27, 1950.

31. Denison Olmstead, "Observations on the Meteors of November 13th, 1833," *American Journal of Science*, 1st ser. 25 (1834).

32. Baden Powell, "A Catalogue of Observations of Luminous Meteors," *Report of the British Association* (1852).

33. C. A. Chant, *Journal of the Royal Astronomical Society of Canada* 7 (1913).

34. William Henry Pickering, "The Meteoric Procession of February 9, 1913," *Popular Astronomy* 30 (1922).

35. C. C. Wylie, "Those Flying Saucers," *Science*, July 31, 1953.

36. Melbane, Alexander D., "The Great Fireball Procession of 1913," *Science*, December 11, 1953.

37. Carl Sagan and Frank Drake, "The Search for Extraterrestrial Intelligence," *Scientific American*, May, 1975.

38. Carl Sagan "The Planetary Perspective," Joseph Priestly Award Address, Dickinson College, April 3, 1975.

39. John A. O'Keefe, "Tektites and the Cyrillid Shower," *Sky and Telescope* 21, no. 1 (January, 1961).

40. A. R. G. Owen, *Can We Explain the Poltergeist?* (Garrett Publications, New York, 1964).

41. "A Vain Bid to Save a Driver," *Liverpool Echo*, April 7, 1938.

42. "Burned to Death," *Liverpool Echo*, January 2, 1939.

43. Vincent H. Gaddis, *Invisible Horizons* (Ace Books, New York, 1965).

44. "Viking Soil Sampler Straightens Out Hitch," *Harrisburg Patriot*, July 26, 1976.

Index

Ablaze, 314

Abrams, Dr. Albert, 87, 150, 156, 162, 167

Adamski, George, 189

aeroform, 56, 266, 273

aeroforms, 2, 22, 58, 64, 67, 84-85, 110, 130, 135, 192, 238, 258, 279, 319

aerospace, 9, 15, 21, 23, 65, 91, 156, 253

Ahriman, 142-149, 243-244, 248

Ahrimanic powers, 143-146, 148-150, 153, 165, 167, 172, 187-188, 190, 216-217, 243-245, 247-249, 251, 256

Aldrin, Buzz, 246

Alexander the Great, 84

aliens, 22, 28, 74, 151, 190, 218-219, 246, 255, 333

Allen, William G., 285

Alpha Series, 53, 65, 258

Alpine Journal, 70

Altai-Himalaya, 74

America, 2, 20, 99-100, 102, 106, 119, 125, 153, 157, 162, 165, 188, 205-206, 211, 214, 243-245, 249-250, 253, 316, 319, 326

American Journal of Science, 316, 334

Amoeba constablea, 314, 316

amoebae, 56, 73, 280

"angels"—radar anomolies, 8, 94-95

angels, 134

animals, extraterrestrial, 69

Anthroposophical Society, xiii, 34, 102-104, 108, 128, 156

Anthroposophical spiritual science, 80, 144

Anthroposophy, 34, 101, 104-105, 108-109, 111, 128

Apollo, 38, 53, 119, 212-213, 284

Apollo Lunar Landing Module, 119

Apollo moon missions, 212

apportation, 132

Arizona, 19, 192-193, 222, 224-225

Arlesheim, 107

Armageddon, 214

armor, 12, 201, 203, 207

armor, muscular, 201, 203, 207

Arnold, Kenneth, 22, 82, 218, 230, 235, 314

Arnold, Larry, v, 75, 303, 314

Ashtar, 60, 61

astral body, 39, 76, 234

astronomy, 105, 107, 144, 195, 295, 325, 334

astrophysics, 105, 144, 179, 195

Atomic explosives, 16

atomic power, 16

Attack on Everest, The, 69, 333

authority, 2-3, 63, 78, 89, 100-102, 117, 118, 132, 153, 198, 227, 248, 331

azoic rocks, 115, 117, 204

Baker, Dr. Elsworth F., 213

Bastian, H. Charlton, 204

Battle Mountain, 72

Bean, Alan, 246, 284

Bearden, Thomas E., 253

Beck, Bob, 94

Beginnings of Life, The, 204

Bermuda Triangle, 138, 139

Bessor, John P., 314, 333

Big Medicine, 151, 153-154, 156-157, 164-165, 210, 221

biodynamic, 106-107

biodynamic gardening and farming, 106

bioenergetic, xii-xiii, 9, 12, 42-44, 51, 53, 61, 82, 89, 100, 126, 148, 151, 168, 170, 173-174, 184, 186-187, 190, 199-200, 207, 209, 211, 230, 232, 234-235, 238, 248

bioenergetic beacon, 51
bioenergetic communications, 151, 186-187
bioenergetic continuum, 174, 187
Bioenergetics, 151
bioenergy, 62, 64, 89, 151, 190, 201, 230-231, 246, 251, 253
bioforms, 54, 114, 122, 149, 280
bioforms, plasmatic, 114
Biological Radio Communications, 251, 253
biology, 22, 34, 46, 74-75, 82, 85, 107, 126, 167, 195, 199, 203, 315, 319, 326-327
Biometric evaluation, 166
biophysics, 203, 218, 224, 252
Biophysics Laboratory, 252
Bioplasma: The Fifth State of Matter?, 252
bioplasmic body, 157, 179
bioplasmic energy, 44, 157, 165
Birth Ray, 180
Blavatsky, Madame, 156
blood, 73, 116, 158, 163, 172-174, 185, 187, 193, 205, 243, 257, 315-316, 334
Boccone, Luciano, v, xv, 75, 83, 258, 293-294, 300-301, 304-306
Book of the Damned, 67, 326, 333
Borderland Sciences Research Foundation, xv, 5, 67, 76
Boys Downstairs, The, 136, 138-144, 167, 217, 221, 226, 234
Boys From Topside, The, 132
Brady, Mildred Edie, 210
Bravo, 63, 86, 96, 258, 268-276
Bravo Series, 96, 258
breathing of the planet Earth, 71
British National Laboratories, 157
brownite, 218, 221
BSRA, 76, 130, 133, 135
BSRF, 67, 76, 115
Buddhists, 178
Burr, Dr. Harold Saxton, 89, 98

Cage, John M., 81
Cambridge Research Laboratories, 94
Canadian government, 13, 109, 132
cancer, 107, 174, 202, 206, 209, 222
Cancer Biopathy, The, 202, 209
Cannon, Dr. Alexander, 36
Cathie, Bruce, 190
cave, 117-119
CFP, 55
changes of consciousness, 111
character analysis, 196, 207
character structure, 2, 17, 92, 196
Chascomús, Argentina, 137
Chichester, Sir Frances, 67
Chidlaw, Gen. Benjamin, 138
children, ix, 1, 17, 27, 52, 68, 103-105, 107, 145, 154, 158-159, 197-199, 214, 245, 256
Chladni sound figures, 122
Christ, 105, 118, 142-143, 146-147, 178, 215, 234, 247-248, 255-256
Christ event, 143
Christ mystery, 105
Churchill, Winston, 227, 247
cloudbuster, 122, 173, 192-194, 207, 212, 220-222, 224-225, 230, 232, 240, 258, 277-279, 281-283, 297, 301
cloudbusters, 192, 220, 227, 230
Cluj, Rumania, 61, 228
Collins, Michael, astronaut, 246
Colorado University project, 15, 26
common functioning principle, 55, 64, 75, 140, 237, 242
Communism, 149, 251
Condon, 2, 26, 90
Condon, Dr. Edward, 26
Contact With Space, 214, 217, 222, 235
control, illegal—of humans, 140, 250
Cooper, Gordon, astronaut, 246
Corliss, William, 325
Cornell University, 54
cosmic numbers, 171
Cosmic Orgone Engineering (CORE),

219, 222
cosmogony, 105, 144
cosmology, 92, 105-106, 246
Council of the Seven Lights, 35
Course of My Life, The, 102
Cox, Adrian, 69
Crabb, Riley, 67, 91
Creatures from the Stratosphere, 81
Creighton, Gorden, 137
critters, v, 55, 63-65, 67-69, 71, 73-75, 77-85, 87, 89-90, 96, 114, 117, 122, 134, 148, 188-189, 201, 227, 233-235, 237, 252, 258, 264, 288-290, 295, 297-298, 314-316, 319, 326, 331, 333
Crosse, Andrew, 115
Curie, Madame, 205
D'Arsonval movement, 47, 229
Da Vinci, Leonardo, 36
de la Warr, George, 162
Deadly Orgone Radiation, 219
Delgado, Dr. Jose, 149, 184
dematerialization, 16, 253, 305
Detroit Air National Guard, 10
devil, 13, 28, 141-142, 216, 242, 249, 255
disappearance, 16
discovery of the orgone, xii, 2-3, 12, 20, 28, 42, 102, 106, 147, 151-152, 157, 193-195, 202, 205-206, 217, 232, 252, 297
Discovery of the Orgone, The, 202
Divination of Disease, The, 153
DOR, 219, 221-225
Doyle, Sir Arthur Conan, 67
Drown instruments, 152, 157, 161, 178
Drown rates, 171, 181-182
Drown, Cynthia, 154, 158
Drown, Dr. Ruth B., xiii, 38, 87, 98, 115, 129, 133, 135, 149-166, 168, 170-176, 178-194, 198, 200, 205-207, 210-211, 227, 254-255, 297
Drown, Homer, 154, 158

E.R.A, 157
Eden, Jerome, 86
Edge of the Unknown, The, 67
Edwards, Frank, 13, 77-78, 132, 240
Einstein, Professor Albert, 5, 208-209, 211, 228
electromagnetic radiation, 4-5, 37-38, 48, 54, 79-81, 135, 157, 215
electromagnetism, 80
Electronic Reaction of Abrams, 157
eloptic energy, 162
EM, 135, 227-231, 234-235, 237, 252
emotional plague (EP), 91, 95, 165, 209-210, 214
Engle, Senator Clair, 94
EP, 209-210
ETH, 24, 96, 97, 131
Ether and its Vortices, The, 5
Ether Ship Mystery and Its Solution, The, 67, 131, 134-135
Ethereans, v, 128-138, 140-141, 146, 148, 186, 187
etheria, 131, 133-136
etherian physics, xii, 59, 61-62, 68, 80, 82, 85, 106, 110, 112, 118, 120-123, 126, 128-129, 136, 141, 176, 194, 216, 219-220, 226, 228-229, 234-236, 238, 242, 255, 304
etheric double, 36, 39, 41, 52, 136, 157, 179, 232, 234
Etheric Formative Forces in the Cosmos, Earth and Man, 255
ethers, v, 30, 46-47, 49, 51, 53, 55-56, 110, 113, 120-124, 126, 131, 173, 178, 208, 215, 224, 228, 235, 242-243, 253
eurythmy, 103, 107
Everest 1933, 69
exobiology, 22, 54, 72, 82, 84, 326
extraterrestrial hypothesis, 24, 96, 131
eye beam, 35, 40, 43, 232
Faraday, Michael, 51
Fate Magazine, 43, 78
Fichte, 100

flying saucer, 6, 20, 22, 25-26, 28, 56, 60, 63, 66, 69, 73, 75, 78, 118, 137-138, 233, 236, 254, 278

Flying Saucer Conspiracy, The, 6, 20, 26, 138

Flying Saucer Review (*FSR*), 20, 25, 56, 63, 69, 78, 118, 137, 233, 236

Flying Saucers Are Real, The, 26, 192, 219

Flying Saucers From Outer Space, 26

Flying Saucers Uncensored, 18

Flying Saucers: Serious Business, 13, 77, 132, 240

foo fighters, 17, 18, 84

Food and Drug Administration, 164, 210, 213, 225

Formative Forces in Crystallization, 174

Fort, Charles, ix, 67, 81, 234, 326, 333

Fortune, Dion, 176

Freud, Dr. Sigmund, xiii, 167, 195-200, 202

Friedman, Stanton, 253

Frontiers of Consciousness, 254

Fry, Daniel, 109

FSR, 20, 25, 233

Function of the Orgasm, 193, 202-203

Gaddis, Vincent, 315

Galileo, 144

Gardner, Robert C., 139, 150

Gheorghita, Florin, 298-299

Giant Rock, California, 28-29, 31, 50, 59, 264, 266, 268-269, 273

glasnost, 244, 251

Glenn, Colonel John, 65, 246

Goddard, Robert, 74

Gods, Demons and Space Chariots, 242

Goethe, Johannes, 34, 38, 83, 99-103, 124, 167

Goetheanum, 103, 106, 108

golden mean, 147-178

Gorbachev, 251

Graham, Billy, 108, 250

GRCU, Gruppo di Ricerche Clipeologiche ed Ufologiche, 75, 83, 258, 294-295, 301, 305, 307-308

Haeckel, Ernst, 102, 104

Hall, Manly Palmer, 36, 179

heat, 19, 23, 49-50, 55, 61, 114, 116, 121-123, 127, 135, 148, 207, 209, 227, 231, 237, 273, 304, 319

heat pictures, 116

heat rays, 19, 23, 49

Heindel, Max, 36

Hieronymus, Thomas Galen, 162, 166, 254

Highest State of Consciousness, The 254

Hitler, Adolf, 203, 247-248

Hohokam Indians, 118

Holmes, Dr. Ernest, 165

Horror of the Heights, The, 67

humanoid(s), 78, 130, 238

Humanoids, The, 78

Huntley, Chet, 244

Hynek, Prof. J.A., 299

I Rode a Flying Saucer, 60, 66

Incident at Exeter, 257

Inconnue, 81

infrared, vii, x, 5, 33, 35, 47-52, 54-55, 61-64, 69, 81, 83, 86, 96, 109, 259, 262-266, 268-281, 290, 293, 295-299, 301-302, 304-309, 311

Ingals, Professor C. E., 54

Inner Circle, The, 129, 136, 137

Inyushin, V. M., 252

Johnson, Kendall, 37

Johnson, Lyndon, (LBJ), 250

Journal of Borderland Research, 91

Journal of Orgonomy, 213

Journal of the Drown Radio Therapy, 164

Kabbalah Unveiled, The, 175, 181

Kant, Immanuel, 113

Kant-La Place theory, 113-114, 116-117, 121

karma, 105, 166, 251
Kazhinskiy, Dr. B., 252
Keel, John A., ix, 255, 298
Kepler, Johannes, 114, 124
Keyhoe, Major Donald, 6-7, 20, 26-27, 90, 138-139, 218-219, 234, 333
Kirlian photography, 37, 44, 159, 252
KRW (kreiselwelle waveform), *see also spinning wave*, 202, 207-208, 235
La Place, Pierre, 113
Lapp, Dr. Ralph E., 26
larynx, relation to hearing, 186
Lascaux, 119
Law of Orgonotic Potential, 42
Layne, Meade, 57, 67-68, 73, 76, 81, 115, 123, 128-135, 137-138, 140, 175
Lehrs, Dr. Ernst, 34-35, 40, 80, 99, 104-105, 127
levitation, 238
Lewis, Dr. E., 38
libido, 198-203
life energy, xiii, 12, 25, 39, 51, 75, 81, 117, 155, 157, 159, 168, 172, 201, 203-206, 216, 221, 224, 231, 235, 250, 284
life fields, 81, 202
life-negative, 80, 90, 103, 209, 245, 248-249
life-positive, xi, 23, 27, 98, 100-101, 172, 195, 214, 245, 248
Little Orgonon, Arizona, 224-225
Littlefield, Dr. Charles W., 115
Loom of Creation, The, 127
LSD, 68, 112
Lucifer, 142-143, 247
Lucifer and Ahriman, 142-143
Lundberg, Ferdinand, 140
MacArthur, General Douglas, 140, 214-215, 221
macrobacteria, xii, 84
Man or Matter, 34-35, 40, 99, 104, 127
Mark Probert Controls, 129-131, 134-135
Mass Psychology of Fascism, The, 203
Mathers, MacGregor, 175, 181
McCampbell, James M., 235, 254, 257
McCullough, Robert A., 191-193, 217-219, 222-225
mechanism, xi, 2, 15, 17, 30, 81, 113, 173, 204, 328
mechanistic science, xi, 1-3, 9-10, 13, 16-17, 19, 23, 100-101, 105-106, 167, 194, 206, 215, 224, 227
medical orgone therapy, 207
melanor, 218-219, 221
Menzel, Dr. Donald, 18, 69
Merlin Weather Engineering, 226, 295
Metcalfe, Eugene, 19
Meteorological Study of Radar Angels, 94
Michel, Aimé, 54, 118-119
microwave(s) 3, 5-6, 49, 54, 79-81, 230-231
Mills, Peter, 211
Milner, Dr. Dennis, 127
Mitchell, Captain Edgar, 244-245
Mitchell, General Wm. "Billy", 93
Moncla, Lt. Felix Jr., 20
Moray, T. Henry, 254
Morley-Martin Experiments, The, 115
Morley-Martin, British biochemist, 115, 204
Moss, Dr. Thelma, 37
Mothman Prophecies, The, 255
Murder of Christ, The, 215
Mystical Qabalah, The, 176
Nansei-shoto, 6-8
NASA, 22, 38, 53, 65, 212, 245-246, 284, 300, 309-310, 312, 333
natural forces, 51, 147
Natural History of the Heavens, 113
Naumov, Edward, 253
Neue Illustrierte Woehenschau, 81

New Republic, 210
New School for Social Research, 206
Newton, Sir Isaac, 101
NICAP, 20, 26, 57, 90, 118, 138, 227
Nietzsche, 102, 104
Nixon, Richard M., 249-250
nonhumans, 149
Norman, Eric, 242
Oberth, Dr. Hermann, 43
official science, vii, xi-xiii, 1-2, 5-6, 8, 12, 14-15, 22, 30, 33, 38-39, 43, 54-55, 64, 75, 79, 87, 98, 103, 112, 115, 120, 125, 129-130, 135, 142, 144-146, 151, 168, 194, 204, 212, 253, 327
oedipus complex, 198
Oranur Experiment, 159, 213, 215, 221-223
orgasm theory, 199
orgone accumulator, 42, 66, 121, 127, 164, 202, 205-212, 231-232, 242
orgone energy, xii-xiii, 3, 5, 12-13, 16, 19-20, 23, 28, 38, 40, 42-44, 48, 52, 59-63, 71, 75, 80, 86, 88, 102-103, 106, 117, 121, 125, 147-148, 151-152, 157, 170-173, 178, 192-195, 201-202, 205-213, 215, 217-221, 226-228, 230-233, 235-237, 239-240, 242, 252, 258, 263, 284-285, 297, 299, 331
Orgone Energy Bulletin, The, 209, 218
Orgone Energy Weather Engineering, 193, 226, 240
Orgone Institute, 193, 217, 219
orgone motor, 80, 211, 212
orgonomy, 40, 193-194, 196, 213, 232, 297, 299
Orgonon, 213, 217-225
ORUR, 223-225
Ostrander, Sheila, 38, 252-253
Outline of Occult Science, 142
Paasche, J., 252
Paleolithic UFO Shapes, 118
Palmer, Ray, 71, 82
Pasteur, Louis, 204

Pedley, George, 236
Pennisi, Albert, 236
Perry, Pierre, 19
Pfeiffer, Dr. Ehrenfried, 106, 174
Phenomenes Spatiaux, 238
Philosopher's Stone, The, 164
Philosophy of Spiritual Activity, The, 101
Planck, Max, 151
Plank, Vernon G., 94
plasmoids, 297-298, 306-308
Poppelbaum, Dr. Hermann, 106
primary energy, 28, 30, 33, 60, 62, 190, 211, 217, 240, 279, 282-283
Probert, Irene, 129
Probert, Mark, 129-131, 134-136
Proceedings of the College of Universal Wisdom, 35
Project Magnet, 13, 109
psychedelic revolution, 31, 146
psychic control, viii, 149, 184, 188, 214, 216, 221-222, 244, 246-248, 254
Psychic Discoveries Behind the Iron Curtain, 38, 137, 252
psychoanalysis, xiii, 167, 195-198, 200, 202-203
Qabala, 175-176, 178-179
Qabalist, 128-129, 176
Qabalists, 128, 179
Queensland (UFO Nests), 236-237, 239
radar, 3, 5-12, 16, 20-21, 39-41, 43-44, 49-50, 54-55, 60, 69, 73, 75, 78-81, 94, 148, 188, 217, 234, 275
radionics, 152-154, 157, 159, 162, 184, 215
radionics, application to UFOs, 189-190
radiovision photography, 170, 179
Radiovision—Scientific Milestone, 166, 191
Ravenscroft, Trevor, 247
Ravitz, Dr. L. J., 89
ray weapons, 20

Reich, Dr. Eva, 192-193, 297
Reich, Dr. Wilhelm, v, xii-xiii, 1-2, 5, 12, 19, 38, 44, 51, 59, 62, 75, 85, 87-89, 93, 98, 100, 102-103, 106, 115-117, 120-122, 125-127, 133, 135, 147, 149, 151-152, 157, 159, 164-165, 167-168, 172-173, 190-196, 199-228, 231-232, 235-236, 238, 240, 251, 254-256, 263, 277, 282, 284-285, 297, 300-301, 331
Reincarnation of Animal and Plant Life from Protoplasm Isolated from the Mineral Kingdom, The, 115
Religion and Science Merged, 35
Rich and the Super-Rich, The, 140
Rio Vista, California 77-78
Roerich Expedition, 74
Rudolf Steiner Schools, 100, 102, 107, 166
Russia(ns), 44, 130, 149, 157, 164, 195, 214, 216, 244-246, 249-253
Ruttledge, Hugh, 69, 333
Saga magazine, 10
Sagan, Dr. Carl, 46, 56, 327, 331, 334
Sanderson, Ivan T., 73-74, 82-83, 96, 114, 334
Sandy Cape, Tasmania, 73
Satan, 142-143
Saucerian Press, Inc., 137
Schirmer, Patrolman Herbert, 228
Schroeder, Lynn, 38, 252-253
Schultz, Edward S., 135
Schwenk, Dr. Theodor, 106
Science magazine, 334
scientism, 1, 24, 82, 89, 92, 126
scientist, xii-xiii, 1, 13, 21, 28, 34, 38, 40, 43, 73, 81-82, 91-92, 96-97, 101, 104, 109, 112, 114, 117, 119, 125-126, 132, 146, 148, 164, 184, 192, 195, 199, 201-202, 213-214, 218, 221-222
scientists, vii, 1-2, 5-6, 8, 14-16, 20-22, 24, 26, 28, 33, 38, 43, 57, 63, 73, 80, 82, 85, 87-92, 94-97, 109-112, 114-115, 125, 129-130, 132, 135,

140-141, 145, 147-149, 151, 156-157, 174-175, 179, 190, 192, 194, 196, 214, 216, 218, 236, 246, 248, 251-253, 256, 258, 284-285, 299, 326, 331
Second Coming of Christ, 146
Second Law of Thermodynamics, 42, 62, 121, 127, 206-207
Seers, Stan, 236
Sensitive Crystallization Processes: A Demonstration of Formative Forces in the Blood, 174
sequestration, 5, 34, 72, 100, 168, 174
sexuality, infant, 197-200
Shaw, George Bernard, 99, 113
Smart, Edward, 127
Smith, Wilbert, 13, 109, 132
Smythe, Frank, 69-71
Soviet Union, 37-38, 65, 130, 137, 149, 157, 216, 244, 246-247, 250, 252-253, 256-257
space animals, 22, 46, 69, 333
Spacecraft From Beyond Three Dimensions, 5
Spear of Destiny, The, 247
spinning wave, 200, 207, 219
Spiritual Science and Medicine, 104
Spurious Echoes on Radar, A Survey, 94
Stanford University, 157, 162
Star Exercise, 51-53, 71, 277
Steep Rock Echo, The, 240
Steiger, Brad, 10, 297, 333
Stein, Dr. Walter Johannes, 247-248
Steiner, Dr. Rudolf, xii, 34-36, 40, 76, 80, 98-101, 103, 106-109, 114, 120, 127-128, 135, 142, 154, 156, 174, 178, 183, 212, 216, 247-248, 255
Strader, 102-103, 209, 212
Strand, Lt. Col. Howard C., 10, 12
Strange Effects From UFOs, 227
Stringfield, Leonard, 138-139, 150
Strong, Dr. Frederick Finch, 156-158
Stuart, Lyle, 150
Stuart-Menzies, John, 239

T-bacilli, 204-205
telepathy, 27-28, 33, 109, 130, 134, 136, 149, 186, 262
They Live in the Sky, vii, xv, 19, 31, 46, 54, 82, 87, 139, 192, 219, 234, 258, 293, 314, 333
Thomas, Dr. Franklin, xii-xiii, 24, 31-34, 50-52, 55, 64, 99, 133, 152-154, 185, 187-188, 297
Thule group of Satanists, 247-248
Tichit, Germain, 238-239
Tiller, Dr. William, 37, 162
Tomlinson, Dr. Henry, 153
Truth and Science, 100
U.S.A., 37, 164, 217, 244, 247, 295
Uberlandia, Brazil, 229
UFO attacks, 19-21, 130-131, 138-140, 178, 218-226, 227, 233-234, 238
UFO propulsion, 14, 22-23, 27, 30, 60-65, 74, 80, 92-93, 184, 199, 208, 211-213, 216-219, 221, 224, 230-242, 262, 263, 284, 287, 288
Ufology: New Insights from Science and Common Sense, 257
UFOs, biological, xii, 12-13, 53-56, 64, 69-75, 85-87, 91, 95, 199, 227, 233, 237-238
UFOs—A New Look, 57
UFOs, contactees, 60, 252, 254-255
UFOs, effects of radar upon, 78-81
UFOs, etheric, 131-133
UFOs—Here and Now, 10
UFOs, interest in water, 239-242
UFOs, scientific reaction to, 87-98
Unger, Dr. Georg, 106
Uninvited Visitors, 73-74, 82-83, 96, 114
ur example, 72
USA, ii, vii, 149, 216, 229
USAF, 10, 19, 20, 22, 28, 30, 43, 49, 60, 64, 69, 91, 93-94, 96, 132, 138, 224, 234, 274-275, 300
Van Tassel, George, 28-30, 35, 50, 59-62, 66, 109, 112, 134, 186, 262-263, 266
Velikovsky, Dr. Immanuel, 87, 92, 115, 144
ventlas, 60, 119
visual ray (eye beam), 35, 40-44, 58, 60-62, 88, 196, 232, 252
Wachsmuth, Dr. Guenther, 80, 106, 120-121, 125
Waldorf Schools, 100, 103-104, 107, 255
Walker, Clint, 240, 242
Wassilko-Serecki, Countess Zoe, 81
Watergate, 140, 250
weather control, 52, 121, 125, 192, 220, 282
Welles, Orson, 29
weltanschauung, 8, 14
White, John, 254, 257, 333
Whritenour, Joan, 10
Wilkins, Harold T., 18, 139
Wood, Don (Jr.), 71-75, 238, 316-318
Woods, Dr. James O., 46, 51, 54, 61, 64, 87, 123, 188, 258, 263, 273-275, 277, 279, 311
Worlds in Collision, 92
x-rays, 162
Yada Di'Shi-ite, 129
Young, Dr. Arthur, 165

BOOKS MENTIONED OR RELATING TO
THE COSMIC PULSE OF LIFE

THE FLYING SAUCERS ARE REAL, by Donald Keyhoe. Was the first in-depth, authoritative look at flying saucers and to this day is considered one of the best books on the subject ever written. Well researched with documented facts, containing none of the disinformation and hype that has crept into the field over time. The author was a retired Marine Corps Major who became an aviation writer, and was therefore perfect for researching and writing this book. Using his own knowledge and information from friends in the military, he came to the conclusion that UFOs are from somewhere else and the U.S. military was engaged in a cover-up. Shows how and why the veil of government secrecy was put into place and enforced. ISBN 978-1-58509-264-2 * 156 pages * $15.95

THE BOOK OF THE DAMNED, by Charles Fort. Time travel, UFOs, mysterious planets, stigmata, rock-throwing poltergeists, huge footprints, bizarre rains of fish and frogs—nearly a century after being published, the strange phenomena in this book remains largely unexplained by modern science. Fort spent his time collecting reports of strange events, sent to him from publications around the globe. This was his first book on unusual and unexplained events and to this day, remains the most popular. If you agree that truth is often stranger than fiction, then this book is for you. ISBN 978-1-58509-278-9 * 228 pages * $18.95

THE COMING OF THE GUARDIANS: An Interpretation of the "Flying Saucers" as Given from the Other Side of Life, by Meade Layne. Where do UFOs really come from? Layne holds an inter-dimensional, etheric-based theory. Those from other realms he terms as "Etherians" and the UFOs themselves are sometimes living creatures, referred to as "aeroforms". Much of the information is corroborated or comes directly from intelligences that were channeled through Mark Probert, the most authentic and amazing medium of his time. Do not dismiss this kind of material without first considering its message. Covers propulsion systems, vibrational frequencies, and the sudden appearance of these craft more clearly than any scientifically based book one could find. ISBN 978-1-58509-525-4 * 100 pages * $14.95

THE MAGIC BAG, by Mark Probert. His classic work. Introduction by Meade Layne. Reveals great wisdom and amazing truths from "The Inner Circle"—the group of spiritual entities that were allowed to speak through Mark. Covers time and space, the nature of the cosmos, creation, the Etherians, and much more. The few people still alive who knew him will attest to the fact that there was no better trance medium than Probert. ISBN 1-58509-258-4 * 196 pages * $12.95

TO ORDER CALL THE BOOK TREE AT 1-800-700-8733 with credit card or send check or money order to PO Box 16476, San Diego, CA 92176. Include $4.50 up to first two books, $1 each thereafter. CA residents add 7.75% tax.

Printed in the United States
126497LV00002B/10/P

9 781585 091157